Finite elements

Finite elements

An introduction to the method and error estimation

Ivo Babuška
University of Texas at Austin

John R. Whiteman
Brunel University

Theofanis Strouboulis
Texas A&M University

OXFORD
UNIVERSITY PRESS

Great Clarendon Street, Oxford OX2 6DP

Oxford University Press is a department of the University of Oxford.
It furthers the University's objective of excellence in research, scholarship,
and education by publishing worldwide in

Oxford New York

Auckland Cape Town Dar es Salaam Hong Kong Karachi
Kuala Lumpur Madrid Melbourne Mexico City Nairobi
New Delhi Shanghai Taipei Toronto

With offices in

Argentina Austria Brazil Chile Czech Republic France Greece
Guatemala Hungary Italy Japan Poland Portugal Singapore
South Korea Switzerland Thailand Turkey Ukraine Vietnam

Oxford is a registered trade mark of Oxford University Press
in the UK and in certain other countries

Published in the United States
by Oxford University Press Inc., New York

British Library Cataloguing in Publication Data
Data available

Library of Congress Cataloging in Publication Data

Babuška, Ivo.
Finite elements : an introduction to the method and error estimation /
Ivo Babuška, John R. Whiteman, Theofanis Strouboulis.
p. cm.
ISBN 978–0–19–850670–6
1. Finite element method. 2. Estimation theory. 3. Error analysis (Mathematics)
I. Whiteman, J. R. (John Robert) II. Strouboulis, Theofanis. III. Title.
QA276.8.B33 2011
518′.25—dc22 2010033235

Typeset by SPI Publisher Services, Pondicherry, India

ISBN 978–0–19–850669–0 (Hbk.)
978–0–19–850670–6 (Pbk.)

1 3 5 7 9 10 8 6 4 2

This book is dedicated to Tinsley Oden and to the late Olek Zienkiewicz, collaborators, colleagues and friends over many years

Preface

There are many mathematical and engineering books on the finite element method and, at a recent count, more than 200 000 papers addressing finite elements and their applications; (for details of the annual count over the last 40 years see, e.g., the website of Jaroslav Mackerle). One has therefore to ask the questions 'Why write yet another book on the finite element method, what is the philosophy and what are the aims that justify another book, and what is the audience for the book?'

In spite of the existence of the large number of finite element books, it can be said that they are all in the main devoted primarily either to the mathematical theory of finite element methods, or to engineering applications, but usually not to both. So, in response to the above questions we feel that there is a need for a book that presents the main theoretical ideas of the finite element method and the analysis of its errors in an accessible way, and that demonstrates the interrelationship between the theory and the computed numbers. The book should show how the theory contributes to the understanding of computed results and builds confidence in the method and the correctness of a finite element program. Further, the book should require the minimum of pre-requisites for understanding the basic theory presented, it should as far as possible be self-contained, and it should address questions that are relevant to numerical methods for partial differential equations. Finally, it should address the numerical computation of typical simple engineering problems, showing the numerical features, analysing them, explaining them and drawing conclusions.

Within the above specification the present book originated from a suggestion by the second author, following publication in 2001 of the monograph by the other two authors (Babuška and Strouboulis, (2001)), that there is a need for a book introducing the verification aspects of the task of reliability of computed quantities of interest; i.e. to check when using *a priori* and *a posteriori* error estimation that they are sufficiently close to the exact values. Verification addresses the correctness of methods and of computer programs, which are related to understanding both the methods and the features of the computational results. This book is aimed at final-year undergraduates and at graduate students in mathematics, science and engineering, and also at practicing engineers and applied mathematicians to whom the numerical results are tailored.

In order to try to achieve the above goals we focus in this book mainly on problems of heat transfer that lead to scalar elliptic differential equations. The ideas are explained in detail initially for one-dimensional problems, as these provide the simplest context, and then the discussion of their extension to two-dimensional problems aims to illustrate the magnitude of the step change in the technicalities in proceeding to higher-space dimensions.

We start the book by discussing the need for reliable computations, and then define the problems to be tackled in physical contexts from first principles. In the third chapter we describe the finite element method, presenting it as a recipe, and give finite element solutions to a range of problems of interest. In the first instance, we present many numbers related to example benchmark problems and to simple 'engineering' problems in one- and two-space dimensions, simply to show what finite element calculations will produce and to arouse the curiosity of the reader as to why the numbers do what they do. In subsequent chapters we use the numbers to demonstrate the theory, noting that the numbers require the theory. We prove theorems relevant to all essential parts of the finite element method; convergence, *a priori* and *a posteriori* error estimation, computation of quantities of interest and their *a priori* and *a posteriori* errors. This approach, both dual and interacting, is a feature that we believe makes this book unique. By using this book to master the above the student can view it as an introduction to more advanced books on finite element error estimation.

It should be stressed that this book is not intended to provide instruction on using commercial finite element programs, but rather to give the necessary knowledge and understanding, and therefore confidence, that persons wishing to perform finite element computations need. We also emphasize that we concentrate initially on the one-dimensional setting not because of its practicality, but rather to ease understanding. We hope that this book can form the basis for a one-semester first course on finite element methods for students of 'applied mathematics' in its broadest sense.

A book of this nature cannot be produced without the help and support of many people. First our thanks go to Dr Realino Hidajat who, in a masterly manner, carried out the computations for all the examples in the book. The steadfast and reliable support of Realino was invaluable. Other contributions to the computations came from Shyan Mohan Keralvarma and many other students at Texas A&M University, as well as from Dr Chandra Shekar Upadhyay. Similarly, our thanks go to students over the years taking the Brunel University Masters numerical analysis and modelling courses on whom early versions of some of the chapters were 'tried out', and who supplied many helpful comments.

Secondly, we would like to thank Maureen Senatore who has, with great competence, typed the many versions of our manuscript. Her cheerful and gracious collaboration has been a key component of getting the book into print, and has been far beyond the call of duty. As always, we owe a debt of gratitude to many colleagues in ICES at the University of Texas at Austin, in BICOM at Brunel University, and in the Department of Aerospace Engineering at Texas A&M University. The second author would also like to thank ICES at the University of Texas at Austin for continuing financial support over a long period, in the form of Distinguished Visiting Fellowships that have enabled him to visit Austin many times in order to progress this manuscript.

Ivo Babuška, Austin, Texas, USA
John Whiteman, Uxbridge, UK
Theofanis Strouboulis, College Station, Texas, USA
February 2010

Contents

1
Introduction

1.1 The finite element method

The finite element method (FEM) is today without doubt one of the major numerical tools in computational science. This position of pre-eminence has arisen as a result of a number of complementary factors. These include: the requirement for engineers, scientists, financiers and clinicians to be able to simulate and solve numerically ever more complex problems; the results on the theory of finite elements that have been produced by mathematicians since the 1960s, particularly regarding error estimation; and last but by no means least, the inexorable increase in computer power over the last half-century and the corresponding increase in the expectations of users. The result of these simultaneous factors has been that theory for finite elements is now well founded for a very large range of applications, and that much commercial software is now available for implementing finite element techniques. Thinking specifically of engineering applications there is perhaps another important factor that comes under the heading of 'technology pull', by which the ability to simulate industrial processes has enabled great advances in these processes to take place, with the result that processing ability now drives design. Instances of this can be found, for example, in the glass industry, where the ability of glass makers to produce windshields and rear windows for cars in ever more exotic shapes is allowing manufacturers correspondingly to develop cars with superior aerodynamic characteristics, resulting in a saving in the fuel consumed, or in the aerospace industry for the design of aircraft, among many others.

The rise in the importance of computational simulation has brought with it the need to assess the *reliability* of computer predictions. The reliability of a simulation is established via two processes. These are known as **validation** and **verification** (V&V), see, e.g., Babuška *et al.* (2007). The validation process addresses the quality of the model in representing the physical context, whilst the verification process addresses the quality of the numerical solution of the mathematical model. It is within the verification process that error analysis for finite element methods has been perceived as playing a crucial role, and this is one of the reasons behind our decision to write a book on finite element methods and error estimation. However, the role of error estimation is far wider than simply to contribute to verification, and it is in fact crucial to validation and the assessment of reliability of a model.

1.2 Mathematical model

The mathematical model is the starting point for computational simulation. The model is a mathematical problem, which has a structure and includes input and output information. For example, it could be a boundary value problem for a partial differential equation, the structure of which is the mathematical form of the differential equation, together with the domain in which the equation holds and the boundary conditions on the boundary of that domain. The forms of the predictions arising from such a problem are, for example, various functionals. The inputs to the mathematical problem in this case are the coefficients in the governing differential equation, the shape of the domain and the boundary data. The outputs are the computed specifics that are produced to approximate the output functionals, for example stresses or fluxes.

A mathematical model of a practical situation inevitably contains uncertainty, which arises in a number of ways. For example, the uncertainty may be **aleatory**, that is that it is inherent to the physical event and arises as a result of 'nature'. Alternatively, the uncertainty may be **epistemic**, and arises due to lack of knowledge. Here, one thinks, for example, of a physical experiment in which the apparatus will introduce unknown errors into its readings. In the specific case of a model consisting of a boundary value problem, the governing differential equation is selected by the analyst and is chosen on the basis of, for example, physical laws. The input is determined from experiments that give values for the coefficients in the differential equations and for the boundary data. The output from the mathematical model must therefore contain uncertainty, which arises from the uncertainty in the input.

1.3 Validation and verification

As has been said above, *validation* relates to the question of whether the mathematical model correctly models the physical event being considered. As any mathematical model is a representation of the physical event as conceived by the modeller, it can only be an approximation to the actual happening. Validation is the process whereby it is determined that the model meets necessary conditions for acceptance from the perspective of the goal of the simulation. Validation is in fact based on a validation pyramid, consisting of a set of validation problems of differing complexity. These problems are analyzed both computationally and experimentally so that the corresponding results can be compared. If the difference is not acceptable, then the model on which the computation is based must be revised; see Babuška *et al.* (2007).

Returning to verification, as has been said, this is the process by which it is ascertained that the computed solution of the mathematical model is sufficiently accurate. Verification has two distinct parts, the first of which is the verification that the theoretical solution of the numerical scheme has an acceptable error in the data of interest; error estimation plays a major role in this. The second part is the verification that the computer program that is applied to the numerical scheme does what it is supposed to and again does not introduce unacceptable error. In this book we shall assume that the 'programming' part is always satisfactory. However, a few words of warning on this aspect of the simulation process are appropriate here. Verification of a program is usually done by applying it to a number of simple problems for which the solution is known, and it is assumed that the effect of round-off is benign. This usually turns out to be satisfactory, but there have been notable exceptions! For example, there is the case of the Ariane 5 rocket that in 1996 failed due to round-off errors in the calculations. (see Mordicai Ben-Aki (1997) for references). It may of course be that a mishap occurs simply because of human error! For example, the loss of the Mars Climate Orbiter in 1999 in the Mars atmosphere occurred as a result of the unintended mixing up of imperial and metric units in the model.

More scientifically, many accidents occur because the initial mathematical model is incorrect (validation difficulties!). We mention a few examples. The collapse of the Tacoma Narrows bridge across the Puget Sound in Washington State, USA in 1940 arose from the fact that the model did not properly take into account the aerodynamic forces and the effects of the Von Karman vortices on the bridge, nor did it correctly describe the behaviour of the bridge cables. Similarly, the collapse in Connecticut, USA of the roof of the Hartford Civic Center in 1978 was due to the use of an inappropriate linear model. In 2003, the loss of the US Columbia Shuttle occurred as a result of a piece of the fuel tank heat shield breaking off. After this was observed, computations were undertaken that indicated that the damage was not serious. However, the model used was not adequate for analysing the effect of a large piece of debris and for its location.

Failures can also be caused by inadequate numerical treatment. For example, in 1991 the Sleipner offshore platform made of reinforced concrete sank in Gandsfjord, Norway during ballast test operations. The reason for this was that the finite element analysis gave a 47% underestimation of the shear forces at a critical part of the structure.

1.4 The finite element method, error analysis and estimation and its role in the processes of verification and validation

As has been said above, due to its versatility, and for the reasons stated in Section 1.1, the FEM is widely applied to obtain numerical solutions to mathematical models. As we shall see in the later chapters of this book it is frequently the case that *a priori* error analysis will indicate when and at what rate the FE solution will converge to the true solution of the mathematical problem. Further, it is also frequently possible to derive *a posteriori* error bounds and computable error estimates for the FE solutions. Armed with this machinery, not only does the modeller have a computed solution to the problem, but he/she also has the calculated estimates of the error that can be used adaptively to produce a FE solution of acceptable accuracy; this is the process of *verification* of the numerical algorithm.

But the FE solution and accompanying error estimates have a further use in the *validation* process. Here, the originally chosen model of a physical event can be viewed as a **coarse** model of the event. A **fine** model, incorporating more physical detail of the event, can then be chosen, and by substituting coarse FE solutions into this, residuals can be calculated. If the residuals are large, then the appropriateness of the coarse model is suspect and the model must be refined. This is the process of hierarchical modelling; see, e.g., Oden and Prudhomme (2002) and Shaw *et al.* (2010).

1.5 The purpose of this book and its layout

Our views on the importance of the FEM in the V&V process have convinced us that there is a need for an expository book that treats from scratch reliability, *a posteriori* error analysis and adaptivity for the finite element method. The result is that in the context of second-order boundary value problems in one- and two-space dimensions we have tried to write a book that does this, which is both clear and introductory, and that concentrates on the basics of the methods and analysis.

The form of this book is as follows. This first chapter is introductory and is intended to be motivational. In the subsequent chapters the sequence each time is to do everything first in the one-dimensional context, and then to move to two dimensions. In Chapter 2 we derive in the contexts of elasticity and heat transfer the second-order partial differential equations and boundary value problems to which we shall apply the FEM, together with the machinery of function spaces, weak formulations and norms. We thought it necessary to include these derivations of the problems so as to indicate the quantities that are important in these contexts. The finite element method and its methodology are introduced in Chapter 3, together with the concept of error and the *best approximation* theorem for the finite element error. In Chapter 4 we consider the finite element error over a single element, and relate this to interpolation error over the element. The results of Chapter 4, together with the best approximation theorem for the finite element error, allow us in Chapter 5 to derive global *a priori* error estimates for the finite element solutions. In Chapter 6 we address both the computation and error in functionals of the solutions of problems and the effect of superconvergence. In Chapter 7 we move to calculable error estimates and derive *a posteriori* bounds for the calculated finite element solutions. At this point, we have reached the stage where we have set up the machinery for the verification of the FE solution, which is the aim of the book.

1.6 Literature

There is a vast literature on the FEM, so much so that it is daunting to the student wishing to study the method *ab initio*. We therefore list here a number of books to which the student should have access in order to study further specific points.

Mathematics

Ainsworth and Oden (2000), Babuška and Strouboulis (2001), Becker and Carey (1981), Braess (1997), Brenner and Scott (1994), Ciarlet (1979), Oden and Reddy (1976), Wait and Mitchell (1985), Whiteman (ed.) (1973, 76, 78, 82, 85, 88, 91, 94, 97, 2000).

Engineering

Bathe (1996), Becker and Carey (1981), Demkowicz (2006), Ern and Guermond (2004), Hinton and Owen (1977), Hughes (1987), Ladeveze and Pelle (2005), Martin and Carey (1973), Oden (1972), Reddy (1984), Schwarz (1988), Szabo and Babuška (1991), Zienkiewicz (1977), Zienkiewicz and Taylor (2005).

2 Formulations of the problems

Summary

- Mathematical models are formulated for problems of linear elasticity in one dimension and for steady-state heat flow in one and two dimensions, in terms of second-order linear differential equations together with boundary conditions.
- The connection between these classical (strong) formulations and the principle of virtual work is demonstrated, and weak formulations of the problems are defined in suitable spaces.
- Mathematical properties of the bilinear and linear forms occurring in the weak formulations and of the solutions of these formulations are presented.
- Various norms are defined.
- One- and two-dimensional 'engineering-type' heat-transfer problems, which will be used throughout the book to demonstrate the use of the finite element method and its performance, are defined.

Throughout this book we shall describe finite element methods, and discuss their analysis and effectiveness, in the context of linear elasticity in one dimension and heat conduction in one and two dimensions. The actual implementations of the methods will be in the context of *integral* forms of these problems. Thus, in this chapter we introduce first a one-dimensional problem involving the linear elastic deformation of a bar. This enables us to derive the *classical* problem in terms of the equation of equilibrium and the boundary conditions, to extend this into the *weak* (integral) form, and to discuss the place of this in the contexts of *virtual work* and *energy*. It is then shown that this exact classical problem can also be reinterpreted as the model for heat flow in the bar, which again extends to the weak formulation as above. The case of two-dimensional heat flow is a natural extension of the one-dimensional model, and the classical and weak forms of this are also given.

The association of the heat-conduction problem with 'energy' is not completely obvious, unlike the case of linear elasticity. However, because in the two-dimensional setting we wish to describe finite element methods in the context of problems involving only *scalar* solutions, we have decided that our discussion of the methods will be only in the context of heat conduction. The weak formulations of all the problems are the starting points for applying the finite element techniques. These concepts also motivate the choice of *norms* in our later finite element analyses.

Although we have chosen only these simple linear elastic and heat-conduction problems, these possess many of the ingredients of 'real-life' practical industrial problems for which computational modelling must be undertaken. In the context of these simple problems we shall demonstrate modern facits of finite element methods such as superconvergence, *a posteriori* error estimation and adaptivity.

2.1 One-dimensional deformation of an elastic bar and one-dimensional heat conduction

2.1.1 Classical differential equation formulation of the bar problem

We begin by considering an elastic bar $\mathcal{B}$ of length l with axis in the x_1-direction, $0 \leqq x_1 \leqq l$, where $(0,l) \equiv I$ and with rectangular cross-sectional area $A(x_1)$. Thus, we may write

$$\mathcal{B} \equiv \{x_1, x_2, x_3 \mid 0 < x_1 < l,\ (x_2, x_3) \in A(x_1)\}. \tag{2.1}$$

We assume that the width and height of the bar are very much smaller than l, and that $A(x_1)$ changes slowly with x_1; i.e. that $\frac{d}{dx_1}(A(x_1)) << 1$.

We are interested in the deformation of the bar under the action of an axial load in the x_1-direction. More precisely, if $a \equiv (x_1^a, x_2^a, x_3^a)$ is a point of the bar before deformation, then this moves to become $b \equiv (x_1^b, x_2^b, x_3^b)$ after deformation, where

$$x_1^b = x_1^a + u_1(x_1, x_2, x_3); x_2^b = x_2^a + u_2(x_1, x_2, x_3); x_3^b = x_3^a + u_3(x_1, x_2, x_3),$$

and the u_1, u_2, u_3 are *displacements*. It is further assumed that $u_2 = u_3 = 0$, that any cross-section of the bar remains a planar cross-section under the deformation, and that $u_1 \equiv u(x_1)$ is small with respect to the length of the bar. This means that the deformation of the bar can be visualized as a *one-dimensional effect* characterized by the displacement function $u(x), 0 \leqq x \leqq l$, where $x \equiv x_1$. (For the *one-dimensional* problem we shall write x for x_1, and $x_j, j = 1, 2, 3, \ldots,$ will be specific values of the variable x. These should *not* be confused with variables denoting other space directions in higher-space dimension contexts).

Consider now the slice $S(x, \Delta)$ of the bar between the cross-section $BB'C'C$ at $x + \frac{\Delta}{2}$ and $AA'D'D$ at $x - \frac{\Delta}{2}$, as shown in Fig. 2.1, where each cross-section is perpendicular to the x-axis. When the bar is deformed, we assume that the cross-sections relocate, respectively, to positions $x + \frac{\Delta}{2} + u(x + \frac{\Delta}{2})$ and $x - \frac{\Delta}{2} + u(x - \frac{\Delta}{2})$, and that each remains both planar and perpendicular to the x-axis. The distance between the two cross-sections, which was Δ before the deformation, becomes $\Delta + u(x + \frac{\Delta}{2}) - u(x - \frac{\Delta}{2})$ and hence the *change* in this distance is $\left(u\left(x + \frac{\Delta}{2}\right) - u\left(x - \frac{\Delta}{2}\right)\right)$. The relative value of this change with respect to Δ is called *strain* and is denoted by $\varepsilon(x)$. For small Δ (neglecting higher-order terms) we may write

$$\left(u\left(x + \frac{\Delta}{2}\right) - u\left(x - \frac{\Delta}{2}\right)\right) / \Delta \approx \frac{du}{dx}(x) \equiv \varepsilon(x). \tag{2.2}$$

Throughout this chapter, when we are seeking to model the effects of elastic deformation, we shall be concerned with the dimensions of the quantities involved,

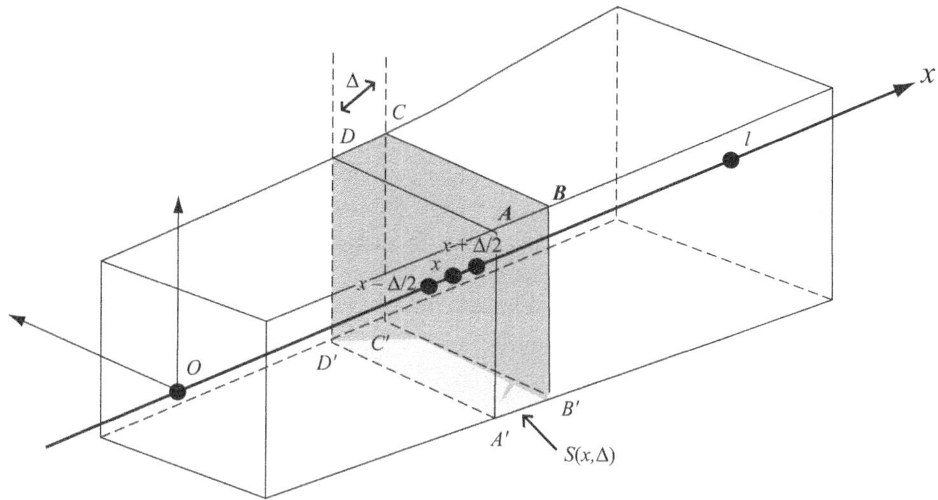

Fig. 2.1 Bar $\mathcal{B}$, $[0, l]$ with rectangular cross-section $A(x)$, and slice $S(x, \Delta)$ having cross-sections $AA'D'D$ and $BB'C'C$.

and we thus from now on introduce the relevant *dimensions*. If $[L]$ denotes the unit of length, then $u(x)$ and Δ each have dimension $[L]$, whilst the strain is dimensionless.

Suppose that the bar is in its state of deformation subjected to a distributed axial load $f(x)$ per unit length, so that the slice $S(x, \Delta)$ is subject to an axial force $P_1(x) = f(x)\Delta$. This load has the units of force $[F]$ so that $f(x)$ has the dimensions, force per unit length, $[F]\,/\,[L]$.

The bar will often be embedded in a surrounding medium that causes resistance to displacement in the form of a distributed axial support. The resistance due to the support is modelled in terms of springs with rigidity coefficients $c(x), c(x) \geqq 0$. The reactive axial force acting on the slice due to the surrounding medium is proportional to the displacement and thus for the slice can be written as

$$P_2(x) \equiv p(x)\Delta = -c(x)u(x)\Delta, \tag{2.3}$$

where the minus sign on the right-hand side of (2.3) results from the fact that $P_2(x)$ acts *against* the displacement, and $p(x)$ is the reaction per unit length of the surrounding medium on the bar. The dimension of $c(x)$ is thus $[F]\,/\,[L^2]$ and that of $p(x)$ is $[F]/[L]$. If the force on the cross-section at x is $F(x)$, then the force/unit area on the cross-section, known as the *stress*, $\sigma(x)$, is

$$\sigma\,(x) = \frac{F(x)}{A(x)}, \tag{2.4}$$

and the dimension of stress is $[F]\,/\,[L^2]$.

From Hooke's law for a linear elastic material we have that the stress $\sigma(x)$ is related to the strain by the *Young's modulus* $E(x)$ of the material, so that

$$\sigma(x) = E(x)\,\varepsilon(x) = E(x)\frac{du}{dx}(x). \tag{2.5}$$

In engineering, the Young's modulus of elasticity E is described as the force per unit area that leads to relative change of length between adjacent cross-sections. The dimensions of $E(x)$ are thus again $[F]\,/\,[L^2]$. We now wish to derive the differential equation that models the equilibrium of the bar, $0 < x < l$. Considering now the equilibrium of the slice $S(x,\Delta)$, of Fig. 2.1, where $F(x)$ is the force acting on the cross-section we have

$$F\left(x+\frac{\Delta}{2}\right) - F\left(x-\frac{\Delta}{2}\right) + \int_{x-\Delta/2}^{x+\Delta/2} f(x)dx + \int_{x-\Delta/2}^{x+\Delta/2} p(x)dx = 0.$$

Assuming that $f(x)$ and $p(x)$ are continuous for $x \in I$ we can apply the mean value theorem and let $\Delta \to 0$ to obtain

$$\frac{dF}{dx}(x) + f(x) + p(x) = 0,$$

and hence, using (2.3),

$$-\frac{dF}{dx}(x) + c(x)\,u(x) = f(x), x \in I. \tag{2.6}$$

Combining (2.4) and (2.5) we can see that

$$F(x) = A(x)\sigma(x) = A(x)E(x)\frac{du}{dx}(x), \tag{2.7}$$

and hence from (2.6) we obtain

$$\pounds u(x) \equiv -\frac{d}{dx}\left(A(x)E(x)\frac{du}{dx}(x)\right) + c(x)u(x) = f(x), x \in I, \tag{2.8}$$

which is the *equilibrium equation* for any point x of the rod.

Note that a necessary condition for the equilibrium equation (2.8) to be correct is that all the terms have the same dimensions. In the case of (2.8) each term has dimension $[F]/[L]$. This means that the symbols in (2.8) represent specific physical quantities, the dimensions of which can be expressed in terms of force $[F]$ and length $[L]$. If we consider the displacement function $u(x)\cdot[L]$, cross-sectional area $A(x)\cdot[L^2]$, Young's modulus $E(x)\cdot[F]/[L^2]$, axial load $f(x)\cdot[F]/[L]$ and reactive rigidity $c(x)\cdot[F]/[L^2]$, the $u(x)$, $A(x)$, $E(x)$, $f(x)$ and $c(x)$ are all *dimensionless* quantities. (Clearly the units of any problem must all belong to a single system; e.g. metric, British engineering, ...).

In order to complete the formulation of the bar problem we must associate with (2.8) boundary conditions at the ends of the bar. The following different cases will be considered.

a) Displacement boundary conditions (Dirichlet conditions).

The simplest boundary condition is that in which the displacement u is prescribed at a boundary point. In this case the displacement at the end points $x = 0, l$ could be given as

$$u(0) = g_1 \qquad u(l) = g_2, \tag{2.9}$$

where g_1 and g_2 are known and have dimension $[L]$. Conditions of this type are called Dirichlet, kinematic or essential boundary conditions.

b) Force boundary conditions (Newton and Neumann conditions)

In this case, we assume that applied forces are prescribed at the end points $x = 0, l$. For the case at $x = 0$ we now consider the slice $S(\frac{\Delta}{2}, \Delta)$ at $x = \frac{\Delta}{2}$ and located at the left-hand end of the bar as in Fig. 2.2, with loadings as indicated.

The $f(x)$ and $c(x)$ are as before. At $x = 0$, we assume that the bar is subjected to both an applied force G_1 of dimension $[F]$ and to an applied spring with rigidity constant α_1 of dimension $[F]/[L]$. For the slice to be in equilibrium we must therefore have that

$$F(\Delta) - \alpha_1 \, u(x) \mid_{x=0} + G_1 + \int_0^{\Delta} (f(x) - c(x) \, u(x)) \, dx = 0.$$

Now, in the limit as $\Delta \to 0$ we have that

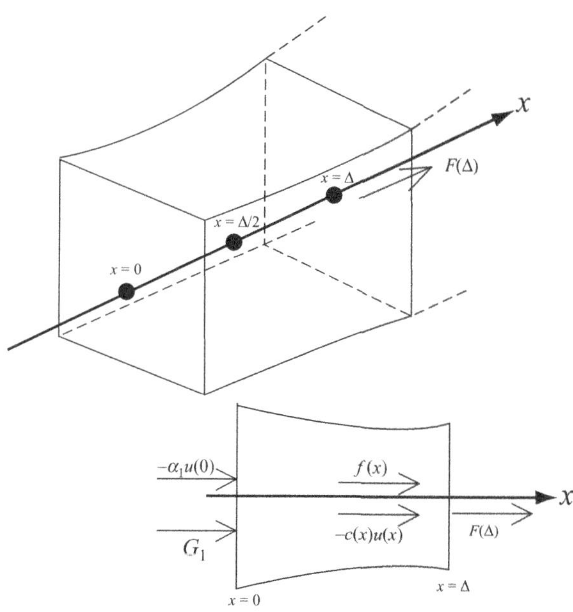

Fig. 2.2 Slice $S(\frac{\Delta}{2}, \Delta)$ at the left-hand end of the bar with the force boundary conditions at $x = 0$. The lower figure is the centre vertical section of the slice in the direction of the axis.

$$\lim_{\Delta \to 0} \int_0^{\Delta} (f(x) - c(x)\, u(x))\, dx = 0,$$

so that

$$F(0) = \alpha_1\, u(0) - G_1. \tag{2.10a}$$

But from (2.7) we know that

$$F(0) = (A(x)\, E(x)\, \frac{du}{dx}(x)) \mid_{x=0}, \tag{2.10b}$$

and so by combining (2.10a) and (2.10b) we obtain the (Newton-type) boundary condition

$$(A(x)\, E(x)\, \frac{du}{dx}(x)) \mid_{x=0} = \alpha_1 u(0) - G_1. \tag{2.11}$$

In order to conform with the usual notation from the literature for elliptic boundary value problems, we adopt the notation that $A(x)\, E(x) \equiv a(x), 0 < x < l$. Our *Newton (or Robin)* boundary condition at $x = 0$ is thus,

$$-(a(x)\, \frac{du}{dx}(x)) \mid_{x=0} + \alpha_1\, u(0) = G_1. \tag{2.12a}$$

If $\alpha_1 = 0$, then (2.12a) becomes the *Neumann* boundary condition

$$-(a(x)\, \frac{du}{dx}(x)) \mid_{x=0} = G_1. \tag{2.12b}$$

For the case of a force-type boundary condition at $x = l$, we consider the slice of the bar $S(l - \frac{\Delta}{2}, \Delta)$ at $x = l - \frac{\Delta}{2}$, located at the right-hand end of the bar, as in Fig. 2.3 with loadings as indicated.

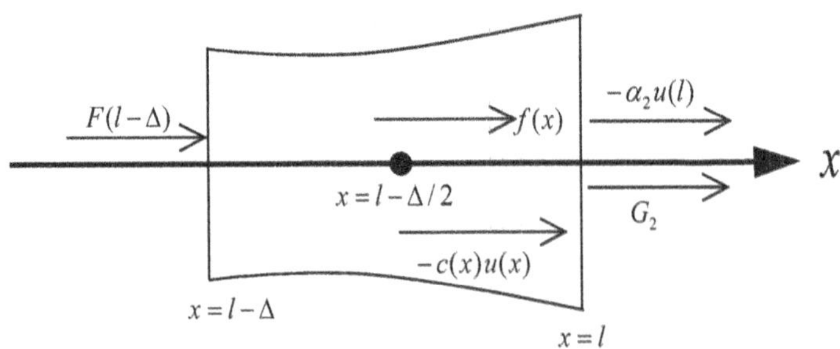

Fig. 2.3 Verticle section of the slice $S(l - \frac{\Delta}{2}, l)$ at the right-hand end of the bar, with force boundary conditions at $x = l$.

Correspondingly, for the slice to be in equilibrium we must have that

$$F(l-\Delta) - \alpha_2\, u(l) + G_2 + \int_{l-\Delta}^{l} (f(x) - c(x)\, u(x))\, dx = 0,$$

where at $x = l$ the bar is subjected to both an applied force G_2 and an applied spring with rigidity constant α_2. Again taking limits we find that

$$F(l) = \alpha_2\, u(l) - G_2,$$

and since

$$F(l) = -(a(x)\,\frac{du}{dx}(x))\mid_{x=l},$$

we obtain that

$$(a(x)\,\frac{du}{dx}(x))\mid_{x=l} + \alpha_2\, u(l) = G_2. \qquad (2.13)$$

We can now formulate a specific problem for the bar $x \in \bar{I}$. Find u that satisfies

$$\pounds u(x) \equiv -\frac{d}{dx}\left(a(x)\frac{du}{dx}(x)\right) + c(x)\, u(x) = f(x), 0 < x < l, \qquad (2.14a)$$

$$u(0) = g_1, \qquad (2.14b)$$

$$(a(x)\frac{du}{dx}(x))\mid_{x=l} + \alpha_2\, u(l) = G_2, \qquad (2.14c)$$

where (2.14c) is a Newton boundary condition because of the presence of $\alpha_2 \neq 0$.

Here, we have adopted the convention of denoting by u the *exact* solution of the problem, in order to distinguish it from other functions, such as approximations to the solution, which occur in later contexts.

Problem (2.14 a–c) will be our *model* one-dimensional problem and, unless we indicate otherwise, this will be the problem that we shall consider in one-dimensional contexts.

We note the two different signs for the du/dx terms in (2.12) and (2.13), because both these derivaties are in the x-direction. If we consider the derivatives in the outward directions at $x = 0, l$ and we denote these by $\left(\frac{du}{dx}\right)_0$, we have that

$$\left(\frac{du}{dx}\right)_0 = -\frac{du}{dx} \text{ at } x = 0,$$

$$\left(\frac{du}{dx}\right)_0 = \frac{du}{dx} \text{ at } x = l.$$

Further, we define the *conormal derivative* $\left(\dfrac{d\mathrm{u}}{dn_c}\right)$ as

$$\frac{d\mathrm{u}}{dn_c} = a\left(\frac{d\mathrm{u}}{dx}\right)_0,$$

which physically represents the outward flux. We shall henceforth use the conormal derivative notation, see, e.g., (2.49c).

The analogy of this will be seen in the two-dimensional context, where the derivatives at points on the boundary will be in the direction of the *outward* normal to the boundary.

Note that we have chosen as our *model* one-dimensional problem the configuration of one *displacement* (type (a)) and one *force* (type (b)) boundary condition, as these illustrate the main features of problems of this type. Clearly, other combinations of displacement and force boundary conditions are possible.

Equations (2.14 a–c) are formulated in a one-dimensional setting where I is the region of the problem, the boundary of I is $(\{0\},\{l\})$ and $\bar{I} \equiv I \cup (\{0\},\{l\})$. This notation is a specific case for the one-dimensional context of a more general notation suited to problems in two and three dimensions.

In general, the region in which a problem is defined is $\Omega \subset \mathbb{R}^n, n = 1,2,3$, which has boundary $\Gamma = \Gamma_{\mathrm{D}} \cup \Gamma_{\mathrm{N}}$. Dirichlet boundary conditions are specified on Γ_{D}, whilst Newton or Neumann conditions are specified on Γ_{N}. If we consider the one-dimensional problem (2.14 a–c) in this more general notation, we see that $\Omega \equiv I, \Gamma_{\mathrm{D}} \equiv \{0\}$, $\Gamma_{\mathrm{N}} = \{l\}, \Gamma = (\{0\},\{l\})$ and $\bar{\Omega} \equiv \bar{I} = (0,l) \cup (\{0\},\{l\})$.

Remark 2.1 *Very often, in any problem involving a differential equation the quantity of interest is not the solution itself, but rather some associated quantity. For example, for the deformation of the bar this quantity might be strain* $\varepsilon(x) \equiv d\mathrm{u}(x)/dx$, *or the force* $F(x) \equiv a(x)\,\varepsilon(x)$ *at a prescribed point, or their maxima over* I.

Remark 2.2 *The formulation (2.14) of course only has sense if each of the symbols contained in the equations is meaningful; e.g. we must be able to differentiate* $\mathrm{u}(x)$. *Hence, we seek the solution of the problem in a specific class of functions; e.g. if* $a(x)$ *has continuous derivatives on* $[0,l]$, *then we seek the solution in the class of functions that are continuous and have continuous first and second derivatives on* $[0,l]$. *Under these conditions* u *is called the strong solution of the problem.*

Remark 2.3 *In undertaking any* **computation** *concerning a problem involving a differential quation, such as the bar problem, which seeks to serve some practical purpose, it is necessary that the* **goal** *of the computation be clearly specified right from the beginning; i.e. the quantitiy of primary interest must be specified. This will obviously influence the whole foundation of our analysis and numerical schemes. Also, because we are able to find the solution only approximately, we have to define the manner in which the* **error** *is measured and when the numerical solution will be acceptable.*

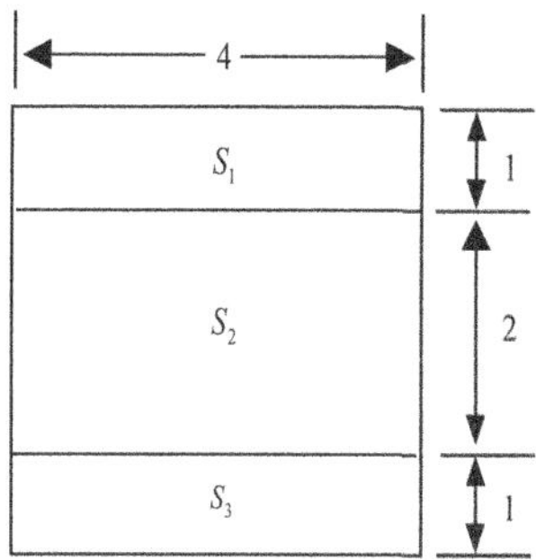

Fig. 2.4 Cross-section of three layer bar.

Example 2.1 Consider a bar with square cross-section that consists of three layers S_1, S_2, S_3 as shown in Fig. 2.4, and for which, in British standard units, the Young's moduli of elasticity of the layers are:

$$\text{for } S_1;\ E_1 = 12 \times 10^6 \text{ lb/(inch)}^2;$$

$$\text{for } S_2;\ E_2 = 5 \times 10^5 \text{ lb/(inch)}^2;$$

$$\text{for } S_3;\ E_3 = 20 \times 10^6 \text{ lb/(inch)}^2.$$

The bar is not embedded in a surrounding medium so that $c(x) = 0$. We are here using British standard units in which $[L] = \text{inch}$, $[F] = \text{lb/(inch)}^2$. Assuming that planes of cross-section remain planar under the deformation, we have from (2.7) that

$$F(x) = (4\ E_1 + 8\ E_2 + 4\ E_3)\frac{d\mathfrak{u}}{dx} = (132 \times 10^6)\frac{d\mathfrak{u}}{dx},$$

and hence that

$$\pounds\mathfrak{u} = -(132 \times 10^6)\frac{d^2\mathfrak{u}}{dx^2}(x).$$

Assume further that in (2.14) the length l of the bar is 12 inches, $f = 3$ lb/inch, $g_1 = 5$ inches and $G_2 = 0$. Then, the complete problem is

$$-\left(132 \times 10^6\right)\frac{d^2\mathfrak{u}}{dx^2}(x) = 3,\ 0 < x < 12,$$

$$\mathfrak{u}(0) = 5$$

$$\left(132 \times 10^6\right)\frac{d\mathfrak{u}}{dx}(x)\ |_{x=12} = 0,$$

and the goal is to determine $\mathfrak{u}(12)$.

Exercise 2.1 Reformulate the problem of Example 2.1 in metric units using the Newton as the force unit. Are the two problems formulated in different units equivalent,

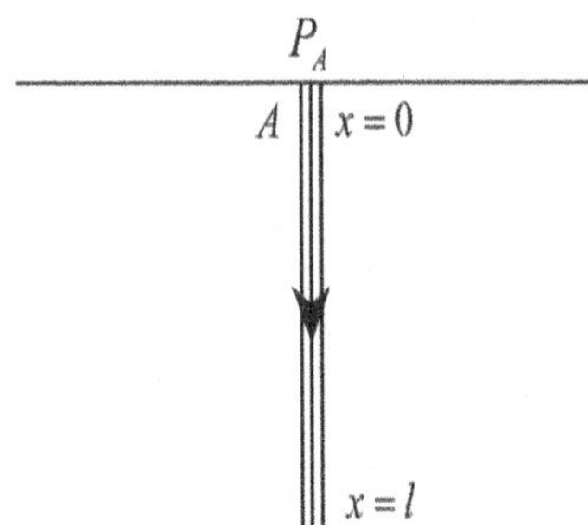

Fig. 2.5 Pile of length $l = 120$ inches and diameter 5 inches embedded in clay.

in the sense that knowing u for one it is possible to evaluate immediately u for the other? Evaluate the solution u in either set of units.

Example 2.2 Consider a wooden pile of circular cross-section with diameter $d = 5$ inches and length $l = 120$ inches embedded in clay, see Fig. 2.5. We assume that the reactive force of the clay is proportional to the surface area of the pile so that, from (2.3), $P_2(x) = -c(x)\mathrm{u}(x)\pi d\Delta$, and we take $c(x) = 4 \times 10^4$ pounds/inch^2. We further assume that the Young's modulus of the wood is $E(x) = 3 \times 10^5$ pounds/inch^2, that the force P_A acting on the top of the pile at A is 2×10^4 pounds and that the reaction on the bottom of the pile is small, due to the small diameter of the cross-section, and can thus be neglected. We wish to find the vertical displacement of the pile at A. The force acting on the cross-section of the pile is

$$F(x) = \frac{\pi}{4}5^2 \times 3 \times 10^5 \frac{d\mathrm{u}}{dx}(x) = \left(\frac{75}{4}\pi \times 10^5\right)\frac{d\mathrm{u}}{dx}(x).$$

The reaction of the clay on the cross-section is $-20\pi \times 10^4 \mathrm{u}(x)$. As there are no other reactions on the pile we have from (2.14a) that

$$\pounds\mathrm{u}(x) \equiv \left(-\frac{75\pi}{4} \times 10^5\right)\frac{d^2\mathrm{u}}{dx^2}(x) + \left(2\pi \times 10^5\right)\mathrm{u}(x) = 0,\ x \in (0, l),$$

and hence (computing to two places of decimals) that

$$-\frac{d^2\mathrm{u}}{dx^2}(x) + (0.11)\mathrm{u}(x) = 0,\ x \in (0, l).$$

In order to complete the problem we need boundary conditions at $x = 0, l$. Due to the loading at $x = 0$ in the positive x-direction the rod will be in compression, so that the boundary condition at $x = 0$ has the form of (2.12b) and is

$$-\frac{d\mathrm{u}}{dx}(x) = 3.39 \times 10^{-3},\ x = 0.$$

Correspondingly, at $x = l$ we have

$$\frac{d\mathrm{u}}{dx}(x) = 0,\ x = l.$$

For this problem it is straightforward to obtain the analytical solution. We know that

$$u(x) = C_1 e^{0.33x} + C_2 e^{-0.33x}, 0 < x < l,$$

where the C_1 and C_2 are coefficients to be determined. Now

$$-\frac{du}{dx}(x) \mid_{x=0} = 0.33\ C_1 + 0.33\ C_2 = 3.39 \times 10^{-3},$$

so that

$$-C_1 + C_2 = 1.027 \times 10^{-2}.$$

Similarly

$$-\frac{du}{dx}(x) \mid_{x=l} = 0.33\ C_1 e^{39.6} - 0.33\ C_2 e^{-39.6} = 0,$$

from which it follows that $C_1 = 0$, and hence that $C_2 = 1.027 \times 10^{-2}$. Thus, $u(0) = 1.027 \times 10^{-2}$, so that under the loading the top of the pile is compressed in the x-direction by 1.027×10^{-2} inches.

Exercise 2.2 Suppose that in Example 2.2 it is impossible to neglect the reaction at the end $x = l$, due for example to the pile being in contact with solid rock. Assume that the reactive force at $x = l$ is proportional to the displacement with coefficient $\alpha_2 = 3 \times 10^7$ pounds/(inch)^2. Derive the boundary condition at $x = l$ and then compute $u(l)$ for this case.

Exercise 2.3 Assume that in (2.14a) $l = 1$ and $a(x)$, $c(x)$ and $f(x)$ are symmetric with respect to $x = \frac{1}{2}$, and that instead of (2.14b) and (2.14c) we have the boundary conditions

$$u(0) = u(1) = 0.$$

Show that the boundary value problem can be solved on the interval $[0, \frac{1}{2}]$ with boundary conditions

$$u(0) = 0$$

$$a(x)\frac{du}{dx}(x) \mid_{x=\frac{1}{2}} = 0.$$

What can you say about $u(x)$, $x > \frac{1}{2}$? (**Hint** Show that the solution is symmetric with respect to $x = \frac{1}{2}$.)

2.1.2 The principle of virtual work and weak formulation

In this section we demonstrate the connection between our classical problem (2.14) and the principle of *virtual work*. This can be achieved by multiplying (2.14a) by an arbitrary (differentiable) test function $v(x)$ and then integrating over I.

In order to interpret this formulation physically, as is done in mechanics, we take $v(x)$ to be a *virtual* displacement; i.e. it is the difference between any two *kinematically admissible* displacement fields. By a kinematically admissible displacement we mean one that satisfies all geometric (kinematic) conditions; i.e. it is sufficiently smooth to give meaning to the equation (2.14a) and it satisfies any *prescribed* geometric (i.e. Dirichlet) boundary conditions (e.g. (2.14b)).

For the case where we have the Dirichlet boundary condition $u(0) = g_1$ prescribed at $x = 0$, if $u(x)$ is kinematically admissible then $u(0) = g_1$ and $u(x) + v(0)$ is also kinematically admissible only if and only if $v(0) = 0$. Hence, the space of virtual displacements consists of all functions that are sufficiently smooth and that are zero at $x = 0$. (i.e. where the Dirichlet conditions are prescribed).

Note that the force-type boundary condition (2.14c) (known as Newton type, or Neumann type when $\alpha_1 = 0$) is not kinematic because it does not involve any geometric condition of displacement.

We return to (2.14), where first we assume that $g_1 = 0$; the general case $g_1 \neq 0$ will be addressed later.

Multiplying (2.14a) by any differentiable function $v(x)$ with $v(0) = 0$ (i.e. an admissible function) and integrating over I we get

$$-\int_0^l \left(\frac{d}{dx}\left(a(x)\frac{du}{dx}(x) \right) v(x) + \int_0^l c(x)\, u(x)\, v(x) \right) dx = \int_0^l f(x)\, v(x)\, dx,$$

which, after integration by parts, becomes

$$\int_0^l \left(a(x)\frac{du}{dx}(x)\frac{dv}{dx}(x) + c(x)\, u(x)\, v(x) \right) dx$$

$$- \left[a(x)\frac{du}{dx}(x)v(x) \right]_0^l = \int_0^l f(x)\, v(x)\, dx. \tag{2.15}$$

Recalling (2.14c) and using the fact that $v(0) = 0$ we may write (2.15) as

$$\int_0^l \left(a(x)\frac{du}{dx}(x)\frac{dv}{dx}(x) + c(x)\, u(x)\, v(x) \right) dx + \alpha_2\, u(l)\, v(l)$$

$$= \int_0^l f(x)\, v(x) dx + G_2\, v(l). \tag{2.16}$$

Further, if we define

$$B(u, v) \equiv \int_0^l \left(a(x)\frac{du}{dx}(x)\frac{dv}{dx}(x) + c(x)\, u(x)\, v(x) \right) dx + \alpha_2\, u(l)\, v(l), \tag{2.17}$$

$$F(v) \equiv \int_0^l f(x)\, v(x) dx + G_2\; v(l), \tag{2.18}$$

for u, v admissible functions for problem (2.14) with $g_1 = 0$ (i.e. $u(0) = v(0) = 0$), then $B(u, v)$ in (2.17) has the physical meaning of the *virtual* work of the internal stresses associated with u corresponding to the virtual displacement $v(x)$. Similarly, $F(v)$ in

(2.18) is the virtual work resulting from the external forces. Then, (2.16), which is written in the form

$$B(\mathrm{u}, v) = F(v), \tag{2.19}$$

is the well-known principle of virtual work. Clearly $B(u, v)$ is symmetric, i.e. $B(u, v) = B(v, u)$.

In summary, this says that the virtual work of the internal stresses of the exact solution is equal to the virtual work of the external forces corresponding to any kinematically admissible virtual displacement.

So far, our arguments for the virtual work principle have been physical. We now formulate the principle in mathematical terms under precise conditions. In particular, we assume that:

$$a(x) \text{ and } c(x) \text{ are defined for } x \in I \equiv (0, l), \tag{2.20a}$$

and that there exist constants $a_{\min}$, $a_{\max}$ and $c_{\max}$ such that

$$0 < a_{\min} \leqq a(x) \leqq a_{\max} < \infty, \quad x \in I, \tag{2.20b}$$

$$0 \leqq c(x) \leqq c_{\max} < \infty, \tag{2.20c}$$

$$0 \leqq \alpha_2 < \infty, \tag{2.20d}$$

$$f(x) \text{ is square integrable on } \bar{I}; \text{ i.e. } \int_0^l (f(x))^2 \, dx < \infty. \tag{2.20e}$$

We now define two *spaces*

$$\mathcal{U} \equiv \{v = v(x) \mid \|v\|_{\mathcal{U}} < \infty\}, \tag{2.21a}$$

and

$$\mathcal{U}_0 \equiv \{v \in \mathcal{U} \mid v(0) = 0\}, \tag{2.21b}$$

where the energy norm $\|u\|_{\mathcal{U}}$ on $\mathcal{U}$ or $\mathcal{U}_0$ is defined as

$$\begin{aligned} \|u\|_{\mathcal{U}} &\equiv \left[\int_0^l \left(a(x)\left(\frac{du}{dx}(x)\right)^2 + c(x)\,u(x)^2\right) dx + \alpha_2\, u(l)^2\right]^{\frac{1}{2}} \\ &= (B(u, u))^{\frac{1}{2}}. \end{aligned} \tag{2.21c}$$

As we shall see below, every $u \in \mathcal{U}$ is a continuous function on $[0, l]$ and hence $u(0)$ has meaning. We now show that $\mathcal{U}$ and $\mathcal{U}_0$ are linear spaces; i.e. if $u, v \in \mathcal{U}$ (respectively, $\mathcal{U}_0$) then

$$w = u + v \in \mathcal{U}, \text{ (respectively, } \mathcal{U}_0\text{)}, \tag{2.22a}$$

and

$$\alpha u \in \mathcal{U}, \text{ (respectively, } \mathcal{U}_0), \text{ for } \alpha \in \mathbb{R}. \tag{2.22b}$$

In order to show that (2.22) hold, we have to show that $\|u+v\|_{\mathcal{U}} < \infty$ and $\|\alpha v\| < \infty$. As $(u+v)^2 \leqq 2(u^2+v^2)$ and $(d(u+v)/dx)^2 \leqq 2(du/dx)^2 + 2(du/dx)^2$ we see that $w \in \mathcal{U}$. Additionally, $\|\alpha u\|_{\mathcal{U}} \leqq |\alpha| \; \|u\|_{\mathcal{U}}$ is obvious, so that (2.22) hold. The same argument leads to (2.22) for $u, v \in \mathcal{U}_0$; see Appendix A.1.1 for the definition of a linear space.

Because functions $v \in \mathcal{U}$ are continuous on $[0, l]$, we define the hyperplane (which is not a linear space)

$$\mathcal{U}_{g_1} \equiv \{v \in \mathcal{U} \mid v(0) = g_1\}. \tag{2.23}$$

Recalling the more general notation for the region Ω and its boundary Γ, which were defined previously, we could equally well define $\mathcal{U}_0$ and $\mathcal{U}_{g_1}$ as:

$$\mathcal{U}_0 \equiv \{v(x) \in \mathcal{U} \mid v(x) = 0, \; x \in \Gamma_D\}, \tag{2.24a}$$

$$\mathcal{U}_{g_1} \equiv \{v(x) \in \mathcal{U} \mid v(x) = g_1(x), \; x \in \Gamma_D\}. \tag{2.24b}$$

Note that if $g_1(x) = 0, \; x \in \Gamma_D$, then $\mathcal{U}_{g_1}$ becomes $\mathcal{U}_0$.

Remark 2.4 *The space* $\mathcal{U}$ *is called the* ***energy space,*** $\|\cdot\|_{\mathcal{U}}$ *is the* ***energy norm*** *and has dimension* $([F][L])^{\frac{1}{2}}$ *; it is called the energy norm because its physical meaning is related to the energy of the system. The energy space is a Hilbert space, see Appendix A.2.3.*

As we shall see below $\|\cdot\|_{\mathcal{U}}$ of (2.21c) has all the properties of a norm. The continuity of $v \in \mathcal{U}$, as used above, follows immediately from the fact that

$$\begin{aligned} |\, u(y) - u(x) | &= \Big| \int_x^y \frac{du}{ds}(s)ds \Big| \\ &\leqq |\, y - x\,|^{\frac{1}{2}} \left(\int_0^l \left(\frac{du}{ds}\right)^2 \right)^{\frac{1}{2}} \\ &\leqq \frac{1}{(a_{\min})^{\frac{1}{2}}} \|u\|_{\mathcal{U}} \; |\, y - x\,|^{\frac{1}{2}}. \end{aligned}$$

Exercise 2.4 Let $q \in \mathbb{R}$ with $q \neq 0$ and define

$$\mathcal{U}_q \equiv \{u \in \mathcal{U} \mid u(0) = q\}.$$

Show that $\mathcal{U}_q$ is linear space only if $q = 0$.

We show below that the expression $\|u\|_{\mathcal{U}}$ in (2.21c) has all the properties of a norm; the notion of the norm of a function $u(x)$ being that it is some measure of the size of u.

In our context, we need norms in order to consider specifically defined measures of functions; for example of problem solutions, of approximations to the solutions and, of course, of errors.

Any norm $\|.\|$ on a linear space X satisfies the following properties, see Appendix A.1.2. For any $u, v \in X$

$$\|u\| \geqq 0 \text{ and } = 0 \text{ if and only if } u = 0, \tag{2.25a}$$

$$\|\alpha u\| = |\alpha| \, \|u\| \text{ for any } \alpha, \tag{2.25b}$$

$$\|u + v\| \leqq \|u\| + \|v\| \text{; triangle inequality.} \tag{2.25c}$$

Exercise 2.5 Let $\mathcal{U} \equiv \{\beta \in \mathbb{R} \mid \|\beta\| = |\beta| < \infty\}$. Show that $\|\beta\|$ satisfies the properties (2.25).

Exercise 2.6 Assume in (2.14) that $a(x) = c(x) = 1 \; \forall x \in I$. Show that, if in (2.14c) $\alpha_2 < 0$ is sufficiently large in magnitude, then (2.25a) for $\|\cdot\| \equiv \|\cdot\|_{\mathcal{U}}$ is not satisfied. (**Hint:** Construct one function $u \in \mathcal{U}$ for which $B(u, u) < 0$.)

Before we show that $\|\cdot\|_{\mathcal{U}}$ has all the properties of a norm, i.e. (2.25), let us analyze the properties of the bilinear form $B(u, v)$ defined in (2.17). For all $u, v, w \in \mathcal{U}$ we have

$$B(u, v) = B(v, u), \tag{2.26a}$$

$$B(u + v, w) = B(u, w) + B(v, w), \tag{2.26b}$$

$$B(\alpha u, v) = \alpha \, B(u, v), \tag{2.26c}$$

$$B(u, u) = \|u\|_{\mathcal{U}}^2 \,, \tag{2.26d}$$

$$|B(u, v)| \leqq \|u\|_{\mathcal{U}} \cdot \|v\|_{\mathcal{U}} \,. \tag{2.26e}$$

Conditions (2.26 a–c) follow immediately from (2.17). Condition (2.26e) is the so-called Schwarz inequality, which is proved for the special case $B(u, v) \equiv \int_0^l u(x) \, v(x) \, dx$ in Appendix A.1.4.

In order to show (2.25a), from (2.26d) we require that $\|u\|_{\mathcal{U}} \equiv (B(u, u))^{\frac{1}{2}} \geqq 0$ and that $\|u\|_{\mathcal{U}} = 0$ if and only if $u = 0$. From the definition of $B(u, v)$ it follows that $\|u\|_{\mathcal{U}} \geqq 0$. If $u = 0$, then obviously $\|u\|_{\mathcal{U}} = 0$. If $\|u\|_{\mathcal{U}} = 0$, then from (2.17)

$$\int_0^l a(x) \left(\frac{du}{dx}(x) \right)^2 dx = 0, \; \int_0^l c(x) \, u(x)^2 dx = 0, \; \alpha_2 \, u(l)^2 = 0.$$

Because $a(x) > a_{\min} > 0$ we see that $du/dx = 0$ and hence that u is a constant. Hence, if $c(x) > 0$ or $u(0) = 0$ or $\alpha_2 > 0$ we see that $u(x) \equiv 0$.

Exercise 2.7 For $u, v \in U$ prove (2.26e), i.e. $|B(u, v)| \leqq \|u\|_{\mathcal{U}} \cdot \|v\|_{\mathcal{U}}$. (Hint: use the fact that for $\lambda \in R$ we have $B(u + \lambda v, u + \lambda v) \geqq 0$ and follow the proof of the Schwarz inequality; see Appendix A.1.4).

We now prove the triangle inequality (2.25c) in the context of $\|u\|_{\mathcal{U}}$ defined as $(B(u,u))^{\frac{1}{2}}$. Using (2.26e) we have that

$$\begin{aligned}\|u+v\|_{\mathcal{U}}^2 &\equiv |B\,(u+v,u+v)| = |B(u,u)+B(v,v)+2B(u,v)|\\ &\leqq \|u\|_{\mathcal{U}}^2+\|v\|_{\mathcal{U}}^2+2\,\|u\|_{\mathcal{U}}\,\|v\|_{\mathcal{U}}\\ &= \left(\|u\|_{\mathcal{U}}+\|v\|_{\mathcal{U}}\right)^2,\end{aligned}$$

as required. Hence, $\|u\|_{\mathcal{U}} = (B(u,u))^{\frac{1}{2}}$ satisfies (2.25) and so is a norm on $\mathcal{U}$.

Our model problem will be as defined in (2.14); i.e. with one Dirichlet and one Newton boundary condition. However, we could alternatively consider a problem with (2.14c) holding at $x = 0, l$, i.e. a Newton condition holding at each end. If we were to do this and to take $\alpha_2 = 0$ in (2.14c) we would have a Neumann problem.

If, in addition, $c(x) = \alpha_1 = \alpha_2 = 0$, giving a purely Neumann problem, $u \in \mathcal{U}$ (not $\mathcal{U}_0$) and $\|u\|_{\mathcal{U}} = 0$, then u is a constant. Hence, $\|u\|_{\mathcal{U}}$ will be a *semi-norm* (modulo a constant function); this means that the constant function has zero norm or that we do not distinguish between functions that differ by a constant. In this case, the virtual work formulation reads

$$B(\mathrm{u},v) \equiv \int_0^l a\frac{du}{dx}\frac{dv}{dx}dx = \int_0^l fvdx + G_1v(0) + G_2v(l) \equiv F(v).$$

Because $v_0(x) = 1$ belongs to $\mathcal{U}$, and $\|v_0\|_{\mathcal{U}} = 0$, we see from $B(\mathrm{u},v) = F(v)$ that a necessary condition for the existence of the solution u is that $F(v_0) = 0$, which is the consistency condition.

Remark 2.5 *If we were being completely rigorous we would understand all the integrals as being defined in the Lebesgue sense. However, this is beyond the scope of this book, and we simply treat the integrals as if they are those that are encountered in the calculus.*

Let us elaborate more on the properties of the energy space $\mathcal{U}_0$. We have

$$|u(x)| \leqq \frac{1}{(a_{\min})^{\frac{1}{2}}}x^{\frac{1}{2}}\,\|u\|_{\mathcal{U}}. \tag{2.27}$$

In fact,

$$\begin{aligned}|u(x)| &= \left|\int_0^x \frac{du}{ds}(s)ds\right| \leqq x^{\frac{1}{2}}\left(\int_0^l\left(\frac{du}{ds}\right)^2(s)ds\right)^{\frac{1}{2}}\\ &\leqq \frac{x^{\frac{1}{2}}}{(a_{\min})^{\frac{1}{2}}}\left(\int_0^l a\,(s)\left(\frac{du}{ds}\right)^2(s)\right)^{\frac{1}{2}}\\ &\leqq \frac{x^{\frac{1}{2}}}{(a_{\min})^{\frac{1}{2}}}\,\|u\|_{\mathcal{U}}.\end{aligned}$$

Now,

$$\int_0^l u^2(s)ds \leqq \frac{1}{a_{\min}} \int_0^l s\, ds \, \|u\|_{\mathcal{U}}^2$$
$$= \frac{l^2}{2a_{\min}} \|u\|_{\mathcal{U}}^2 . \tag{2.28}$$

Further, if $\int_0^l c(x)dx > 0$ or $\alpha_2 > 0$, then we have for $u \in \mathcal{U}$ (not just $\mathcal{U}_0$) that

$$|u(x)| \leqq \mathcal{C} \ \|u\|_{\mathcal{U}}$$
$$\left(\int_0^l u(x)^2 dx\right)^{\frac{1}{2}} \leqq \mathcal{C} \ \|u\|_{\mathcal{U}} , \tag{2.29}$$

where $\mathcal{C}$ depends on a, c, l and α_2.

The bilinear form $B(u, v)$ is *continuous* for $u, v \in \mathcal{U}$. In order to demonstrate this, we note first that with any pair (u, v), $u, v \in \mathcal{U}, B(u, v)$ associates a real number. For $u_1, u_2, v_1, v_2 \in \mathcal{U}$ we compute

$$B(u_1 + u_2, v_1 + v_2) = B(u_1 + u_2, v_1) + B(u_1 + u_2, v_2)$$
$$= B(u_1, v_1) + B(u_2, v_1) + B(u_1, v_2) + B(u_2, v_2),$$

from which

$$|B(u_1 + u_2, v_1 + v_2) - B(u_1, v_1) - B(u_2, v_2)|$$
$$\leqq |B(u_2, v_1) + B(u_1, v_2)| \leqq \|u_2\|_{\mathcal{U}} \ \|v_1\|_{\mathcal{U}} + \|u_1\|_{\mathcal{U}} \ \|v_2\|_{\mathcal{U}} ,$$

using all the properties (2.26), so that slight changes in u_1, v_1 cause the value of $B(u, v)$ to change slightly. This means that the form $B(u, v)$ is continuous in u and v. We have seen that essential to the continuity is property (2.26e), which we designate to be the *continuity* of the bilinear form.

Thinking back to (2.19), just as we defined the bilinear form $B(u, v)$, we may also define the *linear functional* $F(u)$ for $u \in \mathcal{U}$ that has analogous properties to B for fixed $v \in \mathcal{U}$; see Appendix A.3.1.

Let $F(u)$ associate with every $u \in \mathcal{U}_0$ or $\mathcal{U}$ a real number $F(u) \in \mathbb{R}$ and F have the properties of a linear functional

$$F(u_1 + u_2) = F(u_1) + F(u_2),$$
$$F(\alpha u,) = \alpha F(u), \tag{2.30}$$
$$|F(u)| \leqq \mathcal{C} \ \|u\|_{\mathcal{U}} ,$$

where $\mathcal{C}$ is a constant depending on the data of the problem but not on u. If on $\mathcal{U}$ we had only the semi-norm, then we would need that $F(u) = 0$ if $u = 1$ and then condition (2.30) remains valid.

Example 2.3 Let $\int_0^l f^2(x)dx < \infty$. Then, $F(u) = \int_0^l fudx$ is a *linear functional on* $\mathcal{U}_0$. To show this we have only to check that

$$\left|\int_0^l fudx\right| \leqq \left(\int_0^l f^2(x)dx\right)^{\frac{1}{2}} \left(\int_0^l |u|^2\, dx\right)^{\frac{1}{2}} \leqq \left(\int_0^l f^2(x)dx\right)^{\frac{1}{2}} \|u\|_{\mathcal{U}};$$

this follows from (2,28), (2.29) and the Schwarz inequality.

The formulation for (2.14), with $g_1 = 0$,

$$\text{find } \mathrm{u} \in \mathcal{U}_0 \text{ such that}$$

$$B(\mathrm{u}, v) = F(v) \qquad \forall v \in \mathcal{U}_0 \tag{2.31}$$

is called the weak formulation of problem (2.14) and is based on the principle of virtual work and u is the *weak* solution of (2.14).

If $B(u,u) = \|u\|_{\mathcal{U}}^2$ and $F(v) \leqq C\, \|v\|_{\mathcal{U}}$, then by the Lax–Milgram lemma, see Appendix A.3.3, there exists a unique u such that $B(\mathrm{u}, v) = F(v)$, $\forall v \in \mathcal{U}$. In (2.31) we have $\mathcal{U}_0$ and $\|\cdot\|_{\mathcal{U}_0}$ is a norm.

As has been said, if we have a problem with Neumann boundary conditions at both $x = 0$ and $x = l$, and $c(x) = 0$ then $\|\cdot\|_{\mathcal{U}}$ is only a semi-norm. Nevertheless, if the consistency condition is satisfied we still have $F(v) \leqq C\, \|v\|_{\mathcal{U}}$, so that the Lax–Milgram lemma is again applicable. Of course, in this case the solution is unique only up to a constant for which the norm is zero. In this case, we would only seek to determine a quantity of interest, e.g. stress, which did not depend on this abitrary constant. If $c(x) \neq 0$, or $\alpha_1, \alpha_2 \neq 0$ then $\|\cdot\|_{\mathcal{U}}$ is again a norm.

As was mentioned earlier, the case where $g_1 \neq 0$ in (2.14) must of course be considered. This will be done shortly.

We now show that the solution of (2.31) is unique; (i.e. if u_1 and $u_2 \in \mathcal{U}$ solve (2.31), then $u_1 = u_2$). In order to do this we assume on the contrary that the problem has two solutions, denoted by u_1 and u_2. Then $u_1 - u_2 \equiv z \in \mathcal{U}_0$, and for any $v \in \mathcal{U}_0$ we have that $B(z, v) = 0$. Hence, particularly taking $v = z$ we get that

$$B(z, z) = \|z\|_{\mathcal{U}} = 0,$$

and, as a result of the fact that $z \in \mathcal{U}_0$ together with the properties of norms, this gives $z = 0$ as required.

Remark 2.6 *From the uniqueness of the weak solution we see immediately that the classical solution is also the weak solution, but not vice versa because the weak formulation admits a solution for which not all the symbols are defined in the strong sense; e.g. the functions of interest may not be differentiable; for example $a(x)$ in (2.17) can be discontinuous and satisfy (2.20).*

Remark 2.7 *When addressing the uniqueness of the weak solution, we have specified clearly the class of functions for which the uniqueness should hold, in our case functions in $\mathcal{U}_0$.*

We have so far in this section considered the case where in (2.14) the Dirichlet boundary condition (2.14b), has $g_1 = 0$. This led to the virtual work formulation (2.31). We shall now consider the case where $g_1 \neq 0$ in (2.14b). This can be done in the following manner. Let $\bar{u} \in \mathcal{U}$ be such that $\bar{u}(0) = g_1$. We then write $\mathrm{u} = \bar{u} + u_0$ where, from (2.31), $u_0 \in \mathcal{U}_0$ satisfies

$$B(u_0, v) = F(v) - B(\bar{u}, v) \equiv \bar{F}(v), \forall v \in \mathcal{U}_0. \tag{2.32}$$

Note that $\bar{u}$ has to be in $\mathcal{U}$ so that $B(\bar{u}, v)$ is defined and $|B(\bar{u}, v)| \leqq C \|v\|_{\mathcal{U}}$ $\forall v \in \mathcal{U}_0$. Now, $\bar{F}(v)$ satisfies (2.30) and hence u_0 exists and is unique. Thus, instead of solving (2.31), we solve (2.32). Note that $\bar{u}$ is an arbitrary function, so that u_0 depends on $\bar{u}$, whilst $\mathrm{u} = \bar{u} + u_0$ does not.

We remark that for $g_1 \neq 0$ we can define the virtual work statement for problem (2.14) using the notation of (2.22b) as follows:

find $\mathrm{u} \in \mathcal{U}_{g_1}$, such that

$$B(\mathrm{u}, v) = F(v) \qquad \forall v \in \mathcal{U}_0. \tag{2.33}$$

It is easy to show that $\mathrm{u} = \bar{u} + u_0$, with $u_0 \in \mathcal{U}_0$ satisfying (2.32), satisfies (2.33); and if u satisfies (2.33) then $\mathrm{u} - \bar{u} = u_0$ satisfies (2.32).

Remark 2.8 *We have of course said nothing as yet about how we should choose $\bar{u}$. Recalling our aim of obtaining numerical solutions to problems of the type (2.14), the selection of $\bar{u}$ will be done in a way that is suited to the numerical calculations.*

Remark 2.9 *In order to obtain (2.32) one has to be able to produce $\bar{u}$ such that $B(\bar{u}, v)$ has meaning. Whilst the task of finding such a function is straightforward in the one-dimensional case, it is much less so for problems in higher space dimensions!*

We have seen that the boundary condition (2.14b) is explicitly used as a constraint in (2.33) (find $\mathrm{u} \in \mathcal{U}_{g_1}$) and hence it is called an *essential boundary condition.* On the other hand, boundary condition (2.14c) has not been explicitly imposed in the weak formulation, and hence is called a *natural boundary condition.* We thus have to distinguish between these two types of conditions.

We previously defined the energy space $\mathcal{U}$ and the energy norm $\|\cdot\|_{\mathcal{U}}$. On $I \equiv (0, l)$ we now define additional linear spaces that we shall need later. Let

$$C(I) \equiv C^0(I) \equiv \{u(x), x \in I \mid u \text{ is continuous on } \bar{I}\} = [0, l], \tag{2.34a}$$

and

$$\|u\|_{C(I)} = \max_{x \in \bar{I}} |u(x)|, \tag{2.34b}$$

$$C^k(I) = \left\{ u \in C(I) \mid \frac{d^j u}{dx^j}(x) \in C(I), j = 0, 1, \ldots, k \right\}, \tag{2.34c}$$

and

$$\|u\|_{C^k(I)} = \max_{j=0,\ldots,k} \left\| \frac{d^j u}{dx^j} \right\|_{C(I)}. \tag{2.34d}$$

We also define the semi-norm

$$|u|_{C^k(I)} = \left\| \frac{d^k u}{dx^k}(x) \right\|_{C(I)} . \tag{2.35}$$

The spaces $C^k(I)$ are linear spaces and $\|\cdot\|_{C^k(I)}$ has all the properties of a norm.

Finally, we define the spaces and norms

$$H^0(I) = L^2(I) \equiv \left\{ u(x), x \in I \mid \int_0^l u^2(x) dx < \infty \right\}, \tag{2.36a}$$

and

$$\|u\|_{L^2(I)} \equiv \left(\int_0^l u^2(x) dx \right)^{\frac{1}{2}}, \tag{2.36b}$$

$$H^k(I) \equiv \left\{ u \in L^2(I) \mid \int_0^l \left(\frac{d^j u}{dx^j}(x) \right)^2 dx, < \infty, j = 0, \ldots, k \right\}, \tag{2.36c}$$

and

$$\|u\|_{H^k(I)} \equiv \left(\sum_{j=0}^{k} \left\| \frac{d^j u}{dx^j} \right\|_{L^2(I)}^2 \right)^{\frac{1}{2}}, \tag{2.36d}$$

$$|u|_{H^k(I)} \equiv \left(\int_0^l \left(\frac{d^k u}{dx^k} \right)^2 dx \right)^{\frac{1}{2}}, \tag{2.36e}$$

where (2.36e) is the semi-norm modulo a polynomial of degree $k-1$. We note that all the definitions of (2.34)–(2.36) can be used on any subdomain $\tau \subset I$.

The spaces $H^k(I)$ are complete linear (Hilbert) spaces and $\|\cdot\|_{H^k(I)}$ has all the properties of a norm, see Appendix A.1.2. We should, for correctness, consider the derivatives in the generalized sense and the integrals in the Lebesgue sense, see Remark 2.5, but for our purposes, as we shall deal only with functions that are at worst piecewise continuous over I and for which the classical derivatives exist (and are equal to the generalized derivatives), it will be sufficient to use the definitions of derivatives and integrals from the calculus.

In subsequent chapters we shall need a number of relations between the various spaces. We list these now.

For any $u \in \mathcal{U}_0$ there exist constants $C_1$and C_2, independent of u such that

$$C_1 \|u\|_{\mathcal{U}} \leqq \|u\|_{H^1(I)} \leqq C_2 \|u\|_{\mathcal{U}} . \tag{2.37}$$

In this situation, we say that the spaces $\mathcal{U}_0$ and $H_0^1(I)$ are equivalent, where $H_0^1(I) = \{u \in H^1(I) \mid u(0) = 0\}$.

If $u \in H^1(I)$, then $u \in C^0(I)$ and

$$\|u\|_{C^0(I)} \leqq C \|u\|_{H^1(I)} , \tag{2.38}$$

and

$$\|u\|_{L^2(I)} < \mathcal{C}\,|u|^2_{H^1(I)} \;\forall u \in H^1_0(I). \tag{2.39}$$

We shall normally assume that the solution of (2.33) $\mathrm{u} \in \mathcal{C}^0(I)$ and that $\mathrm{u}\,|_{\tau_k} \in C^j(\tau_k), k = 1, \ldots, s,\ j \geqq 1$, so that $\mathrm{u} \in \mathcal{U}$, where the domain I can be divided into non-overlapping subintervals $\tau_k, k = 1, \ldots, s$ such that $\bar{I} \equiv \cup\bar{\tau}_k$ and $\mathrm{u} \in C^j(\tau_k)$; this condition is called the regularity of u on τ_k.

The regularity of $\mathrm{u}(x)$ clearly depends on the regularity of the functions $a(x), c(x), f(x)$ that occur in (2.14a). Specifically, under the assumptions of (2.20), for $k = 1, \ldots, s$ for u satisfying (2.31).

$$\text{if } a(x) \in C^1(\tau_k), c(x) \in C^0(\tau_k), f(x) \in C^0(\tau_k), \text{ then } \mathrm{u}(x) \in C^2(\tau_k), \tag{2.40a}$$

$$\text{if } a(x) \in C^2(\tau_k), c(x) \in C^1(\tau_k), f(x) \in C^1(\tau_k), \text{ then } \mathrm{u}(x) \in C^3(\tau_k). \tag{2.40b}$$

If these local regularity conditions on $a(x), c(x), f(x)$ hold globally over the whole I, then the corresponding regularity conditions as in (2.40 a and b) hold over the whole of I. Namely: for $k = 2, 3$

$$\text{if } a(x) \in C^{k-1}(I), c(x) \in C^{k-2}(I), f(x) \in C^{k-2}(I), \text{ then } \mathrm{u}(x) \in C^k(I). \tag{2.41}$$

Example 2.4 Let us check these conditions for (2.14a), where $c(x) = 0$ and $g_1 = 0$. Thus, $\mathcal{L}(\mathrm{u}) \equiv -\frac{d}{dx}\left(a(x)\frac{d\mathrm{u}}{dx}(x)\right)$ and we consider the equation

$$\mathcal{L}(\mathrm{u}) \equiv -\frac{d}{dx}\left(a(x)\frac{d\mathrm{u}}{dx}(x)\right) = f(x), x \in I.$$

Hence, we have

$$a(x)\frac{d\mathrm{u}}{dx}(x) = \int_0^x f(t)dt + \mathcal{C},$$

giving

$$\frac{d\mathrm{u}}{dx} = \frac{1}{a(x)}\int_0^x f(t)dt + \frac{\mathcal{C}}{a(x)},$$

from which (2.40 a and b) follow.

2.1.3 The principle of minimization of energy

Thinking again of problem (2.33), the expression $\frac{1}{2}B(u, u)$ has the physical meaning of the *strain energy* of the bar, whilst

$$\Pi(u) \equiv \frac{1}{2}\,B(u, u) - F(u) \tag{2.42}$$

has the meaning of *total energy* with dimension $[F][L]$; i.e. the energy due to both the internal forces and the prescribed extended loads on the bar.

We now show that the solution of problem (2.33) minimizes the total energy among all functions $u \in \bar{u} + \mathcal{U}_0$, where again $\bar{u} \in \mathcal{U}$ is such that $\bar{u}(0) = g_1$. Hence, for u the solution of (2.33) we need to show that

$$\Pi(u) \geqq \Pi(\mathrm{u}),$$

where $u = \mathrm{u} + v$ with $v \in \mathcal{U}_0$ the virtual displacement. Now,

$$\begin{aligned}\Pi(\mathrm{u}+v) &= \frac{1}{2}B(\mathrm{u}+v,\mathrm{u}+v) - F(\mathrm{u}+v) \\ &= \frac{1}{2}B(\mathrm{u},\mathrm{u}) + B(\mathrm{u},v) + \frac{1}{2}B(v,v) - F(\mathrm{u}) - F(v),\end{aligned}$$

where we have used the symmetry of $B(\cdot,\cdot)$. Thus,

$$\begin{aligned}\Pi(\mathrm{u}+v) &= \Pi(\mathrm{u}) + (B(\mathrm{u},v) - F(v)) + \frac{1}{2}B(v,v) \\ &= \Pi(\mathrm{u}) + \frac{1}{2}B(v,v) \geqq \Pi(\mathrm{u}),\end{aligned}$$

where we have used the fact that u satisfies (2.33) and that $B(v,v) \geqq 0$ for any $v \in \mathcal{U}_0$. Note that this result can be interpreted as u minimising $\Pi(u)$ over $\mathcal{U}_{g_1}$.

Alternatively, we may assume that u minimizes (2.42) over all kinematically admissible functions, (i.e. over $\bar{u} + \mathcal{U}_0$). This means that if $\mathrm{u} \in \mathcal{U}$ exists and

$$\Pi(\mathrm{u}+v) \geqq \Pi(\mathrm{u}) \qquad \forall v \in \mathcal{U}_0,$$

then

$$B(\mathrm{u},v) = F(v)\ \forall v \in \mathcal{U}_0.$$

This can be written in the form

$$\begin{aligned}&\frac{1}{2}B(\mathrm{u},\mathrm{u}) + B(\mathrm{u},v) + \frac{1}{2}B(v,v) - F(\mathrm{u}) - F(v) - \frac{1}{2}B(\mathrm{u},\mathrm{u}) + F(\mathrm{u}) \\ &= \frac{1}{2}B(v,v) + B(\mathrm{u},v) - F(v) \geqq 0. \end{aligned} \tag{2.43}$$

From this, it follows that

$$B(\mathrm{u},v) - F(v) = 0, \qquad \forall v \in \mathcal{U}_0. \tag{2.44}$$

To show this, assume that the above equation does not hold. Then, there exists a function $\bar{v} \in \mathcal{U}_0, \bar{v} \neq 0$, such that

$$B(\mathrm{u},\bar{v}) - F(\bar{v}) = a \neq 0. \tag{2.45}$$

Taking $v = \lambda\bar{v}$, $\lambda =$ constant, we have that

$$B(\mathrm{u},\lambda\bar{v}) - F(\lambda\bar{v}) = \lambda B(\mathrm{u},\bar{v}) - \lambda F(\bar{v}) = \lambda a.$$

Obviously, we can select the sign of λ such that $\lambda a < 0$. Because

$$\frac{1}{2}B(v,v) = \frac{1}{2}\lambda^2 B(\bar{v},\bar{v}) = \frac{1}{2}\lambda^2 \|\bar{v}\|^2_{\mathcal{U}},$$

we have that

$$\frac{1}{2}B(v,v) + B(\mathrm{u},v) - F(v) = \frac{1}{2}\lambda^2 \|\bar{v}\|_{\mathcal{U}}^2 + \lambda a,$$

which is negative for sufficiently small λ and thus contradicts (2.43). Thus, (2.45) is an incorrect assumption and it follows that (2.44) holds.

Remark 2.10 *As stated previously, in the weak formulation (2.19) it is possible to prove the existence of* u *using the Lax–Milgram Lemma. In the minimizing of the energy we* ***have assumed*** *the existence of the minimizer and have shown that it is the weak solution. The existence of the minimizer, although plausible, is not obvious. Riemann in the ninteenth century used the existence assumption as a fact and with this proved the existence of the solution of the weak problem. Later, Weierstrass pointed out that this was in error; it took about 40 years up to the beginning of the twentieth century, and work by Hilbert, to overcome this difficulty. Essentially, Hilbert proved the existence of the weak solution as the minimizer of the energy.*

Example 2.5 Consider problem (2.14) of the displacement of the rod, where we take $g_1 = 0$. This problem has the interpretation that there are springs of rigidity $c(x)$ attached to the rod on its upper and lower surfaces (representing the reaction of the surrounding medium) and of rigidity α_2 attached to the end $x = l$ (representing an elastic support at $x = l$). In this case, the internal energy of the bar is

$$\frac{1}{2}\int_0^l a(x)\left(\frac{d\mathrm{u}}{dx}(x)\right)^2,$$

the energy of the surface springs is $\frac{1}{2}\int_0^l c(x)\mathrm{u}(x)^2 dx$ and the energy of the end spring is $\frac{1}{2}\alpha_2 \mathrm{u}(l)^2$. Hence, the energy of the system of the rod and springs is $\frac{1}{2}B(\mathrm{u},\mathrm{u})$. The energy of the external forces is $-\int_0^l f(x)\mathrm{u}(x)dx - G(l)\mathrm{u}(l) \equiv -F(\mathrm{u})$. Hence, the total energy is

$$\Pi(\mathrm{u}) = \frac{1}{2}B(\mathrm{u},\mathrm{u}) - F(\mathrm{u}).$$

Exercise 2.8 Consider the special case of problem of the type (2.14) in which $\mathrm{u}(x)$ satisfies:

$$-3\frac{d^2\mathrm{u}}{dx^2}(x) = 2,\ 0 < x < 2,$$
$$\mathrm{u}(0) = \mathrm{u}(2) = 0.$$

Find the exact solution $\mathrm{u}(x)$ of the problem, evaluate the strain energy of the bar and of the external force $f(x)$.

Exercise 2.9 Consider the problem in which $u(x)$ satisfies:

$$-3\frac{d^2u}{dx^2}(x) = 2,\ 0 < x < 2,$$

$$u(0) = 0,\ 3\frac{du}{dx}(x)\mid_{x=2} = 0.$$

Evaluate the total energy $\frac{1}{2}B(u, u)$ for this case and compare this with that of Exercise 2.8.

Exercise 2.10 Show that for Exercises 2.8 and 2.9

$$\Pi(u) = -\frac{1}{2}B(u, u).$$

2.1.4 One-dimensional heat transfer

In Sections 2.1.1 and 2.1.2 we set up the classical and weak forms of the mathematical model of the deformation of a loaded bar. Such formulations can of course be used to describe other physical phenomena, where the concepts of energy and of virtual work have no direct meaning. One such occurrence is that of *steady-state heat conduction* in the same bar as previously but with the interpretation of Fig. 2.1 now as follows.

In that heat-conduction context $u(x)$ in (2.14a) is now the *temperature* in the bar, and this is assumed to be constant on any cross-section, analogously to the displacement in the elasticity context; $u(x)$ has dimension $[T]$. Correspondingly, $F(x)$ is now the *total flux,* which is the amount of heat/unit time going through the cross-section of the bar at x; the dimension of $F(x)$ is $[C]/[t]$, where $[C]$ is the dimension of heat. The flux/unit area of $\sigma(x) = F(x)/A(x)$ and has dimension $[C]/[t][L^2]$. In the heat context $E(x)$ is the *coefficient of thermal conductivity,* which determines the heat conducted by the material of the bar and has dimension $[C]/[L][t][T]$, where $[T]$ is the dimension of temperature. The analogue of Hooke's law in heat flow is the Fourier law that relates $\sigma(x)$ with the *temperature gradient* $\varepsilon(x) \equiv du(x)/dx$ at x. Fourier's law has the form $\sigma(x) = E(x)\varepsilon(x)$, where the dimension of $\varepsilon(x)$ is $[T]/[L]$. Also, in (2.14a) we have $c(x)$, which is the coefficient of *heat transfer* to the surrounding medium, which is again heat/unit length/unit time/degree and that also has dimension $[C]/[L][t][T]$. The function $f(x)$ is now an *internal heat source* at x generating heat/unit length/unit time and having units $[C]/[L][t]$. Finally, as before, we define $a(x) \equiv A(x)\,E(x)$. Equation (2.14a) is the equation of heat balance in the slice $S(x, \Delta)$ of Fig. 2.1, and is generated exactly as the equilibrium equation in elasticity, but now with each quantity having meaning as above. Finally, we note that $B(u, u) = \|u\|^2_{\mathcal{U}}$ has dimension $[C][T]/[t][L]$.

We must now analogously consider the boundary conditions (2.14b) and (2.14c).

a) Dirichlet boundary conditions

Again, the simplest boundary condition is that where the temperature is described at a boundary point. Thus, if we prescribe a known boundary temperature g_1 at, see

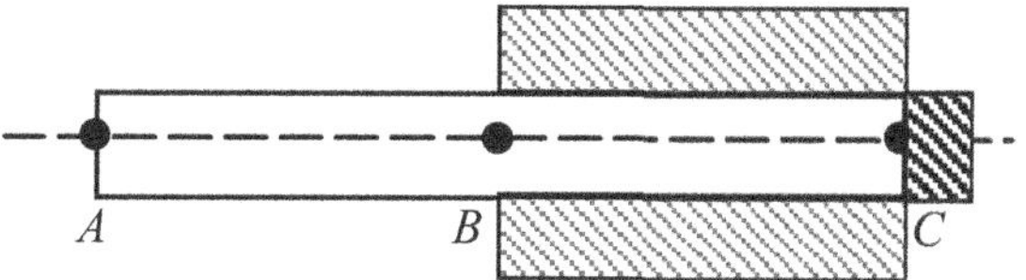

Fig. 2.6 Part insulated bar with heat source at C.

Fig. 2.6, the point A where $x = 0$, we again have as in (2.14b) the Dirichlet boundary condition

$$u(0) = g_1. \tag{2.46}$$

b) Newton boundary conditions

We refer again to Fig. 2.6. If we apply a heat source G_2 to the bar at the point C where $x = l$, and have the bar in contact at $x = l$ with a conductor with coefficient of conductivity α_2 with dimension $[C]/[t][T]$, then we shall again have at $x = l$ the boundary condition

$$(a(x)\frac{du}{dx}(x)) \mid_{x=l} + \alpha_2 \, u(l) = G_2, \tag{2.47}$$

which now has dimension $[C]/[t]$.

Exercise 2.11 Consider the problem of heat transfer in a rod of square cross-section $A(x)$, see Fig. 2.6, in which the part BC of the lateral surface is insulated so that no heat flow takes place across it. There is a heat source of 5000 calories/s at C, the temperature at A is kept at $u_A = 5°\text{C}$ and the environment surrounding the bar has temperature 20°C. The coefficient of thermal conductivity of the bar is 300 calories/cm.s.°C, and the flux/unit area through the surface is 50 calories/cm^2.s. Formulate the differential equation for the temperature, together with the boundary conditions. **Hint:** the heat flux through the surface is $c(u - u_0)$ where u_0 is the temperature of the environment.

2.1.5 Engineering application, one-dimensional heat-transfer problem

In the spirit of our contention that our purpose is to use the finite element method to solve genuine engineering problems reliably and effectively, we complete this section on one-dimensional boundary value problems by defining a *one-dimensional problem from heat transfer*, with which we shall illustrate the effectiveness of the finite element method at appropriate places in subsequent chapters; a similar approach will be followed for two dimensions.

Consider an infinite pipe consisting of two materials with constant properties and of circular cross-section as in Fig. 2.7, through which fluid at temperature T_0 flows and for which the temperature of the surrounding environment is T_1.

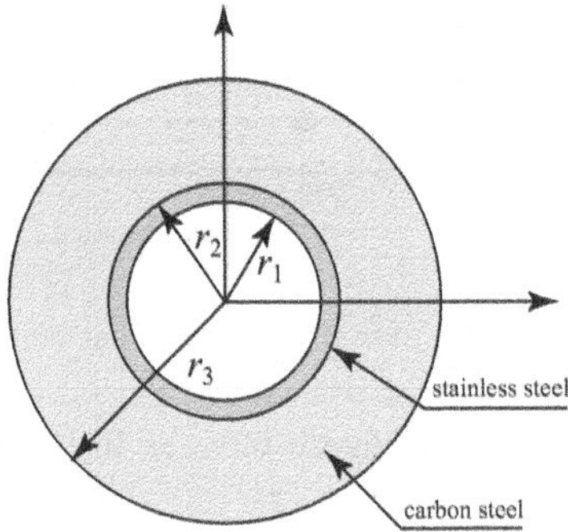

Fig. 2.7 Model problem.

Because the pipe is long the heat flux in the longitudinal direction down the pipe will be neglected and the heat-transfer problem may be formulated in any cross-section of the pipe. Further, since the two materials have constant properties and the pipe is axisymmetric, the temperature $\mathfrak{u}$ in the pipe depends only on r and the problem becomes one dimensional with solution $\mathfrak{u}(r)$. The differential equation for the temperature can be derived exactly as in Section 2.1.1, when the cross-section $A = 2\pi r\Delta$, where $\Delta = 1$ is unit length of the pipe, with r replacing x.

Consider the cross-section of the pipe as in Fig. 2.7, where the pipe consists of two concentric rings, the internal of which is made of stainless steel with thermal conductivity E_1, whilst the outer ring is made of carbon steel with thermal conductivity E_2. The radii of the internal surface of the pipe, of the interface between the two materials, and of the outer pipe surface, are, respectively, r_1, r_2, r_3. We assume that the heat transfer from the outer pipe surface is convective, that the coefficient of convective transfer to the medium surrounding the pipe of temperature T_1 is β, and that the temperature of the fluid in the pipe is T_0.

In this chapter we have adopted the practice of mathematics and have formulated our problems in terms of dimensionless quantities. However, for a practical problem the data have physical meanings and the values of these quantities are usually obtained from material handbooks. In our engineering examples we will work with these physical material quantities.

Returning to the heat transfer from the outer pipe surface, Newton's law describes the convective heat transfer to the environment in terms of the *coefficient of convective heat transfer* β; (which has dimension $[C]/[t][L]^2[T]$). The α_2 of (2.14c) is $A(r_3)\beta$, where $A(r) = 2\pi r\Delta$. Similarly, if γ is the *coefficient of conductivity* of the material, (which in the literature is usually denoted by c) then the 'mathematical' quantity $a \equiv AE$, as in (2.11) and (2.12) is in fact $a = A\gamma$, as $E = \gamma$.

Thus, with the Dirichlet boundary condition $\mathfrak{u} = T_0$ on the inner surface at $r = r_1$, and the Newton boundary condition arising from the convection $\frac{\partial \mathfrak{u}}{\partial n_c} + \alpha_2(\mathfrak{u} - T_1) = 0$ on the outer surface at $r = r_3$, the problem is axisymmetric so that $\mathfrak{u} = \mathfrak{u}(r)$ and, see (2.14), the temperature $\mathfrak{u}(r)$ satisfies the equation,

$$\frac{d}{dr}\left(a(r)\frac{d\mathfrak{u}(r)}{dr}\right) \equiv \frac{d}{dr}\left(A(r)\gamma\frac{d\mathfrak{u}(r)}{dr}\right) = 0, \quad r_1 < r < r_3, \tag{2.48a}$$

together with the boundary conditions

$$\mathfrak{u}(r) = T_0, \quad r = r_1, \tag{2.48b}$$

$$\frac{d\mathfrak{u}(r)}{dn_c} + A(r)\beta(\mathfrak{u}(r) - T_1) = 0, \quad r = r_3. \tag{2.48c}$$

This problem is of type (2.14), when written in terms of the Cartesian coordinate $x, a \leqq x \leqq c$ (where we take $a = r_1, b = r_2, c = r_3$) and our one-dimensional engineering problem is: find $\mathfrak{u} = \mathfrak{u}(x)$ that satisfies

$$-\frac{d}{dx}\left(a(x)\frac{d\mathfrak{u}}{dx}(x)\right) = 0, \quad a < x < c, \tag{2.49a}$$

$$u(x) = T_0, \quad x = a, \tag{2.49b}$$

$$\frac{d\mathfrak{u}}{dn_c}(x) + \alpha_2(\mathfrak{u}(x) - T_1) = 0, \quad x = c. \tag{2.49c}$$

In order to obtain the weak form of problem (2.49) we multiply (2.49a) by a test function $v \in \mathcal{U}_0$ and integrate over [a,c], so that we seek $u \in \mathcal{U}_{g_1}$ such that

$$\begin{aligned} B(u,v) &= \int_a^c a(x)\frac{du}{dx}(x)\frac{dv}{dx}(x)dx + \alpha_2\mathfrak{u}(c)v(c) \\ &= \alpha_2 T_1 v(c) \equiv F(v), \quad \forall v \in \mathcal{U}_0, \end{aligned} \tag{2.50}$$

where

$$\text{for } a \leqq x \leqq b,\ a(x) = x\gamma_{ss} \quad \text{(stainless steel)}$$

$$\text{for } b < x \leqq c,\ c(x) = x\gamma_{cs} \quad \text{(carbon steel)},$$

and, following (2.24),

$$\mathcal{U}_0 \equiv \{v \in \mathcal{U} \mid v(a) = 0\}, \tag{2.51a}$$

$$\mathcal{U}_{g_1} \equiv \{v \in \mathcal{U} \mid v(a) = g_1\}. \tag{2.51b}$$

We now consider specific examples of problem (2.49); which we denote **1D Eng Problem 1** and **1D Eng Problem 2**. For each of these we take

$$T_0 = 500°\text{C}, \quad T_1 = 20°\text{C}, \tag{2.52}$$

the thermal conductivity of stainless steel as

$$\gamma_{\text{ss}} = 0.90 \text{ cal /s cm °C}, \tag{2.53a}$$

the thermal conductivity of carbon steel as

$$\gamma_{\text{cs}} = 2.10 \text{ cal /s cm °C}, \tag{2.53b}$$

the coefficient of convective heat transfer as

$$\beta = 0.55 \text{ cal }/\text{s cm}^2\ {}^\circ\text{C}, \tag{2.53c}$$

and

$$a \equiv r_1 = 3 \text{ cm}, \ c \equiv r_3 = 6.5 \text{ cm}, \tag{2.53d}$$

which are both rational numbers.

For 1D Eng Problem 1 we have

$$b = r_2 = 3.5 \text{ cm}, \tag{2.54}$$

again rational, whilst for 1D Eng Problem 2

$$b = r_2 = r_1 + (r_3 - r_1)\ /\ 2\pi, \tag{2.55}$$

is an irrational number.

Note that the $a(x)$ in (2.50) is discontinuous but that is satisfies (2.20); see also Remark 2.6 so that the weak solution exists, whilst the strong solution does not.

Because problem (2.49) is one dimensional, it is straightforward to compute the solutions of 1D Eng Problems 1 and 2 analytically.

We wish for both 1D Eng Problem 1 and 1D Eng Problem 2 to determine three quantities of interest, (QoI), $Q_i, i \equiv 1, 2, 3$, where

$Q_1([T])$ is the temperature $\mathfrak{u}$ at the outer surface of the pipe,
$Q_2 \equiv Q_2([C]/[t])$ is the amount of heat per unit length $(\Delta = 1)$ transferred to the environment surrounding the pipe,
$Q_3 \equiv Q_3([C][T]/[t])$ is the total energy $\Pi(\mathfrak{u})$ per unit length of the pipe as defined in (2.42).

We see that

$$Q_1 = \mathfrak{u}(c)$$

$$Q_2 = \left|\frac{d\mathfrak{u}}{dn_c}(c)\right|$$

$$Q_3 = \tfrac{1}{2}\left(\int_{r_1}^{r_3} a(x)\left(\frac{d\mathfrak{u}}{dx}\right)^2 + \alpha_2 \mathfrak{u}^2(c)\right) - \alpha_2 T_1 \mathfrak{u}(c).$$

From the analytical solutions it is found that the exact values of the quantities of interest Q_i are:

for 1D Eng Problem 1

$Q_1 = 200.03414, \quad Q_2 = 4043.99611, \quad Q_3 = 153\ 754.27310,$

for 1D Eng Problem 2

$Q_1 = 197.58989, \quad Q_2 = 3939.09362, \quad Q_3 = 151\ 657.12854.$

We note that when we compute approximations to problem (2.49), the computations will have a specific purpose. The computed results have to describe reliably those features of the problem upon which engineering decisions rely. Thus, the goals of the computations are very specific, rather than abstract and general.

Problem (2.49) as formulated essentially has three parts: the structure, i.e. the differential equation and the boundary conditions; the input data $T_0, T_1, \gamma_{ss}, \gamma_{cs}, \beta$; the output data Q_1,Q_2,Q_3. In terms of these three components we need to consider whether the formulation will lead to information that is *reliable* in the engineering sense. This we now do.

It has been assumed that there is no heat flow in the longitudinal direction down the pipe. This is of course only approximately true as obviously the fluid cools as it travels down the pipe. Nevertheless, it is reasonable to assume that this change of temperature does not significantly affect the goal of the computation.

For the structure of the problem the assumption of axisymmetry is very accurate in the context of the steel and the temperature of the fluid in the pipe, but is less accurate with respect to the convective heat transfer from the outer surface of the pipe as the surrounding environment is obviously not homogeneous. It must also be realized that the coefficients of thermal conductivity of the steels are obtained from experiments and depend on temperature.

All the data inputs to problem (2.49) introduce uncertainty into the model, and will affect the reliability of the solution. In order to illustrate this let us assume that the values of $\beta, \gamma_{ss}, \gamma_{cs}$ can vary within the following ranges

$$0.4 \leq \beta \leq 0.55, \quad 0.67 \leq \gamma_{\mathrm{ss}} \leq 0.90, \quad 1.50 \leq \gamma_{\mathrm{cs}} \leq 2.10.$$

As has been said above and in Chapter 1 the computed approximations to problems such as (2.49) and the resulting quantities of interest will be the basis for engineering decisions. Because of the uncertainties in the input data $\beta, \gamma_{\mathrm{ss}}$ and γ_{cs}, decision makers have only at their disposal ranges in which the quantities of interest lie. For a discussion of uncertainty see eg. Babuska and Silvia (2010). For the above ranges of input data, and using the analytic solutions to problem (2.49), the ranges of the quantities of interest for 1D Eng Problems 1 and 2 are given in Tables 2.1 and 2.2. These illustrate the corresponding levels of uncertainty in the quantities of interest. Clearly for more complicated problems analytical solutions will not be available, in which case approximations such as finite element solutions are used.

Table 2.1 Ranges of quantities of interest for varying input data. 1D Eng Problem 1.

		γ_{ss}	γ_{cs}	β
Q_1	min = 165.547	0.67	1.5	0.55
	max = 237.022	0.9	2.1	0.4
Q_2	min = 2953.542	0.67	1.5	0.4
	max = 4043.996	0.9	2.1	0.55
Q_3	min = 111 609.755	0.67	1.5	0.4
	max = 153 754.675	0.9	2.1	0.55

Table 2.2 Ranges of quantities of interest for varying input data. 1D Eng Problem 2.

		γ_{ss}	γ_{cs}	β
Q_1	min = 163.4708	0.67	1.5	0.55
	max = 234.4349	0.9	2.1	0.4
Q_2	min = 2897.8831	0.67	1.5	0.4
	max = 3989.0929	0.9	2.1	0.55
Q_3	min = 110 171.064	0.67	1.5	0.4
	max = 151 656.698	0.9	2.1	0.55

2.2 Two-dimensional heat-conduction problem

2.2.1 Classical partial differential equation formulation

Let us now consider the two-dimensional analogues of the one-dimensional problems that we defined in the previous sections.

The two-dimensional analogue of the bar problem is the problem of plane elasticity, for which the solution is the *displacement vector* having components in the x_1- and x_2-directions. We believe that, if we were to introduce a problem with a *vector solution*, this would cause an unnecessary complication to our defined aim, which is to present finite element methods in a simple manner. Other two-dimensional problems of elasticity that have scalar solutions are those of torsion and of membrane deflection. However, these do not have the generality that we seek.

On the other hand, the problem of *two-dimensional heat flow* is a clear analogue of the one-dimensional problem of Section 2.1.4 although, as was said in the introduction to this chapter, the association of heat conduction problems with energy is not obvious. However, we shall take as our two-dimensional model one of heat conduction as follows. Consider a thin plate

$$\tilde{\Omega} \equiv \left\{ (x_1, x_2, x_3) \,\Big|\, |x_3| < \frac{d(\mathbf{x})}{2}, \mathbf{x} \equiv (x_1, x_2) \in \Omega \right\},$$

where $\Omega \subset \mathbb{R}^2$ is a bounded polygonal domain that is the centre plane of the plate and has boundary Γ, and where thickness $d(\mathbf{x}) << 1$, $\mathbf{x} \in \Omega$, and changes slowly with $\mathbf{x}$. We seek the temperature $u(x_1, x_2, x_3)$, $(x_1, x_2, x_3) \in \tilde{\Omega}$ in the plate due to heat sources $f(\mathbf{x})$ in the plate through the thickness, heat transfer from the surfaces of the plate, and boundary conditions on the boundary $\tilde{\Gamma}$ of $\tilde{\Omega}$. It is reasonable to assume that the temperature is constant through the thickness of the plate for any $\mathbf{x} \in \tilde{\Omega}$. With this assumption we are able to consider a two-dimensional problem in Ω in the central plane of the plate with boundary conditions imposed on Γ the boundary of Ω, which is a polygon with vertices $A_1, A_2, i = 1, 2, \ldots, n$ such that $A_{n+1} = A_1$ and consists of the open edges $\Gamma_i \equiv A_i, A_{i+1}, 1,\ 2, \ldots, n$ such that $\Gamma \equiv \cup\bar{\Gamma}_i$, see Fig. 2.8.

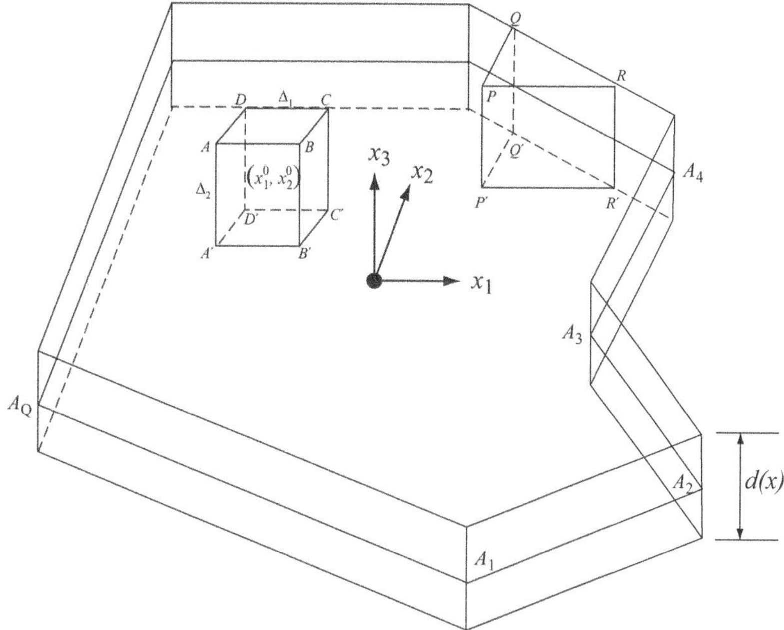

Fig. 2.8 Thin plate $\tilde{\Omega}$ with central plane Ω in the (x_1, x_2)-plane, with thickness $d(\mathbf{x}), \mathbf{x} \in \Omega$ and with Γ the boundary of Ω. The edges of Γ are Γ_i, and the boundary vertices are $A_i, i = 1, 2, \ldots, n$. The plate has upper and lower faces $\left\{\mathbf{x}+\frac{d(x)}{2}\right\}, \mathbf{x} \in \Omega$. The incremental slice $S(x_1^0, x_2^0) \equiv ABCDA'B'C'D'$ and the boundary incremental slice $\hat{S}(\hat{x}_1^0, \hat{x}_2^0) \equiv PQRP'Q'R'$ are also shown in the figure.

In order to model the temperature in $\tilde{\Omega}$ we may thus seek the temperature $u(x_1, x_2) \equiv u(\mathbf{x}), \mathbf{x} \in \Omega$, which has units $[T]$, again subject to heat sources, heat transfer as above and boundary conditions on Γ. As in the bar problem we shall deal with gradients of u, where in the two-dimensional context we have

$$\nabla u \equiv \left[\frac{\partial u}{\partial x_1}, \frac{\partial u}{\partial x_2}\right]^T,$$

which has units $[T]/[L]$.

In order to produce a mathematical model of the heat conduction in the plate we consider a slice $S\left(x_1^0, x_2^0\right) \equiv ABCDA'B'C'D'$ of the plate with a centre at the point $\left(x_1^0, x_2^0\right)$ as shown in Fig. 2.8, where

$$S\left(x_1^0, x_2^0\right) \equiv \left\{(x_1, x_2, x_3) \,\middle|\, \left|x_1 - x_1^0\right| < \frac{\Delta_1}{2}, \left|x_2 - x_2^0\right| < \frac{\Delta_2}{2}, |x_3| < \frac{d(x_1, x_2)}{2}\right\}.$$

By analogy with the one-dimensional case we consider the total flux, i.e. heat/unit time with dimension $[C]/[t]$, passing through plane sections $\left(x_1^0, x_2^0\right)$ that are parallel to the sections $DD'A'A$ and $CC'B'B$ and denote these fluxes by

$$F_1\left(x_1^0,x_2^0\right)=\Delta_2 d\left(x_1^0,x_2^0\right)\sigma_1\left(x_1^0,x_2^0\right), \tag{2.56a}$$

$$F_2\left(x_1^0,x_2^0\right)=\Delta_1 d\left(x_1^0,x_2^0\right)\sigma_2\left(x_1^0,x_2^0\right), \tag{2.56b}$$

where now $\sigma_i, i=1,2$, is the flux/unit area in the direction x_i, which has dimensions $[C]/[L^2][t]$.

Again, in analogy with the one-dimensional situation we use the Fourier law and obtain

$$\begin{bmatrix}\sigma_1\left(x_1^0,x_2^0\right)\\ \sigma_2\left(x_1^0,x_2^0\right)\end{bmatrix}=\begin{bmatrix}E_{11}\,E_{12}\\ E_{21}\,E_{22}\end{bmatrix}\left(x_1^0,x_2^0\right)\begin{bmatrix}\partial u/\partial x_1\\ \partial u/\partial x_2\end{bmatrix}\left(x_1^0,x_2^0\right), \tag{2.57}$$

where $E=\begin{bmatrix}E_{11}E_{12}\\ E_{21}E_{22}\end{bmatrix}$ is a symmetric positive definite matrix; the dimensions of E_{ij} are $[C]/[L][T][t]$. For the case where $E_{11}=E_{22}=E$, and $E_{12}=E_{21}=0$, we have an *isotropic* material. Otherwise, the material is *anisotropic*.

Continuing the analogy with the one-dimensional case we now consider the heat balance in the slice S. Denoting by $f(x_1^0,x_2^0)\Delta_1\Delta_2$ the total amount of heat/unit time created by the source f in S, and the heat loss/unit time to the surroundings through the top and bottom surfaces by

$$-u(x_1^0,x_2^0)\,c(x_1^0,x_2^0)\Delta_1\Delta_2$$

we have the dimensions of f as $[C]/[L^2][t]$ and c as $[C]/[t][L^2][T]$. Thus, for heat balance for the slice $S(x_1^0,x_2^0)$ we have, using (2.56) and (2.57)

$$\begin{aligned}&F_1(x_1^0+\frac{\Delta_1}{2},x_2^0)-F_1(x_1^0-\frac{\Delta_1}{2},x_2^0)+F_2(x_1^0,x_2^0+\frac{\Delta_2}{2})\\ &-F_2(x_1^0,x_2^0-\frac{\Delta_2}{2})+f(x_1^0,x_2^0)\Delta_1\,\Delta_2\\ &-c(x_1^0,x_2^0)\,u(x_1^0,x_2^0)\Delta_1\,\Delta_2=0.\end{aligned} \tag{2.58}$$

Taking limits as $\Delta_1,\Delta_2\to 0,(x_1^0,x_2^0)\to(x_1,x_2)$ we therefore get

$$\frac{\partial}{\partial x_1}\left(d(\mathbf{x})\left(E_{11}\frac{\partial u}{\partial x_1}+E_{12}\frac{\partial u}{\partial x_2}\right)\right)+\frac{\partial}{\partial x_2}\left(d(\mathbf{x})\left(E_{21}\frac{\partial u}{\partial x_1}+E_{22}\frac{\partial u}{\partial x_2}\right)\right)$$

$$-c(\mathbf{x})\,u(\mathbf{x})+f(\mathbf{x})=0,$$

or

$$-\nabla\cdot(d(\mathbf{x})[E(\mathbf{x})]\nabla u(\mathbf{x}))+c(\mathbf{x})u(\mathbf{x})=f(\mathbf{x}). \tag{2.59}$$

As in one dimension we shall write

$$d(\mathbf{x})[E(\mathbf{x})]=[\mathcal{A}(\mathbf{x})]=\begin{bmatrix}a_{11}\,a_{12}\\ a_{21}\,a_{22}\end{bmatrix},$$

so that we get finally the equation of heat balance in the plate

$$-\nabla \cdot ([\mathcal{A}(\mathbf{x})]\nabla u(\mathbf{x})) + c(\mathbf{x})u(\mathbf{x}) = f(\mathbf{x}), \mathbf{x} \in \Omega. \tag{2.60}$$

We need now to consider the boundary conditions that must be applied on the boundary Γ of Ω, where $\Gamma \equiv \bigcup_{i=1}^{n} \Gamma_i, \Gamma_i \cap \Gamma_j = 0$. Let us now denote by $\mathbf{n} = (n_1, n_2) \equiv (\cos nx_1, \cos nx_2)$ the outward unit normal to Γ_i at $\mathbf{x} \in \Gamma_i$, $i = 1, 2, \ldots, n$; note that $\mathbf{n}$ is not defined at A_i.

The boundary conditions for points of Γ_i may be of the type where we prescribe temperature (as in (2.14b)) or of the flux type (as in (2.14c)); note that only one type of boundary condition is defined on each Γ_i. Suppose that the boundary Γ is such that $\Gamma = \Gamma_{\mathrm{D}} \cup \Gamma_{\mathrm{N}}$, where $\Gamma_{\mathrm{D}} \equiv \cup \Gamma_i$ for those Γ_i on which temperature is prescribed and $\Gamma_{\mathrm{N}} \equiv \cup \Gamma_i$ for those Γ_i on which there is a flux-type condition. Analogously to the one-dimensional case we shall often assume that $\Gamma_{\mathrm{D}} \neq 0$.

Then, we have

$$u(\mathbf{x}) = g(\mathbf{x}), \mathbf{x} \in \Gamma_{\mathrm{D}}, \tag{2.61}$$

where g is the boundary temperature prescribed on Γ_{D}.

Alternatively, in order to determine a flux-type boundary condition analogous to (2.14c) at $\mathbf{x} \in \Gamma_N$, we need to determine the total flux $F(\mathbf{x})$ across an incremental area $QRR'Q'$ centered on $(\hat{x}_1^0 \hat{x}_2^0) \in \Gamma_N$, where $PQ = \Delta x_2, PR = \Delta x_1, QR = \Delta s$, see Fig. 2.9. This condition is derived by considering the flux equilibrium for the incremental prism slice $PQRP'Q'R'$. From Fourier's law, in the absence of heat sources and in the steady state the net heat entering (leaving) the prism is zero; i.e. the heat flux entering through $PQQ'P'$ and $PRR'P'$ equals that leaving through $QRR'Q'$.

The flux through $PQQ'P'$ is

$$d(\mathbf{x})\Delta x_2 \left((a_{11}\partial u/\partial x_1)_{(\hat{x}_1^0 - \Delta x_1/2, \hat{x}_2^0)} + (a_{12}\partial u/\partial x_2)_{(\hat{x}_1^0 - \Delta x_1/2, \hat{x}_2^0)}\right),$$

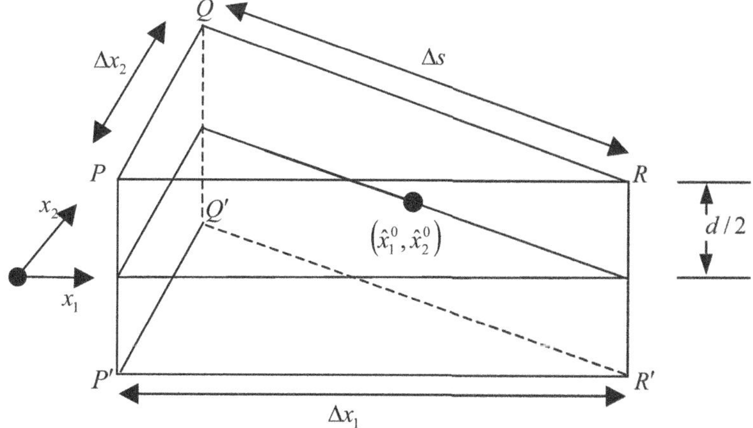

Fig. 2.9 Boundary incremental slice $\hat{S}(\hat{x}_1^0, \hat{x}_2^0) \equiv PQRP'Q'R'$.

whilst that through $PRR'P'$ is

$$d(\mathbf{x})\Delta x_1\left((a_{21}\partial u/\partial x_1)_{(\hat{x}_1^0,\hat{x}_2^0-\Delta x_2/2,)}+(a_{22}\partial u/\partial x_2)_{(\hat{x}_1^0,\hat{x}_2^0-\Delta x_2/2)}\right).$$

The sum of these two fluxes equals the flux across $QRR'Q'$ that has area $d(\mathbf{x})\Delta s$, where $QR=\Delta s$. Thus, in the limit we have that the flux at the point $(\hat{x}_1^0,\hat{x}_2^0)\equiv\mathbf{x}$ on the edge in the direction of the outward conormal to the boundary is

$$\frac{\partial u(\mathbf{x})}{\partial n_c}=[\cos\, nx_1\ \cos\, nx_2]\begin{bmatrix}a_{11}(\mathbf{x})a_{12}(\mathbf{x})\\ a_{21}(\mathbf{x})a_{22}(\mathbf{x})\end{bmatrix}\begin{bmatrix}\partial u(\mathbf{x})/\partial x_1\\ \partial u(\mathbf{x})/\partial x_2\end{bmatrix}.$$

In complete analogy with the one-dimensional case in (2.14c) we have that the Newton boundary condition for $\mathbf{x}\in\Gamma_N$ is

$$\frac{\partial u(\mathbf{x})}{\partial n_c}+\alpha(\mathbf{x})\,u(\mathbf{x})=G(\mathbf{x}),\mathbf{x}\in\Gamma_N, \tag{2.62}$$

where the outward conormal derivative is defined by

$$\frac{\partial u(\mathbf{x})}{\partial n_c}\equiv\sum_{i,j=1}^{2}n_i\,a_{ij}(\mathbf{x})\frac{\partial u(\mathbf{x})}{\partial x_j},$$

where the dimensions of $\alpha(\mathbf{x})$ and $G(\mathbf{x})$ are, respectively, $[C]/[t][T][L]$ and $[C]/[t][L]$.

Our problem for the conduction of heat in the plate is thus:

find $\mathrm{u}(\mathbf{x}),\mathbf{x}\in\Omega$ such that

$$-\nabla\cdot([\mathcal{A}(\mathbf{x})]\nabla \mathrm{u}(\mathbf{x}))+c(\mathbf{x})\mathrm{u}(\mathbf{x})=f(\mathbf{x}),\mathbf{x}\in\Omega, \tag{2.63a}$$

$$\mathrm{u}(\mathbf{x})=g(\mathbf{x}),\mathbf{x}\in\Gamma_{\mathrm{D}}, \tag{2.63b}$$

$$\frac{\partial \mathrm{u}(\mathbf{x})}{\partial n_c}+\alpha(\mathbf{x})\,\mathrm{u}(\mathbf{x})=G(\mathbf{x}),\mathbf{x}\in\Gamma_{\mathrm{N}}. \tag{2.63c}$$

As in the one-dimensional context we must again assume that all the symbols in the formulation have meaning. In particular, we do not take the boundary condition (2.63c) at vertices of Γ_{N}, as the conormal derivative cannot be defined at these points. At a point of intersection of Γ_{N} and Γ_{D} we take the Dirichlet boundary condition of (2.63b). In the next section we shall define the weak formulation of problem (2.63). The solution of the weak problems is unique even when the conormal cannot be defined at vertex points interior to Γ_{N}.

Example 2.6 Consider now a problem of heat transfer of the type (2.63), where Ω is the unit square $\{(x_1,x_2)\mid |x_1|<1,|x_2|<1\},[\mathcal{A}(x)]=\begin{bmatrix}1\,0\\0\,1\end{bmatrix}$, $f=1$ and $\Gamma_D\equiv\Gamma$ with a homogeneous ($g=0$) Dirichlet boundary condition on Γ. It is also assumed that there is no heat transfer to the surrounding medium so that $c(\mathbf{x})=0$. Under these assumptions (2.63a) takes the form

$$-\Delta u(\mathbf{x}) \equiv -\frac{\partial^2 u}{\partial x_1^2}(\mathbf{x}) - \frac{\partial^2 u}{\partial x_2^2}(\mathbf{x}) = 1, \ \mathbf{x} \in \Omega,$$

which is of course the classical Poisson equation involving the Laplacian operation Δ, whilst on Γ the boundary condition (2.63b) takes the form

$$u(\mathbf{x}) = 0, \mathbf{x} \in \Gamma.$$

If, on the other hand, $[\mathcal{A}(x)] = \begin{bmatrix} A & 0 \\ 0 & 1 \end{bmatrix}$ then (2.63a) takes the form

$$-\left(A\frac{\partial^2 u}{\partial x_1^2}(\mathbf{x}) + \frac{\partial^2 u}{\partial x_2^2}(\mathbf{x})\right) = 1, \ \mathbf{x} \in \Omega\,.$$

If we make the transformation of coordinates $x_1 = aq_1$, $x_2 = q_2$, then we have the relations

$$\frac{\partial^2 u}{\partial x_1^2}(\mathbf{x}) = \frac{1}{a^2}\frac{\partial^2 u}{\partial q_1^2}(\mathbf{q}), \ \frac{\partial^2 u}{\partial x_2^2}(\mathbf{x}) = \frac{\partial^2 u}{\partial q_2^2}(\mathbf{q}),$$

so that in the new coordinates q_1, q_2 the differential equation becomes

$$-\left(\frac{A}{a^2}\frac{\partial^2 u}{\partial q_1^2}(\mathbf{q}) + \frac{\partial^2 u}{\partial q_2^2}(\mathbf{q})\right) = 1, \ \mathbf{q} \in \Omega_a,$$

where

$$\Omega_a = \left\{ \{q_1, q_2\} \,\middle|\, |q_1| < \frac{1}{a}, |q_2| < 1 \right\},$$

with boundary Γ_a. If we select $a^2 = A$ we see that the transformed problem on Ω_a is

$$-\left(\frac{\partial^2 u}{\partial q_1^2}(\mathbf{q}) + \frac{\partial^2 u}{\partial q_2^2}(\mathbf{q})\right) = 1, \ \mathbf{q} \in \Omega_a$$

$$u(\mathbf{q}) = 0, \ \mathbf{q} \in \Gamma_a.$$

Exercise 2.12 Consider the anisotropic problem as in the above example, but with $[\mathcal{A}] = \begin{bmatrix} a_{11} & a_{12} \\ a_{21} & a_{22} \end{bmatrix}$. Transform this problem into an isotropic (Poisson) problem on a suitable domain. What is the shape of the mapped domain?

Remark 2.11 *The purpose of Example 2.6 is to show that it is possible to transform an anisotropic problem, characterized by a matrix $[\mathcal{A}] \neq [I]$ defined in a square region Ω into an isotropic (Poisson) problem characterized by the matrix $[\mathcal{A}] = [I]$ but defined on a rectangle. It is further seen from Exercise 2.12 that the mapping approach can be extended to the case of a problem characterized by the general constant matrix $\mathcal{A}$, which is transformed into an isotropic problem defined on a parallelogram. This mapping technique is important because, as we shall see later, the regularity of the solution for the isotropic problem depends on the angles of the domain in which it is* *defined. Hence, the analysis for the regularity of the solution in the isotropic case can be utilized to determine the regularity of the solution of the anisotropic problem.*

2.2.2 Weak formulation

For this two-dimensional heat-conduction problem we wish now to derive the weak formulation that corresponds to (2.63). In order to do this, keeping in mind the boundary conditions (2.63b) on Γ_{D} and (2.63c) on Γ_{N}, we multiply (2.63a) by a test function $v(\mathbf{x})$ and integrate of Ω to get

$$\int_{\Omega}\left(\sum_{i,j=1}^{2} a_{ij}(\mathbf{x})\frac{\partial \mathrm{u}}{\partial x_i}(\mathbf{x})\frac{\partial v}{\partial x_j}(\mathbf{x}) + c(\mathbf{x})\,\mathrm{u}(\mathbf{x})\,v(\mathbf{x})\right)d\mathbf{x}$$

$$-\oint_{\Gamma}\frac{\partial \mathrm{u}}{\partial n_c}(\mathbf{x})v(\mathbf{x})ds = \int_{\Omega} f(\mathbf{x})\,v(\mathbf{x})d\mathbf{x}. \tag{2.64}$$

If we now further assume that $v(\mathbf{x}) = 0, \mathbf{x} \in \Gamma_{\mathrm{D}}$, and we use (2.63c) on Γ_{N}, we obtain

$$\int_{\Omega}\left(\sum_{i,j=1}^{2} a_{ij}(\mathbf{x})\frac{\partial \mathrm{u}}{\partial x_i}(\mathbf{x})\frac{\partial v}{\partial x_j}(\mathbf{x}) + c(\mathbf{x})\,\mathrm{u}(\mathbf{x})\,v(\mathbf{x})\right)d\mathbf{x}$$

$$+\oint_{\Gamma_N}\alpha(\mathbf{x})\,\mathrm{u}(\mathbf{x})\,v(\mathbf{x})\,ds = \int_{\Omega} f(\mathbf{x})\,v(\mathbf{x})\,d\mathbf{x} + \int_{\Gamma_N} G(\mathbf{x})\,v(\mathbf{x})\,ds,$$

which we write as

$$B(\mathrm{u}, v) = F(v), \tag{2.65}$$

where

$$B(\mathrm{u}, v) \equiv \int_{\Omega}\left(\sum_{i,j=1}^{2} a_{ij}(\mathbf{x})\frac{\partial \mathrm{u}}{\partial x_i}(\mathbf{x})\frac{\partial v}{dx_j}(\mathbf{x}) + c(\mathbf{x})\mathrm{u}(\mathbf{x})v(\mathbf{x})\right)d\mathbf{x}$$

$$+\oint_{\Gamma_N}\alpha(\mathbf{x})\,\mathrm{u}(\mathbf{x})\,v(\mathbf{x})\,ds \tag{2.66}$$

and

$$F(v) \equiv \int_{\Omega} f(\mathbf{x})\,v(\mathbf{x})\,d\mathbf{x} + \int_{\Gamma_N} G(\mathbf{x})\,v(\mathbf{x})\,ds. \tag{2.67}$$

We shall of course need to impose on $\mathrm{u}(\mathbf{x})$ the boundary condition (2.63b) for $\mathbf{x} \in \Gamma_{\mathrm{D}}$, assuming that the length of Γ_{D} is positive. Note that the line integrals in (2.64) are always taken in the direction such that the interior of Ω is on the left-hand side.

Completely analogously to the one-dimensional case, we define for the two-dimensional problem the spaces

$$\mathcal{U} \equiv \{v = v(\mathbf{x}) \mid \|v\|_{\mathcal{U}} < \infty\}, \tag{2.68}$$

$$\mathcal{U}_0 \equiv \{v \in \mathcal{U} \mid v(\mathbf{x})\,|_{\mathbf{x}\in\Gamma_D} = 0\}, \tag{2.69}$$

where, as before,

$$\|u\|_{\mathcal{U}} = (B(u,u))^{\frac{1}{2}} \tag{2.70}$$

has all the properties of a norm and $B(u,v)$ has those of a scalar product. Further, if $\Gamma_N = \Gamma$ and $\alpha(\mathbf{x}) = c(\mathbf{x}) = 0$, then if $B(v,v) = 0$ then v is a constant function and hence $\|\text{constant function}\|_{\mathcal{U}} = 0$, which means that $\|\cdot\|_{\mathcal{U}}$ is a norm modulo a constant. For the existence of the solution in this case we must have $F(1) = 0$.

Following the notation of (2.22b) we define

$$\mathcal{U}_g \equiv \{v(\mathbf{x}) \mid v \in \mathcal{U} \text{ and } v(\mathbf{x}) = g(\mathbf{x}), \mathbf{x} \in \Gamma_D\}. \tag{2.71}$$

For the weak form of (2.63) we now let $\bar{u} \in \mathcal{U}$ be such that $\bar{u}(\mathbf{x}) = g(\mathbf{x}), \mathbf{x} \in \Gamma_D$ and thus (analogously to the one-dimensional case, see (2.32)) we define $\mathfrak{u} = \bar{u} + u_0$, where $u_0 \in \mathcal{U}_0$, and we have that

$$B(u_0, v) = F(v) - B(\bar{u}, v) \; \forall v \in \mathcal{U}_0.$$

The weak form of (2.63) is thus: find $\mathfrak{u} \in \mathcal{U}_g$ such that

$$B(\mathfrak{u}, v) = F(v) \; \forall v \in \mathcal{U}_0. \tag{2.72}$$

We note that if $g = 0$ then $\mathfrak{u} \in \mathcal{U}_0$.

As in one dimension we have to formulate the assumptions that we make on $\mathcal{A}(\mathbf{x}), c(\mathbf{x}), f(\mathbf{x}), \alpha(\mathbf{x}), G(\mathbf{x})$ and $g(\mathbf{x})$ in (2.63), in order that the bilinear form (2.66) may have proper meaning and that the solution $\mathfrak{u}(\mathbf{x})$ may exist and be unique. To this end, we shall assume that:

(1) there exist constants $a_{\min}, a_{\max} \in \mathbb{R}$, with $0 \leqq a_{\min} < a_{\max} < \infty$, such that for any $\mathbf{x} \in \Omega$

$$a_{\min} \sum_{j=1}^{2} \xi_i \xi_j \leqq \sum_{j=1,2} \xi_i \xi_j \; a_{ij}(\mathbf{x}) \leqq a_{\max} \sum_{j=1}^{2} \xi_i \xi_j. \tag{2.73}$$

Condition (2.73) shows that the matrix $[\mathcal{A}(\mathbf{x})]$ is positive definite.

(2) There exist constants $\mathcal{C}_{\max}, \alpha_{\max} \in \mathbb{R}$ such that

$$0 \leqq c(\mathbf{x}) \leqq \mathcal{C}_{\max} \; \forall \; \mathbf{x} \in \Omega, \tag{2.74a}$$

(3)

$$0 \leqq \alpha(\mathbf{x}) \leqq \alpha_{\max} \; \forall \; \mathbf{x} \in \Gamma_N, \tag{2.74b}$$

(4)

$$\int_{\Gamma_N} G(\mathbf{x})^2 ds < \infty, \tag{2.74c}$$

(5)

$$\int_{\Omega} f(\mathbf{x})^2 d\mathbf{x} < \infty, \tag{2.74d}$$

(6) there exists $\bar{u} \in \mathcal{U}$ such that

$$\bar{u}(\mathbf{x}) = g(\mathbf{x}), \ \mathbf{x} \in \Gamma_{\mathrm{D}}. \tag{2.74e}$$

Analogously to the one-dimensional case, if in (2.63) with $\alpha = c = 0$ we take Neumann conditions on the entire boundary Γ, the necessary consistency condition is

$$\int_{\Omega} f d\mathbf{x} + \int_{\Gamma} G ds = 0.$$

For the case $\alpha = c = 0$ the solution to (2.72) with $\mathcal{U}_0$ replaced by $\mathcal{U}$ exists when the consistency condition holds and is unique up to a constant. Again we are interested only in fluxes, which are not influenced by this arbitrary constant.

Remark 2.12 *In the two-dimensional setting the bilinear form $B(u, v)$ as in (2.66) and the energy space together with (2.70) guarantees that $B(u, u)$ is a norm, but the functions $\bar{u} \in \mathcal{U}$ are not necessarily continuous. Nevertheless, these functions have well-defined 'traces' on both Γ_{N} and Γ_{D}, so that $\int_{\Gamma_N} \alpha uv ds$ is also well defined. If $\Gamma_{\mathrm{D}} \neq \emptyset$ then $\int_{\Gamma_{\mathrm{N}}} G v ds \leqq C \|v\|_{\mathcal{U}}$, and $\int_{\Omega} f v d\mathbf{x} \leqq C \|v\|_{\mathcal{U}}$.*

The assumption of the existence of $\bar{u} \in \mathcal{U}$ is in general a complex one. Nevertheless, if $\bar{u}$ is smooth on every $\bar{\Gamma}_i \subset \Gamma_{\mathrm{D}}$ and if $\bar{\Gamma}_i \subset \Gamma_{\mathrm{D}}$, $\bar{\Gamma}_j \subset \Gamma_{\mathrm{D}}$, $\bar{\Gamma}_i \cap \bar{\Gamma}_j = A_s$, then the values of g on $\bar{\Gamma}_i$ and $\bar{\Gamma}_j$ have to be the same at A_s.

Under all these conditions the solution of (2.72) exists and is unique. We shall henceforth always assume that these conditions hold.

Remark 2.13 *There is in this two-dimensional context a complete analogue for the principle of minimum energy of Section 2.3, if we define as previously*

$$\Pi(u) = \frac{1}{2} B(u, u) - F(u), \tag{2.75}$$

with $B(\cdot, \cdot)$ and $F(\cdot)$ as in (2.66) and (2.67). In this case the solution of the weak problem (2.65) minimizes the functional $\Pi(u)$, and conversely the minimizer is the weak solution. Whilst mathematically these properties are perfectly clear, the physical interpretation of energy and virtual work has no standard meaning in this heat-diffusion context. The interpretation is, however, completely meaningful in the more complicated context of linear elasticity.

Recalling the one-dimensional case we now introduce for the two-dimensional case the spaces that we shall need for the subsequent analysis. Let

$$C(\Omega) \equiv C^0(\Omega) \equiv \left\{ u(\mathbf{x}), \ \mathbf{x} \in \Omega \mid u(\mathbf{x}) \text{ is continuous on } \bar{\Omega} \right\}, \tag{2.76a}$$

and

$$\|u\|_{C^0(\Omega)} = \max_{\mathbf{x} \in \bar{\Omega}} \mid u(\mathbf{x}) \mid, \tag{2.76b}$$

$$C^k(\Omega) \equiv \left\{ \begin{array}{c} u \in C(\Omega) \mid \frac{\partial^j u(\mathbf{x})}{\partial x_1^{j_1} \partial x_2^{j_2}} \in C(\Omega), \\ j_1 + j_2 = j, \ j = 0, \ldots, k \end{array} \right\}, \tag{2.76c}$$

and

$$\|u\|_{C^k(\Omega)} = \max_{j=0,\ldots,k} \left\| \frac{\partial^j u(\mathbf{x})}{\partial^{j_1} x_1 \, \partial^{j_2} x_2} \right\|_{C^0(\Omega)}. \tag{2.76d}$$

We also define the semi-norm

$$|u|_{C^k(\Omega)} = \left\| \frac{\partial^k u(\mathbf{x})}{\partial x_1^{k_1} \, \partial x_2^{k_2}} \right\|_{C^0(\Omega)} \qquad k_1 + k_2 = k. \tag{2.77}$$

These spaces are linear spaces and $\|\cdot\|_{C^k(\Omega)}$ has all the properties of a norm, while $|u|_{C^k(\Omega)}$ is a semi-norm modulo polynomials of degree $k-1$

We further define the spaces

$$H^0(\Omega) = L^2(\Omega) \equiv \left\{ u(\mathbf{x}), x \in \Omega \mid \int_\Omega u^2(\mathbf{x}) d\mathbf{x} < \infty \right\}, \tag{2.78a}$$

and

$$\|u\|_{L^2(\Omega)} \equiv \left(\int_\Omega u^2(\mathbf{x}) dx \right)^{\frac{1}{2}}, \tag{2.78b}$$

$$H^k(\Omega) \equiv \left\{ u \in L^2(\Omega) \mid \textstyle\int_\Omega \left(\frac{\partial^j u(\mathbf{x})}{\partial x_1^{j_1} \, \partial x_2^{j_2}} \right)^2 d\mathbf{x} < \infty, \; j = j_1 + j_2, \; j = 0, \ldots, k \right\}, \tag{2.78c}$$

and

$$\|u\|_{H^k(\Omega)} = \left(\sum_{j=0}^{k} \left\| \frac{\partial^j u(\mathbf{x})}{\partial x_1^{j_1} \, \partial x_2^{j_2}} \right\|_{L^2(\Omega)}^2 \right)^{\frac{1}{2}}, j = j_1 + j_2. \tag{2.78d}$$

We can also define the semi-norm

$$|u|_{H^k(\Omega)} = \left(\sum_{k_1+k_2=k} \int_\Omega \left(\frac{\partial^k u(\mathbf{x})}{\partial x_1^{k_1} \, \partial x_2^{k_2}} \right)^2 d\mathbf{x} \right)^{\frac{1}{2}}, k = k_1 + k_2. \tag{2.79}$$

As in the one-dimensional context, the spaces $H^k(\Omega)$ are linear spaces and $\|\cdot\|_{H^k(\Omega)}$ has all the properties of the norm, and $|\cdot|_{H^k(\Omega)}$ of the seminorm. Again, we should for correctness consider the derivatives in the generalized sense and the integrals in the Lebesgue sense, but once more it will here be sufficient to consider the definitions from the calculus. Similarly, we actually need the notion $u(\mathbf{x}) \mid_{\mathbf{x}\in\Gamma}$ in a generalized (trace) sense; nevertheless, the usual restriction to Γ will also be sufficient.

There are as before, a number of relations between these spaces that we shall need later. These are of course similar to those for the corresponding spaces in the one-dimensional setting. For completeness, we list these relations again:

If Γ_D has positive length and $u \in \mathcal{U}_0$ there exist constants C_1 and C_2 independent of u such that

$$C_1 \|u\|_{\mathcal{U}} \leqq \|u\|_{H^1(\Omega)} \leqq C_2 \|u\|_{\mathcal{U}} . \tag{2.80}$$

If $H_0^1(\Omega) \equiv \{u(x), x \in \Omega \mid u \in H^1(\Omega) \text{ and } v = 0 \text{ on } \Gamma_D\}$ then the spaces $\mathcal{U}_0$ and $H_0^1(\Omega)$ are equivalent, and for $u(\mathbf{x}) \mid_{\mathbf{x}\in\Gamma}$

$$\int_\Gamma (u(\mathbf{x}) \mid_{\mathbf{x}\in\Gamma})^2 ds \leqq C \|u\|^2{}_{H^k(\Omega)}, k \geqq 1. \tag{2.81}$$

If $u \in H^k(\Omega), k > 1$, then $u \in C^0(\Omega)$ and

$$\|u\|_{C^0(\Omega)} \leqq C \|u\|_{H^k(\Omega)}, k > 1. \tag{2.82}$$

Note that this inequality does not hold for $k = 1$; for example $u = \ln|\ln r| \in H^1 - C^0$. If $u \in H^1(\Omega)$ and Γ_D has positive length, then

$$\|u\|_{L^2(\Omega)} \leqq C |u|_{H^1(\Omega)} < C \|u\|_{\mathcal{U}} \; \forall u \in H_0^1(\Omega),$$

and

$$\|u\|_{L^2(\Gamma_j)} \leqq C \|u\|_{H^1(\Omega)} \leqq C \|u\|_{\mathcal{U}} . \tag{2.83}$$

In Section 2.1.2 we considered the regularity of the solution $u(x)$ of (2.33). The situation in two dimensions for the regularity of $u(\mathbf{x}), \mathbf{x} \in \Omega$, the solution of (2.72), is more complicated. The regularity still depends on that of $f(\mathbf{x}), \mathcal{A}(\mathbf{x})$, $c(\mathbf{x})$, $\alpha(\mathbf{x})$, $g(\mathbf{x})$ and $G(\mathbf{x})$, but of course it now also depends on the shape of the boundary Γ of Ω, which in our case is a polygon, together with the boundary conditions. The major difference from the one-dimensional case is that Γ contains corners, and at such corners $u(\mathbf{x})$ can contain *singularities* that reduce the level of regularity of $u(\mathbf{x})$. Much work has been done, see Grisvard, (1985), Nazarov and Plamenevsky (1994), to derive the local forms of the solutions $u(\mathbf{x})$ in the neighbourhoods of the boundary corners. These techniques invariably require the solving of a transcedental equation to produce the parameters that occur in the local solution forms that characterize $u(\mathbf{x})$ near the corners. These characterisations take the form of a smooth function together with a summation of terms, based on the vertex data, some of which may be singular and that are responsible for the reduced regularity. We shall here merely state some of these characterisations for particular cases, and will end this section by stating a general theorem for the two-dimensional regularity.

In the discussion we shall use the minimum level of generality needed for practical computations for solving (2.72) using the finite element method.

If, see Fig. 2.11, $A_j, j = 1, 2, \ldots, \mathcal{Q}$ are the vertices of Γ, with internal angles α_j, then we define neighbourhoods ω_j of A_j for each j such that $\bar{\omega}_i \cap \bar{\omega}_j = \emptyset$ for $i \neq j$.(see Fig. 2.10)

Let us first address the singularity in the neighbourhood ω_j of a single corner A_j of Γ, with arms Γ_j,Γ_{j-1} and interior angle α_j as shown in Fig. 2.10, and take local polar coordinates (r, θ) centered on the corner. We are interested in solving the equation

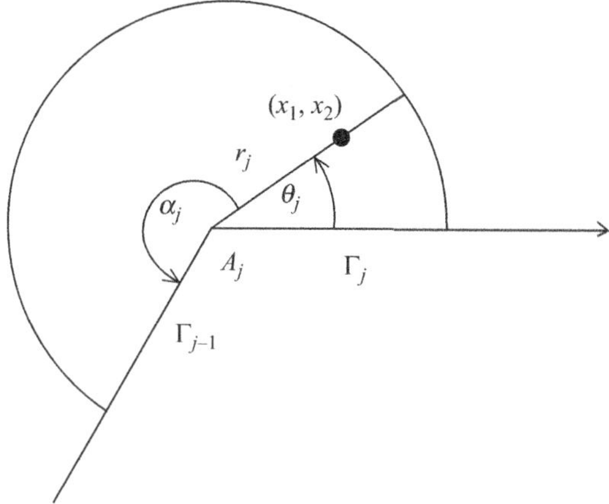

Fig. 2.10 The neighbourhood ω_j of the vertex A_j.

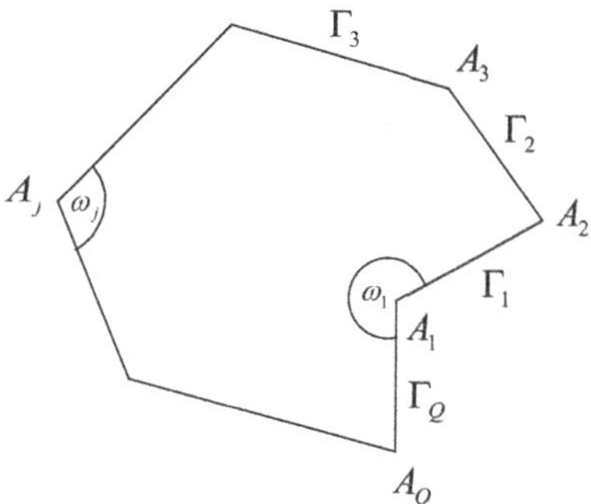

Fig. 2.11 Polygonal region Ω with boundary vertices $A_j, j = 1, 2, \ldots, Q$, vertex subdomains ω_j and interior vertex angles α_j.

$\Delta u = 0$ with zero Dirichlet boundary conditions on Γ_j, and Γ_{j-1}. Writing $\Delta u = 0$ in these local polar coordinates we have that

$$\Delta u(r, \theta) \equiv \frac{\partial^2 u}{\partial r^2} + \frac{1}{r}\frac{\partial u}{\partial r} + \frac{1}{r^2}\frac{\partial^2 u}{\partial \theta^2} = 0, \tag{2.84}$$

with $u(r, \theta) = 0$ on $\theta = 0, (\Gamma_{j-1})$ and $\theta = \alpha\pi, (\Gamma_j)$. From separation of variables the functions

$$r^{\beta_j} \sin \beta_j \theta \text{ and } r^{\beta_j} \cos \beta_j \theta, \tag{2.85}$$

with $\beta_j = \dfrac{\pi}{\alpha_j}$ satisfy (2.84) for any α. Now, $r^\beta \sin\beta\theta$ also satisfies the homogeneous Dirichlet boundary conditions on Γ_{j-1} and Γ_j, so that $r^{\beta_j}\sin\beta_j\theta$ is the leading term in the solution u.

We see that, in general, the singular solution $u(x_i, x_2)$ of (2.63) in the neighbourhood ω_j will be such that with $l = l_1 + l_2$

$$\left|\frac{\partial^{l_1+l_2}u(x_1,x_2)}{\partial x_1^{l_1}\partial x_2^{l_2}}\right| \leq C\, r_j^{\beta_j - l}, \; l = 1,2,3, \; \beta_j - l < 0, \tag{2.86}$$

for the case when β_j is not an integer, and where r_j denotes the distance from (x_1, x_2) to A_j.

If β_j is an integer at a corner A_j then the solution of $\Delta u = 0$ is smooth at A_j, whilst this is not the case for the solution u of $\Delta u = 1$. In this case, the β in (2.86) is replaced by $\beta - \varepsilon$, for any $\varepsilon > 0$ however small, and we have that

$$\left|\frac{\partial^{l_1+l_2}u(x_1,x_2)}{\partial x_1^{l_1}\partial x_2^{l_2}}\right| \leq C(\varepsilon) r_j^{\beta_j - \varepsilon - l}, \; l = 1,2,3, \; \beta_j - \varepsilon - l < 0. \tag{2.87}$$

More specifically, the values of β_j for different boundary coefficient combinations are given in Table 2.3.

For the problem (2.63) we first impose the following restrictions for the data, so that for $j = 1, 2, \ldots, Q$:

(1) The matrix $[\mathcal{A}(\mathbf{x})] = a_{mn}(\mathbf{x})$ is the identity matrix in each $\bar{\omega}_j$. (2.88a)

Table 2.3 Singularity parameters β_j for different boundary conditions and interior vertex angles α_j.

Γ_{j1}	Γ_j	β_j
Dirichlet	Dirichlet	π/α_j
Neumann	Neumann	π/α_j
Newton	Newton	π/α_j
Dirichlet	Neumann	$\pi/2\alpha_j$
Dirichlet	Newton	$\pi/2\alpha_j$
Neumann	Newton	π/α_j

(2)
$$f(\mathbf{x}) \in C^{k-1}(\Omega),\; g(\mathbf{x}) \in C^{k+1}(\Gamma_j),$$
$$G(\mathbf{x}) \in C^k(\Gamma_j), \alpha(\mathbf{x}) \in C^k(\Gamma_j),$$
$$a_{mn}(\mathbf{x}) \in C^k(\Omega),\; c(\mathbf{x}) \in C^{k-1}(\Omega),$$
$$k = 2, 3. \tag{2.88b}$$

(3) On each Γ_j, $j = 1, \ldots, \mathcal{Q}$ of Γ we have either a Dirichlet or a Neumann (Newton) boundary condition.

Note the required regularity on the data in (2.88b) is one level higher than for the one-dimensional case in (2.40).

Although these assumptions are somewhat restrictive, they do allow us to describe the regularity of $u(\mathbf{x}), \mathbf{x} \in \Omega$ relatively easily.

The regularity of the solution $u(\mathbf{x})$ of (2.63) differs between the subdomains ω_j and $\Omega - \cup\bar{\omega}_j \equiv \hat{\Omega}$, i.e. near the boundary vertices and away from them, and has to be considered separately. The regularity in $\hat{\Omega}$ is similar to that which was encountered in the one-dimensional problems, provided that the input data of problem (2.63) satisfies the more stringent regularity conditions. If this data fulfills (2.88b) for $j = 1, 2, \ldots, \mathcal{Q}$, then $u \in C^k(\hat{\Omega})$ and for $1 \leqq l \leqq k$

$$|u|_{C^l(\hat{\Omega})} \leqq \mathcal{C}\left(\|f\|_{C^{k-1}(\Omega)} + \sum_{\Gamma_j \in \Gamma_D} \|g\|_{C^k(\Gamma_j)} + \sum_{\Gamma_j \subset \Gamma_N} \|G\|_{C^{k-1}(\Gamma_j)}\right),$$
$$\equiv \mathcal{C}Q, \tag{2.89a}$$

and on ω_j we have $u = u_1 + u_2$ with

$$\|u_1\|_{C^l(\omega_j)} \leqq \mathcal{C}Q, \tag{2.89b}$$

and

$$\left|\frac{\partial^{l_1+l_2} u_2}{\partial x_1^{l_1} \partial x_2^{l_2}}\right| \leqq Q\mathcal{C}(\varepsilon) r_j^{\beta_j - \varepsilon - l},$$
$$l_1 + l_2 = l \geqq 1 \text{ and } \beta_j - \varepsilon - l < 0, \tag{2.89c}$$

where $\mathcal{C}$ depends on a Ω and ω_j but not on f, g and G.

If in (2.89a) we use $\|f\|_{L^2(\Omega)}$ rather than $\|f\|_{C^{k-1}(\Omega)}$ and $k \geqq 2$, then (2.89a) holds but with $|u|_{C^l(\hat{\Omega})}$ replaced by $\|u\|_{H^l(\Omega)}$, $1 \leqq l \leqq k$. We also have $\|u_1\|_{H^l(\Omega)}$ in (2.89b).

In the neighbourhoods of the vertices A_j the growth of the derivatives can be estimated by (2.89b), where β depends only on the geometry of Γ and the type of boundary conditions. We can now state a theorem on regularity.

Theorem 2.1 *If* $u(\mathbf{x}), \mathbf{x} \in \Omega$ *is the solution of problem (2.72) the weak form of (2.63), defined in the region* Ω *of Fig. 2.11, and the data of (2.63) satisfies (2.88), then for* $j = 1, 2, \ldots, Q$ *and* $k = 2, 3$, *(2.89a) holds, whilst in the neighbourhoods* ω_j *the* u *satisfies (2.89c), where* β_j *is related to the boundary conditions.*

Remark 2.14 *We have assumed for simplicity that in (2.60) the coefficients are such that* $a_{ij} \in C^k(\Omega), c \in C^{k-1}(\Omega), f \in C^{k-1}(\Omega)$. *In many problems this is not the case. Rather, we have* $a_{ij} \in C^k(\Omega_s), c \in C^{k-1}(\Omega_s), f \in C^{k-1}(\Omega_s)$, *where* $\cup\Omega_s \equiv \Omega$ *and the* Ω_s *are the polygons with boundaries* $\partial\Omega_s$. *The boundary segments* $\Gamma_s \subset \Omega$ *are called the interfaces and all the corners* x_t *of the* Ω_s *are potentially points of singularity. Denoting by* ω_t *the neighbourhoods of the* x_t, *we have analogously as before in (2.89a) that*

$$\left| \frac{\partial^{k_1+k_2} u}{\partial x_1^{k_1} \partial x_2^{k_2}} \right|_{C^0(\hat{\Omega})} \leq CQ,$$

where the terms in the Q *associated with the boundary conditions may possibly have vanished,* $\hat{\Omega}_s \equiv \Omega_s - \cup\omega_t$, *and (2.89b) and (2.89c) holds in* $\omega_t \cap \Omega_t$, *where* r_t *is the distance to* x_t. *For example, consider problem (2.63) defined in a polygonal domain* Ω *where* $\mathcal{A} = \mathcal{A}_i$ *in* Ω_i, *and assume that* $u = 0$ *on* Γ, *whilst* $f \neq 0$ *in* Ω. *The boundary conditions may change from Dirichlet to Neumann at points interior to the boundary segments, and these points are also points singularly.*

Example 2.7 Let us consider problem (2.63) in the form

$$-\Delta u(\mathbf{x}) = f(\mathbf{x}), \quad \mathbf{x} \in \Omega,$$

$$u(\mathbf{x}) = 0, \quad \mathbf{x} \in \Gamma,$$

where Ω is the L-shaped domain of Fig. 2.12 and f satisfies condition (2.74d) with $f = 0$ in ω_j. Then, in (2.88b) $\beta_1 = \frac{2}{3}$ in ω_1, whilst $\beta_j = 2, j = 2, \ldots, 6$, so that here all the derivatives of $u(\mathbf{x})$ are bounded. It is only in ω_1 that $u(\mathbf{x})$ is singular.

Example 2.8 Consider again the problem of Example 2.7, but with $f(\mathbf{x}) = 1$, $\mathbf{x} \in \Omega$, Then, not all the derivatives of $u(\mathbf{x})$ are bounded in ω_j, $j = 2, 3, \ldots, 6$, even though the derivative growth condition (2.89) holds in $\hat{\Omega}$.

Exercise 2.13 Consider the problem in which $u(\mathbf{x})$ satisfies

$$-\nabla \cdot [\mathcal{A}(\mathbf{x})] \, \nabla u(\mathbf{x}) = f(\mathbf{x}), \ \mathbf{x} \in \Omega,$$

$$u(\mathbf{x}) = 0, \ \mathbf{x} \in \Gamma,$$

with

$$[\mathcal{A}(\mathbf{x})] = \begin{bmatrix} 3 & 1 \\ 1 & 4 \end{bmatrix},$$

where Ω is the L-shaped domain of the previous two examples and $f(x) = 1$. Show that $u(\mathbf{x})$ satisfies (2.87) in ω_1 and determine the β_1.

Hint: use transformation of coordinates to transform the differential operator into the Laplacian operator Δ.

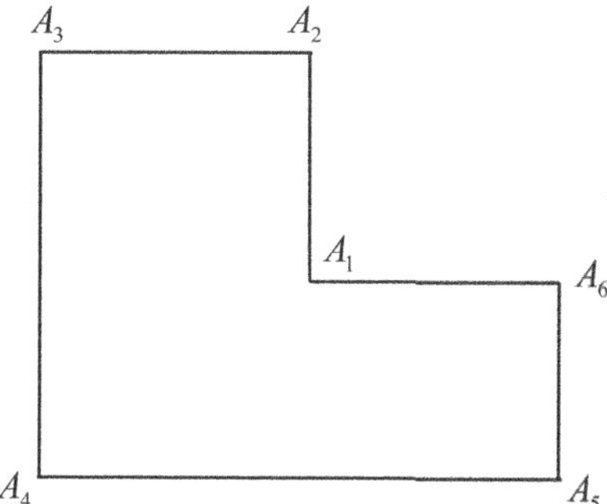

Fig. 2.12 L-shaped domain.

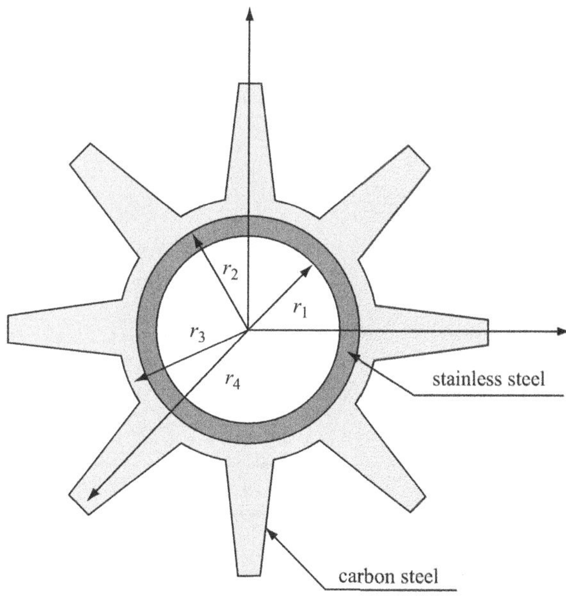

Fig. 2.13 Bimaterial circular pipe with symmetrically positioned fin.

2.2.3 Engineering application; two-dimensional heat-transfer problem

In a manner similar to that of the one-dimensional case of Section 2.1.3., we now define a *two-dimensional heat-transfer problem* (**2D Eng Problem**) with which we shall demonstrate the effectiveness of the finite element method in subsequent chapters.

Consider again a long pipe consisting of materials with constant properties, but with cross-section as shown in Fig. 2.13. The pipe consists of two concentric rings with, as before, the radii of the inner surface, the interface and the outer surface being, respectively, r_1, r_2 and r_3, the inner being made of stainless steel and the outer of carbon steel, but now the pipe structure contains cooling fins that are positioned symetrically at 45° intervals and for which the tips are at distance r_4 from the pipe axis.

It is again assumed that the heat transfer from the outer pipe surface and fins is convective, that the coefficient of convective transfer to the surrounding medium is β, that the temperature of the fluid in the pipe is T_0, whilst that of the surrounding environment is T_1.

The aim of our analysis is, as before, to determine QoI:

Q_1 the temperature at points A and B of the fins.
We have $\mathrm{Q}_1(A)$ and $\mathrm{Q}_1(B)$, see Figs. 2.14 and 2.15.

Q_2 the amount of heat going from the pipe into the surrounding environment per unit length, $\mathrm{Q}_2 = 16 \int_{(3)}^{(6)} \frac{\partial u}{\partial n_c}(\mathbf{x})ds$.

Q_3 the total energy of the pipe per unit length, $\mathrm{Q}_3 = \frac{1}{2}B(u,u) - F(u)$.

Once again, because the pipe is long the flux in the longitudinal direction will be neglected and the problem can be formulated in two dimensions in the region of Fig. 2.13. Because of the symmetry of the problem, we need to consider only one sixteenth of the region of Fig. 2.13, which we give in Fig. 2.14. This region we denote by $\tilde{\Omega}$, taking polar coordinates (r, θ) centred on 0, and denote the relevant boundary points by $1, 2, \ldots, 8$ as shown.

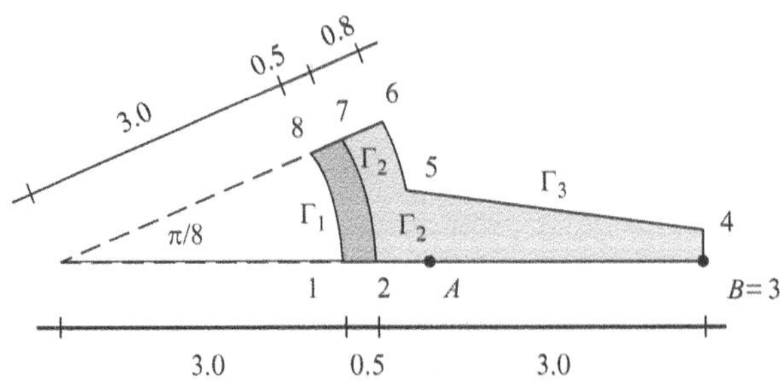

Fig. 2.14 Problem in '1/16' region Ω.

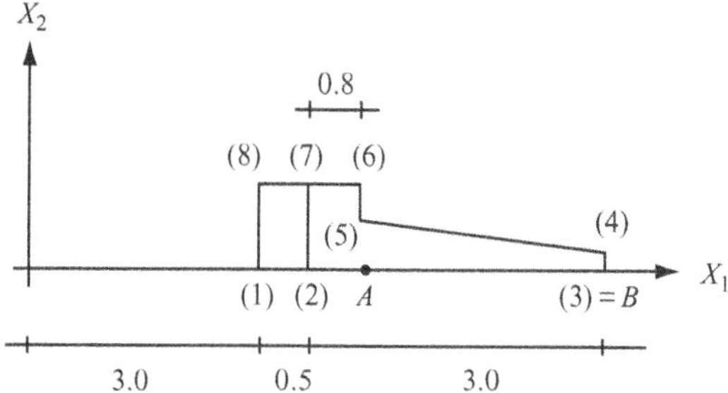

Fig. 2.15 Problem in cartesian co-ordinates (x_1, x_2).

The temperature $u(r,\theta)$ in $\tilde{\Omega}$ satisfies

$$\frac{\partial}{\partial r}\left(\gamma r \frac{\partial u}{\partial r}(r,\theta)\right) + \frac{\gamma}{r}\frac{\partial^2 u}{\partial \theta^2}(r,\theta) = 0, \tag{2.90a}$$

$$0 < \theta < \pi/8,\ r_1 < r < r_4,$$

together with the boundary conditions

$$u(r,\theta) = T_0,\ (r,\theta) \in \overline{18} \tag{2.90b}$$

$$\frac{\partial u}{\partial n_c}(r,\theta) + \beta\left(u(r,\theta) - T_1\right) = 0,\ (r,\theta) \in \overline{3456}, \tag{2.90c}$$

$$\frac{\partial u}{\partial n_c}(r,\theta) = 0,\ (r,\theta) \in \overline{123} \cup \overline{678}. \tag{2.90d}$$

As previously in Section 2.1.3 we write (2.90) in Cartesian coordinates (x_1, x_2), taking now $x_1 = r, x_2 = \theta$ and we thus define the domain Ω with straight sides as in Fig. 2.15 and vertices (1), (2), ..., (8).

Note that we assume that in the (x_1, x_2)-plane the domain is a polygon, whereas the fins in Fig. 2.14 have curved edges.

We define the boundary segments $\Gamma_D \equiv \overline{(1)(8)}, \Gamma_{N_1} \equiv \overline{(1)(2)(3)}, \Gamma_{N_2} \equiv \overline{(3)(4)(5)(6)},\ \Gamma_{N_3} \equiv \overline{(6)(7)(8)}$. In Cartesian coordinates problem (2.91) becomes that of finding $\tilde{u}(x_1, x_2)$ such that (note that we shall still use the symbol u rather than $\tilde{u}$), find $u(x_1, x_2)$ such that

$$\frac{\partial}{\partial x_1}\left(\gamma x_1 \frac{\partial u}{\partial x_1}(x_1,x_2)\right) + \frac{\gamma}{x_1}\frac{\partial^2 u}{\partial x_2^2}(x_1,x_2) = 0,\ (x_1,x_2) \in \Omega \tag{2.91a}$$

$$u(x,0) = T_0,\ (x_1,x_2) \in \Gamma_D \tag{2.91b}$$

$$\frac{\partial u}{\partial n_c}(x_1,x_2) + x_1\beta\left(u(x_1,x_2) - T_1\right) = 0,\ (x_1,x_2) \in \Gamma_{N_2} \tag{2.91c}$$

$$\frac{\partial \mathfrak{u}}{\partial n_c}(x_1, x_2) = 0, (x_1, x_2) \in \Gamma_{N_1} \cup \Gamma_{N_3}. \tag{2.91d}$$

Following the values of Section 2.1.3 we take

$$r_1 = 3.0\,\text{cm}, \; r_2 = 3.5\,\text{cm}, \; r_3 = 4.3\,\text{cm}, \; r_4 = 6.5\,\text{cm}, \tag{2.92}$$

so that

$(1) \equiv (3.0, 0)$, $(2) \equiv (3.5, 0)$ $A \equiv (4.3, 0)$ $B \equiv (3) \equiv (6.5, 0)$,
$(4) \equiv (6.5, 0.030769)$, $(5) \equiv (4.3, 0.093023)$, $(6) \equiv (4.3, 0.392699)$,
$(7) \equiv (3.5, 0.392699)$, $(8) \equiv (3.0, 0.392699)$.

Taking

$$\mathcal{U}_0 \equiv \{v \in \mathcal{U} \mid v(\mathbf{x}) = 0, \; \mathbf{x} \in \Gamma_{\mathrm{D}}\}, \tag{2.93a}$$

$$\mathcal{U}_{g_1} \equiv \{v \in \mathcal{U} \mid v(\mathbf{x}) = g_1, \; \mathbf{x} \in \Gamma_{\mathrm{D}}\}, \tag{2.93b}$$

the weak form of problem (2.91) is : find $\mathfrak{u} \in \mathcal{U}_{g_1}$ such that

$$B(\mathfrak{u}, v) = \int_\Omega \left(\gamma x_1 \frac{\partial \mathfrak{u}}{\partial x_1} \frac{\partial v}{\partial x_1} + \frac{\gamma}{x_1} \frac{\partial \mathfrak{u}}{\partial x_2} \frac{\partial v}{\partial x_2} \right) d\mathbf{x}$$

$$+ \int_{\Gamma_{N_1} \cup \Gamma_{N_2}} x_1 \beta \left(u(x) - T_1 \right) v ds = 0, \quad \forall v \in \mathcal{U}_0, \tag{2.94}$$

where

$$\gamma(x) = \begin{cases} \gamma_{ss}(\mathbf{x}), & r_1 < x_1 < r_2, \\ \gamma_{cs}(\mathbf{x}), & r_2 < x_1 < r_4. \end{cases} \tag{2.61}$$

Returning now to the aim of our computations for this problem, we are interested in the quantities of interest $Q_i, i = 1, 2, 3$.

Figure 2.15 together with Remark 2.14 would suggest that the solution $\mathfrak{u}$ is singular in the neighbourhoods of all the points (1), ..., (8). This is, however, not the case because Fig. 2.15 stems from Fig. 2.13 for which symetry was applied, and in fact only (4) and (5) are points of singularity.

3
Finite element methods

Summary

- **A description of the general Galerkin method for boundary value problems with mixed boundary conditions is first given.**
- **The finite element method is presented as a specific form of the Galerkin technique, and is given in the context of the one- and two-dimensional boundary value problems of Chapter 2.**
- **The basics of the finite element method (in the Galerkin setting) based on piecewise polynomial functions defined over (mesh) partitions of the problem domain are then described, first for one-dimensional and then for two-dimensional problems.**
- **The sequence followed is: use of piecewise linear and piecewise quadratic functions; use of mappings to generate the element stiffness matrices and load vectors; assembly of the global stiffness equations; application of the Dirichlet boundary conditions to produce the constrained global stiffness equations.**

As this chapter is devoted to the basics of the method, we do **not** discuss error analysis. However, thinking ahead, and by way of introduction, we **state** the best approximation property of the finite element solution and discuss **errors** in a series of remarks.

3.1 Introduction

In this chapter we describe the finite element method for boundary value problems of the type presented in Chapter 2. The method will be applied to the weak (virtual work) formulations, see (2.33) and (2.72), of these problems. Our strategy is to seek approximations $u^{(n)}(\mathbf{x})$ to $\mathrm{u}(\mathbf{x})$, which can be associated with an n-dimensional vector $\mathbf{c} \equiv (c_1, \ldots, c_n)^T \in \mathbb{R}^n$, and the finite element method will be the recipe for obtaining $u^{(n)}(\mathbf{x})$ with the help of $\mathbf{c}$ so that the resulting $u^{(n)}(\mathbf{x})$ is **close** to $\mathrm{u}(\mathbf{x})$. Because computers can deal only with finite numbers, and finite numbers of numbers, the approach must be **finite-dimensional.**

We first describe the Galerkin method for producing representations of the solutions of the weak problems in finite-dimensional subspaces of the energy space. The finite element method is a special case of the Galerkin method, in which the finite-dimensional spaces are constructed using bases of continuous piecewise polynomial functions defined on **meshes** that are partitions of the domain Ω into non-overlapping subdomains, **elements**.

It will be shown later that the finite element solution is the function from the finite element space that minimizes the error in the *energy norm*. The significance of this is profound, and will be discussed at length later, as it underlies the process of determining error estimates, convergence rates and error estimators, as well as adaptive techniques and error control for finite element methods. We merely state this property in this chapter.

Whilst our treatment of Galerkin methods is general, we treat one-and two-dimensional **finite element methods** separately, as we consider that this makes the understanding of the underlying techniques simpler.

3.2 The Galerkin method

In Chapter 2 we defined boundary value problems in bounded domains $\Omega \subset \mathbb{R}^n, n = 1, 2$, where Γ is the boundary of Ω. We shall now assume that in the two-dimensional case Γ is a polygon, and that $\Gamma \equiv \Gamma_D \cup \Gamma_N$ with again $\Gamma_D \cap \Gamma_N = \emptyset$.

Referring back to (2.23) and (2.24) and (2.68)–(2.71) we recall that

$$\mathcal{U} \equiv \{v(\mathbf{x}), \mathbf{x} \in \Omega \mid \|v\|_{\mathcal{U}} < \infty\}, \tag{3.1a}$$

$$\mathcal{U}_0 \equiv \{v(\mathbf{x}) \in \mathcal{U} \mid v(\mathbf{x}) = 0, \mathbf{x} \in \Gamma_D\}, \tag{3.1b}$$

$$\mathcal{U}_g \equiv \{v(\mathbf{x}) \in \mathcal{U} \mid v(\mathbf{x}) = g(\mathbf{x}), \mathbf{x} \in \Gamma_D\}, \tag{3.1c}$$

and we define

$$\mathcal{U}_{\Gamma_D} \equiv \{v(\mathbf{x}) \in \mathcal{U}, \mathbf{x} \in \Gamma_D\}. \tag{3.1d}$$

Note that (3.1c) makes the assumption that $g(\mathbf{x}) \in \mathcal{U}_{\Gamma_D}$.

The norm $\|v\|_{\mathcal{U}} = (B(v,v))^{\frac{1}{2}}$ was defined in (2.21c) for the one-dimensional case and in (2.72) for the two-dimensional case.

We now consider again problems of the type (2.33) (one-dimensional) and (2.72) (two-dimensional), where we seek $\mathrm{u} = \bar{u} + u_0$ with $u_0 \in \mathcal{U}_0$, and $\bar{u} \in \mathcal{U}_g$ such that

$$B(\mathrm{u}, v) = F(v) \ \ \forall v \in \mathcal{U}_0, \tag{3.2}$$

where we assume that $\bar{u} \in \mathcal{U}$ is such that

$$\bar{u}(\mathbf{x}) = g(\mathbf{x}), \mathbf{x} \in \Gamma_D.$$

Substituting $\mathrm{u} = \bar{u} + u_0$ we have that

$$B(u_0, v) = F(v) - B(\bar{u}, v) \ \forall v \in \mathcal{U}_0. \tag{3.3}$$

We emphasize again that $B(\cdot,\cdot)$ is a symmetric bilinear form defined on $\mathcal{U} \times \mathcal{U}$ and the Neumann boundary condition on Γ_N produces a term that is contained in $F(v)$. Further, $\|u\|_{\mathcal{U}} \equiv (B(u,u))^{\frac{1}{2}}$ has all the properties of the norm on $\mathcal{U}$ if $c \neq 0$ or $\Gamma_D \neq 0$ or $\alpha_2 \neq 0$ see (2.21c). Otherwise, $\|u\|_{\mathcal{U}}$ is only a semi-norm modulo a constant, i.e. $\|\text{constant}\|_{\mathcal{U}} = 0$. In this case, a necessary and sufficient condition for the existence of the solution is F (constant)$= 0$. The solution is then unique up to an arbitrary additive constant.

Remark 3.1 *The solution* $\mathfrak{u}$ *is independent of the manner in which* $g(\mathbf{x})$, $\mathbf{x} \in \Gamma_{\mathrm{D}}$ *is extended into* $\bar{u} \in \mathcal{U}$, *and many such extensions exist. Obviously,* u_0 *in (3.3) depends on this extension, but two extensions* $\tilde{g}$ *and* $\hat{g}$ *of* $g \in U_{\Gamma_{\mathrm{D}}}$ *produce the same solution* $\mathfrak{u}$.

In order to produce a numerical solution to (3.2) we introduce the finite-dimensional spaces as in (3.7), ${}^{N}S$ of dimension N, ${}^{N}S_0$ of dimension N_1, and ${}^{N}S_{\mathrm{D}}$ and ${}^{N}S_{\mathrm{D}}^{\Gamma}$ of dimension N_2, where ${}^{N}S_{\mathrm{D}}^{\Gamma}$ is the space of traces of ${}^{N}S$ on Γ_{D}, and ${}^{N}S_{\mathrm{D}}^{\Gamma} \subset \mathcal{U}_{\Gamma_{\mathrm{D}}}$. We shall assume that the data functions are in ${}^{N}S$.

Let $\{\Phi_i(\mathbf{x})\}_{i=1}^{N}$ be the *basis* functions of ${}^{N}S$; i.e. $\Phi_i \in \mathcal{U}$, $i = 1, 2, \ldots, N$ and every $u \in {}^{N}S$ can be uniquely written in the form

$$u(\mathbf{x}) = \sum_{i=1}^{N} \bar{c}_i \Phi_i(\mathbf{x}), \tag{3.4}$$

where the $\bar{c}_i$ are parameters, and we have that $u = 0 \Leftrightarrow \bar{c}_i = 0, i = 1, \ldots, N$.

Let us now introduce the sets of numbers

$$\mathcal{N} = \{1, \ldots, N\}, \mathcal{N}_1 \subset \mathcal{N},\ \mathcal{N}_2 \subset \mathcal{N},\ \mathcal{N}_1 \cup \mathcal{N}_2 = \mathcal{N},\ \mathcal{N}_1 \cap \mathcal{N}_2 = 0,$$

so that

$$\Phi_i \subset {}^{N}S_0,\ \ i \in \mathcal{N}_1, \tag{3.5}$$

$$\Phi_i(\mathbf{x}) \neq 0,\ \mathbf{x} \in \Gamma_D, i \in \mathcal{N}_2. \tag{3.6}$$

We shall assume that $\Phi_i(\mathbf{x})$, $\mathbf{x} \in \Gamma_D$, $i \in \mathcal{N}_2$ create the basis of ${}^{N}S_D^{\Gamma}$; i.e. these $\Phi_i(\mathbf{x})$ are linearly independent.

We then have:

$${}^{N}S \equiv \left\{ v \in \mathcal{U} \mid v(\mathbf{x}) = \sum_{i \in \mathcal{N}} c_i \Phi_i(\mathbf{x}) \right\}, \tag{3.7a}$$

$${}^{N}S_0 \equiv \left\{ v \in \mathcal{U}_0 \mid v(\mathbf{x}) = \sum_{i \in \mathcal{N}_1} c_i \Phi_i(\mathbf{x}) \right\}, \tag{3.7b}$$

$$\begin{aligned} {}^{N}S_{\mathrm{D}} &= \{ u \in {}^{N}S \mid u \notin {}^{N}S_0 \} \\ &= \left\{ v \in {}^{N}S \mid v(\mathbf{x}) = \sum_{i \in \mathcal{N}_2} c_i \Phi_i(\mathbf{x}) \right\}, \end{aligned} \tag{3.7c}$$

$${}^{N}S_{\mathrm{D}}^{\Gamma} \equiv \left\{ v \in \mathcal{U}_{\Gamma_{\mathrm{D}}} \mid v(\mathbf{x}) = \sum_{i \in \mathcal{N}_2} c_i \Phi_i(\mathbf{x}) \,|_{\mathbf{x} \in \Gamma_D} \right\}. \tag{3.7d}$$

We shall assume that $g(\mathbf{x}) \in {}^{N}S_{\mathrm{D}}$ and hence we can uniquely write

$$g(\mathbf{x}) = \sum_{i \in \mathcal{N}_2} d_i \Phi_i(\mathbf{x}), \mathbf{x} \in \Gamma_{\mathrm{D}}, \tag{3.8}$$

where the d_i are determined directly from the known Dirchlet boundary function g on Γ_D. Now, the numerical solution $u^N \in {}^N S$ to (3.2) will be expressed in the form

$$^N u(\mathbf{x}) = \sum_{i \in \mathcal{N}} \bar{c}_i \Phi_i(\mathbf{x}), \tag{3.9}$$

with

$$\bar{c}_i = d_i \text{ for } i \in \mathcal{N}_2. \tag{3.10}$$

Hence, our goal is to determine the vector $\mathbf{c} \equiv (c_1, c_2, \ldots, c_N)^T$. We now illustrate this with an example.

Note that the assumption that $g(\mathbf{x}) \in {}^N S_D$ is made for simplicity in order to avoid the influence of the error in the boundary condition when $g \notin {}^N S_D$.

Example 3.1 Let us consider a problem of type (2.63) in which $\mathfrak{u}(\mathbf{x})$ satisfies

$$-\Delta \mathfrak{u}(\mathbf{x}) = 1, \ \mathbf{x} \in \Omega$$

$$\mathfrak{u}(\mathbf{x}) = g(\mathbf{x}) \equiv x_2^2 - 2x_1^2, \ \mathbf{x} \in \Gamma,$$

where Ω is the unit square,

$$\Omega \equiv \{(x_1, x_2) \mid |x_1| < 1, |x_2| < 1\},$$

with boundary $\Gamma \equiv \Gamma_D$. The energy space $\mathcal{U}$ is in this case

$$\mathcal{U} \equiv \left\{ u = u(\mathbf{x}) \mid \int_\Omega \left(\left(\frac{\partial u}{\partial x_1} \right)^2 + \left(\frac{\partial u}{\partial x_2} \right)^2 \right) d\mathbf{x} < \infty \right\},$$

and

$$\mathcal{U}_g \equiv \{ u(\mathbf{x}) \in \mathcal{U} \mid u(\mathbf{x}) = g(\mathbf{x}), \ \mathbf{x} \in \Gamma \}.$$

Clearly, there exists $\bar{u} \in \mathcal{U}$ such that $\bar{u}(\mathbf{x}) = g(\mathbf{x}), \mathbf{x} \in \Gamma$, since we can take $\bar{u}(\mathbf{x}) = x_2^2 - 2x_1^2$.

The weak formulation of the problem is:

$$\text{find } \mathfrak{u} \in \mathcal{U}_g \ \text{ such that}$$

$$B(\mathfrak{u}, v) \equiv \int_\Omega \left(\frac{\partial \mathfrak{u}}{\partial x_1}(\mathbf{x}) \frac{\partial v}{\partial x_1}(\mathbf{x}) + \frac{\partial \mathfrak{u}}{\partial x_2}(\mathbf{x}) \frac{\partial v}{\partial x_2}(\mathbf{x}) \right) dx = \int_\Omega v(\mathbf{x}) d\mathbf{x}, \ \forall v \in \mathcal{U}_0,$$

or equivalently

$$\text{find } u_0 \in \mathcal{U}_0 \text{ such that}$$

$$B(u_0, v) = \int_\Omega v(\mathbf{x}) d\mathbf{x} - B(\bar{u}, v) \ \ \forall v \in \mathcal{U}_0.$$

We now define the space ${}^N S$ by the basis functions $\Phi_i(\mathbf{x}) \in \mathcal{U}, i = 1, \ldots, N, N = 9$, where

$$\Phi_1 = 1, \quad \Phi_2 = x_1, \quad \Phi_3 = x_2, \quad \Phi_4 = x_1^2, \quad \Phi_5 = x_1 x_2, \quad \Phi_6 = x_2^2,$$
$$\Phi_7 = \left(1 - x_1^2\right)\left(1 - x_2^2\right), \quad \Phi_8 = \left(1 - x_1^2\right)\left(1 - x_2^2\right) x_1,$$
$$\Phi_9 = \left(1 - x_1^2\right)\left(1 - x_2^2\right) x_2.$$

Then,

$${}^N S \equiv \left\{ u \mid u(\mathbf{x}) = \sum_{i=1}^{9} \bar{c}_i \Phi_i(\mathbf{x}), \bar{c}_i \in \mathbb{R} \right\},$$

and the space

$${}^N S_0 \equiv \left\{ u \mid u(\mathbf{x}) = \sum_{i=7}^{9} \bar{c}_i \Phi_i(\mathbf{x}), \bar{c}_i \in \mathbb{R} \right\}$$

has dimension 3 and hence $N_1 = 3$. Further,

$${}^N S_{\mathrm{D}} \equiv \left\{ u(\mathbf{x}) \mid u(\mathbf{x}) = \sum_{i=1}^{6} \bar{c}_i \Phi_i(\mathbf{x}), \bar{c}_i \in \mathbb{R} \right\},$$

$${}^N S_{\mathrm{D}}^{\Gamma} \equiv \left\{ u(\mathbf{x}), \mathbf{x} \in \Gamma \mid u(\mathbf{x}) = \sum_{i=1}^{6} \bar{c}_i \Phi_i(\mathbf{x}), \bar{c}_i \in \mathbb{R} \right\}.$$

Note that ${}^N S_{\mathrm{D}}^{\Gamma}$ is defined only on Γ via the traces of the Φ_i. Thus, $N_1 = 3$, $N_2 = 6, \mathcal{N}_1 \equiv \{7, 8, 9\}, \mathcal{N}_2 = \{1, 2, 3, 4, 5, 6\}$, and $d_1 = d_2 = d_3 = d_5 = 0$, $d_4 = -2$ and $d_6 = 1$.

On account of (3.10) we may write

$${}^N u(\mathbf{x}) = \sum_{i \in \mathcal{N}_1} c_i \Phi_i(\mathbf{x}) + \sum_{i \in \mathcal{N}_2} d_i \Phi_i(\mathbf{x}), \tag{3.11}$$

and our task is to determine $c_i \in \mathcal{N}_1$.

Writing $\mathbf{c} = \{c_i\}_{i \in \mathcal{N}_1}$ and $\mathbf{d} = \{d_i\}_{i \in \mathcal{N}_2}$ we now construct a system of linear equations for $\mathbf{c}$. For this we shall use (3.2) and (3.9) and (3.10) to determine $\mathbf{c}$ so that

$$B({}^N u, v) = F(v) \;\; \forall v \in {}^N S_0, \tag{3.12a}$$

or

$$B\left(\sum_{i \in \mathcal{N}_1} c_i \Phi_i, \Phi_j \right) = F(\Phi_j) - B\left(\sum_{i \in \mathcal{N}_2} d_i \Phi_i, \Phi_j \right), \; j \in \mathcal{N}_1. \tag{3.12b}$$

Note that in producing (3.12a) we have interpreted (3.2) in the context of finite-dimensional space; and in particular for all $v \in {}^N S_0$. If (3.12b) holds for each $\Phi_j \in {}^N S_0$, then it holds for all $v \in {}^N S_0$.

Hence, denoting $\bar{\mathbf{f}} \equiv \{f_j\}_{j\in\mathcal{N}}$, $\hat{\mathbf{f}} \equiv \{f_j\}_{j\in\mathcal{N}_1}$, where $f_j \equiv F(\Phi_j)$, we obtain from (3.11) and (3.12) the linear system

$$[K]\mathbf{c} = \mathbf{f} - [K_{12}]\mathbf{d} \equiv \hat{\mathbf{f}}, \tag{3.13}$$

where $[K]$ is called the *constrained stiffness matrix of order* $N_1 \times N_1$, $\mathbf{f}$ is the *constrained load vector,* and $\hat{\mathbf{f}}$ is the *augmented load vector.*

In (3.13) we have that

$$[K] \equiv [B(\Phi_i, \Phi_j)],\ i, j \in \mathcal{N}_1, \tag{3.14a}$$

and

$$[K_{12}] \equiv [B(\Phi_i, \Phi_j)],\ i \in \mathcal{N}_1, j = \mathcal{N}_2. \tag{3.14b}$$

We note also that $[K]$ in (3.14a) is an $N_1 \times N_1$ matrix, whilst $[K_{12}]$ in (3.14b) is an $N_1 \times N_2$ matrix.

In order to complete the system (3.13) we must calculate the vector $\mathbf{d}$ that arises from the Dirichlet boundary conditions on Γ_{D}. One way of doing this is on Γ_{D} to use (3.8) to evaluate

$$\sum_{i\in\mathcal{N}_2} d_i \int_{\Gamma_{\mathrm{D}}} \Phi_i(\mathbf{x})\Phi_j(\mathbf{x})ds = \int_{\Gamma_{\mathrm{D}}} g(\mathbf{x})\Phi_j(\mathbf{x})ds,\ j \in \mathcal{N}_2. \tag{3.15}$$

We are finally able to substitute the calculated values of $\mathbf{c}$ from (3.13) and $\mathbf{d}$ from (3.15) into (3.11) to obtain ${}^{N}u(\mathbf{x})$ the solution of (3.12a). Note that we assumed in (3.8) that the d_i exist and are unique. Hence, they can be obtained from (3.15).

It is often the practice, for ease of programming, instead of setting up (3.13) to set up the unconstrained stiffness matrix $[\bar{K}]$ of dimension $N \times N$ where

$$[\bar{K}]_{N\times N} \equiv [\bar{k}_{ij}]_{N\times N} = \begin{bmatrix} [K_{11}] & [K_{12}] \\ [K_{21}] & [K_{22}] \end{bmatrix}_{N_1+N_2}, \tag{3.16}$$

with $[K_{21}] = [K_{12}]^T$, which is associated with the case in which the boundary value problem being approximated (2.31) or (2.72)) has no prescribed Dirichlet boundary conditions; i.e. $\Gamma_{\mathrm{D}} = \emptyset$. Similarly, one sets up the vector $\bar{\mathbf{f}}_{N\times 1}$. The K_{22} is actually never used, and nor are the elements of $\bar{\mathbf{f}}$ that do not appear in $\mathbf{f}$. However, as noted above, from the programming point of view it is simpler to derive the whole of $[\bar{K}]$ and the whole of $\bar{\mathbf{f}}$. Of course, one could decide not to compute the data that are not needed.

Remark 3.2 *The reason for the (complicated) notation using $\mathcal{N}_1$ and $\mathcal{N}_2$ will be seen later in the two-dimensional setting where there is no 'natural' ordering of the elements and nodal points comparable to that which exists for one-dimensional problems.*

Example 3.2 Considering again Example 3.1 with the nine basis functions $\Phi_i(\mathbf{x}), i = 1, \ldots, 9$, we now compute the unconstrained matrix $[\bar{K}]$ of dimension $N \times N, N = 9$.

By simple computations we obtain

$$
[\bar{K}] = \begin{bmatrix}
0 & 0 & 0 & 0 & 0 & 0 & 0 & 0 & 0 \\
 & 4 & 0 & 0 & 0 & 0 & 0 & 0 & 0 \\
 & & 4 & 0 & 0 & 0 & 0 & 0 & 0 \\
 & & & \frac{16}{3} & 0 & 0 & \frac{-32}{9} & 0 & 0 \\
 & & & & \frac{8}{3} & 0 & 0 & 0 & 0 \\
 & & & & & \frac{16}{3} & \frac{-32}{9} & 0 & 0 \\
 & & & & & & \frac{256}{45} & 0 & 0 \\
 & & & & & & & \frac{3328}{1575} & 0 \\
 & & & & & & & & \frac{3328}{1575}
\end{bmatrix}.
$$

Similarly, for the unconstrained load vector $\bar{\mathbf{f}}$ of dimension $N \times 1, N = 9$ we obtain

$$
\bar{\mathbf{f}} = \begin{bmatrix} 4 \\ 0 \\ 0 \\ \frac{4}{3} \\ 0 \\ \frac{4}{3} \\ \frac{16}{9} \\ 0 \\ 0 \end{bmatrix}.
$$

By partitioning $[\bar{K}]$ as in (3.16) we see that

$$
[K_{11}] = \begin{bmatrix} \frac{256}{45} & 0 & 0 \\ 0 & \frac{3328}{1575} & 0 \\ 0 & 0 & \frac{3328}{1575} \end{bmatrix}, [K_{12}] = \begin{bmatrix} 0 & 0 & 0 & \frac{-32}{9} & 0 & \frac{-32}{9} \\ 0 & 0 & 0 & 0 & 0 & 0 \\ 0 & 0 & 0 & 0 & 0 & 0 \end{bmatrix},
$$

and the augmented load vector is

$$
\hat{\mathbf{f}} = \begin{bmatrix} \frac{304}{27} \\ 0 \\ 0 \end{bmatrix}.
$$

Hence, $\mathbf{c}$ satisfies

$$
\begin{bmatrix} \frac{256}{45} & 0 & 0 \\ 0 & \frac{3328}{1575} & 0 \\ 0 & 0 & \frac{3328}{1575} \end{bmatrix} \begin{bmatrix} c_7 \\ c_8 \\ c_9 \end{bmatrix} = \begin{bmatrix} \frac{304}{27} \\ 0 \\ 0 \end{bmatrix},
$$

and the solution of the problem of Example 3.1 is:

$$
{}^{N}u(\mathbf{x}) = +\frac{215}{108}\Phi_7(\mathbf{x}) - 2\Phi_4(\mathbf{x}) + \Phi_6(\mathbf{x}).
$$

We see that the process of finding the numerical solution ${}^{N}u(\mathbf{x})$ of the Galerkin problem (3.12b) has the following three steps:

Step 1: *Construction of the stiffness matrices* $[K]$ *and* $[K_{12}]$, *the load vector* $\mathbf{f}$, *the vector* $\mathbf{d}$ and the vector $\hat{\mathbf{f}}$.

Step 2: *The solving of the linear equations (3.13) for* $\mathbf{c}$.

Step 3: *The recovery of* ${}^N u(\mathbf{x}), \mathbf{x} \in \Omega$, *and also other data of interest using (3.11).*

To these three steps we may add:

Step 4: *The computation of* error estimates *for the computed data of the problem (3.13). We define here the error to be the difference between the exact (unknown) solution of (3.2) and the numerical solution* ${}^N u$ *of (3.11).*

Step 5: *If the error in Step 4 is not satisfactory, to modify the numerical scheme to improve the numerical solution and hence reduce the error to a required tolerance.*

Note that the constructions in Step 1 often involve the use of numerical integration that has to be done in a manner that introduces sufficiently small errors.

In this book we shall concentrate mainly on Steps 3–5. However, for both the one- and two-dimensional contexts we shall first describe the finite element form of the Galerkin method above. We shall not linger on Step 2, which involves the solving of the linear equations (3.13). Because the matrix $[K]$ is sparse, symmetric and positive definite, many methods are available for solving systems of this type efficiently. Today, Gaussian elimination techniques are available in practice for routinely solving systems up to order 10^6 or more. It is not our purpose to discuss such techniques.

If the Neumann condition is prescribed on the entire boundary Γ, i.e. $\Gamma_{\mathrm{D}} = \varnothing$, and the coefficient $c = 0$ in the bilinear form $B(\cdot,\cdot)$, the matrix $[K]$ is singular with exactly one eigenvalue being zero, and the eigenvector is a *constant*. (Note that this belongs to ${}^N S$ by assumption). This means that $[K]\bar{\mathbf{c}} = \mathbf{0}$ if and only if $\sum c_i \Phi_i = \bar{u} = constant$. The problem $[K]\mathbf{c} = \mathbf{f}$ has a solution if $F(constant) = 0$, which is the necessary and sufficient condition for the existence of the original problem.

In order to obtain a non-singular matrix $[K]$ we impose an additional constraint to determine the arbitrary constant in the solution. For example, let $x_0 \in \Gamma$, then the constraint is $\sum c_i \Phi_i(\mathbf{x}_0) = 0$ or, if $\int_\Omega \Phi_j d\mathbf{x} \neq 0$, the condition could be that $c_j = 0$. With either of these constraints the matrix $[K]$ becomes regular.

Remark 3.3 *Once the constraint is imposed, the solution* ${}^N u$ *also exists when* $F(constant) \neq 0$, *i.e. the consistency condition is not satisfied. In this case, whilst* ${}^N u$ *exists, it cannot be close to the solution* $\mathfrak{u}$ *of the original problem.*

3.3 One-dimensional finite element method

In the one-dimensional setting, we seek to use the finite element method to approximate the solution $\mathfrak{u}(x), x \in [0,l] \equiv \bar{I}$ of the weak form of the bar problem (2.31). As has been said, the finite element method is a specific form of the Galerkin method for constructing ${}^N S$ and ${}^N S_0$, and in this one-dimensional setting is based on piecewise polynomial functions of x defined over partitions of I. We shall consider separately the method using piecewise linear and piecewise quadratic basis functions.

3.3.1 The finite element method with piecewise linear functions

On $\bar{I} = [0, l], \Gamma = \{0\} \cup \{l\}$ with $\Gamma_\text{D} = \{0\}$ we define a **mesh** Δ,

$$0 = x_1^\Delta < x_2^\Delta < \ldots < x_{M+1}^\Delta = l,$$

which partitions I into elements $\tau_q^\Delta = (x_q^\Delta, x_{q+1}^\Delta)$, $q = 1, \ldots, M$. Note that τ_q^Δ is an open element, whilst $\bar{\tau}_q^\Delta \equiv [x_q^\Delta, x_{q+1}^\Delta]$ is closed. We shall use the term *mesh* to denote the collection of elements.

The points with coordinates x_q^Δ are the *vertices* of the mesh, and are denoted by $V_q^\Delta, q = 1, \ldots, M+1$. The terminology vertex actually relates to the two-dimensional case.

We denote by $h_q^\Delta \equiv |\tau_q^\Delta| = x_{q+1}^\Delta - x_q^\Delta$ the element sizes, and let $h^\Delta \equiv \max h_q^\Delta$, $q = 1, \ldots, M$. Where no confusion arises, we shall for simplicity omit the Δ when using the symbols x_q, h_q and τ_q.

For our model problem (3.2), (2.33) we now define the spaces

$$S_\Delta^{[1]} \equiv \left\{ v(x), v \in C^0(\bar{I}) \mid v|_{\tau_q} \in \mathbb{P}^{[1]}(\bar{\tau}_q) \right\}, \tag{3.17a}$$

$$S_{\Delta,0}^{[1]} \equiv \left\{ v(x), v \in S_\Delta^{[1]} \mid v(x) = 0, x \in \Gamma_D \right\}, \tag{3.17b}$$

$$S_{\Delta,D}^{[1]} \equiv \left\{ v(x) \mid v \in S_\Delta^{[1]}, v \notin S_{\Delta,0}^{[1]} \right\}, \tag{3.17c}$$

$$S_{\Delta,g_1}^{[1],} \equiv \left\{ v(x), v \in S_\Delta^{[1]} \mid v(x) = g_1, x \in \Gamma_D \right\}. \tag{3.17d}$$

Note that in (3.17d) we have, as in (2.23), a hyperplane rather than a space. However, we shall speak about a space when no misunderstanding can occur.

Note that now the d in (3.8) in this case consists of a single value because $N_2 = 1$. The dimension of $S_\Delta^{[1]}$ is $M+1, C^0(I)$ is the space of all continuous function on $\bar{I}, \mathbb{P}^{[1]}(\tau_q)$ is the space of linear polynomials on τ_q, and $v \mid \tau_q$ means the restriction of v on τ_q. Clearly, $S_\Delta^{[1]} \subset \mathcal{U}$ and $S_{\Delta,0}^{[1]} \subset \mathcal{U}_0$. We identify the spaces $S_\Delta^{[1]}, S_{\Delta,0}^{[1]}, S_{\Delta,D}^{[1]}$ and $S_{\Delta,g_1}^{[1]}$, respectively, with the spaces ${}^N S$ and ${}^N S_0$ of (3.7), ${}^N S_\text{D}$ and ${}^N S_\text{D}^\Gamma$, with $N = M+1$. This association between the ${}^N S$ spaces and the S_Δ spaces will continue throughout.

The finite element problem for (3.2) is in this case defined as

$$\text{find } u_\Delta^{[1]} \in S_{\Delta,g}^{[1]}, \text{ such that}$$

$$B(u_\Delta^{[1]}, v) = F(v) \; \forall v \in S_{\Delta,0}^{[1]}. \tag{3.18}$$

Now, $v(x) \in S_\Delta^{[1]}$ is completely defined by its values at $x_q, q = 1, \ldots, M+1$. The vertices $V_q, q = 1, \ldots, M+1$ of the mesh are in this *piecewise linear* case called *nodal points* that we denote by z_k (in this case $z_q \equiv V_q$), and we can write

$$v(x) = \sum_{q=1}^{M+1} c_q \Phi_q^{[1]}(x), \tag{3.19}$$

where

$$\Phi_q^{[1]}(x) = \begin{cases} (x_{q+1} - x)/h_q, \; x \in \bar{\tau}_q, \\ (x - x_{q-1})/h_{q-1}, \; x \in \bar{\tau}_{q-1}, \\ 0 \qquad x \notin \tau_q \cup \bar{\tau}_{q-1}, \end{cases} \tag{3.20}$$

for $q = 2, \ldots, M$,

$$\Phi_1^{[1]}(x) = \begin{cases} (x_2 - x)/h_1, \; x \in \bar{\tau}_1, \\ 0 \qquad x \notin \bar{\tau}_1, \end{cases}$$

and

$$\Phi_{M+1}^{[1]}(x) = \begin{cases} (x - x_M)/h_M, \; x \in \bar{\tau}_M, \\ 0 \qquad x \notin \bar{\tau}_M \end{cases}$$

with the bars denoting closures of the elements.

In the case of $S_{\Delta,0}^{[1]}$, we have $\Phi_q^{[1]}(x) = 0, x \in \Gamma_D$, see Fig. 3.1. Clearly, $\Phi_q^{[1]}(x)$ are continuous functions on $\bar{I}$ and satisfy

$$\Phi_q^{[1]}(z_j) = \begin{cases} 1 \;\; j = q, \\ 0 \;\; j \neq k. \end{cases} \tag{3.21}$$

We now return to (3.12b) and for the piecewise linear case define, corresponding, respectively, to ${}^N S, {}^N S_0$ and ${}^N S_{\Gamma_D}$ in (3.7),

$$S_\Delta^{[1]} \equiv \left\{ v \mid v = \Sigma_{q=1}^{M+1} c_q \Phi_q^{[1]} \right\} \tag{3.22a}$$

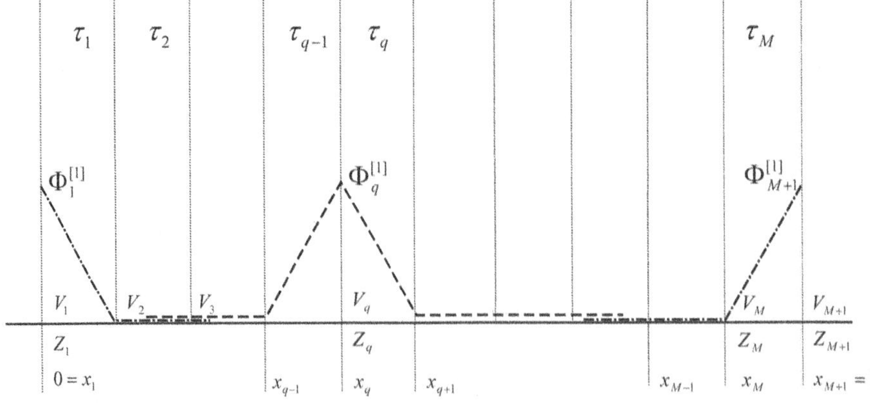

Fig. 3.1 The mesh Δ and the hat functions $\Phi_q^{[1]}$.

$$S^{[1]}_{\Delta,0} \equiv \left\{ v \mid v = \Sigma_{q=2}^{M+1} c_q \Phi_q^{[1]} \right\} \tag{3.22b}$$

$$S^{[1]}_{\Gamma_D} \equiv \left\{ v \mid v = d_1 \Phi_1^{[1]},\ d_1 = g_1 \right\}. \tag{3.22c}$$

In (3.22) the upper index in square brackets indicates the degree of polynomial used in each element, 1, whilst Δ indicates the mesh.

Again, as the context allows we shall as appropriate omit these indices in order to simplify the notation. Note that in (3.22) $\Gamma_D = x_1$.

Because of their distinctive shape the basis functions $\Phi_q^{[1]}, q = 1, \ldots, M+1$ are usually referred to as hat or pyramid functions. Note the specific forms of $\Phi_1^{[1]}$ and $\Phi_{M+1}^{[1]}$. The $\Phi_q^{[1]}$ are of course linearly independent.

The complete set of basis functions $\Phi_q^{[1]}(x), q = 1, \ldots, M+1$ associated with the mesh Δ is illustrated in Fig. 3.1. Using this complete set we construct the unconstrained stiffness matrix $[\bar{K}]_{(M+1)\times(M+1)}$ as in (3.16).

If we had $\Gamma_D = \emptyset$, i.e. there was no part of the boundary Γ with a Dirichlet boundary condition, then we would have $[K] = [\bar{K}]$. If, as is often the case $\Gamma_D \neq \emptyset$ then, having computed $[\bar{K}]$, we have to identify the node numbers $j \in \mathcal{N}_2$ associated with the Dirichlet boundary condition; in the present case the single node number 1. We say that $[K_{12}]$ is the submatrix associated with the rows corresponding to the nodes $j \in \mathcal{N}_2$ and the columns associated with $j \in \mathcal{N}_1$; in our case $[K_{12}] \equiv [\bar{k}_{12}, \ldots, \bar{k}_{1,M+1}]^T$ and $K = [\bar{k}_{ij}]\, i, j = 2, \ldots, M+1$.

Correspondingly, having computed $\mathbf{f}$, we must compute the augmented form $\hat{\mathbf{f}}$, by using K_{12} and (3.13).

Remark 3.4 *An important property of the hat functions $\Phi_q^{[1]}(x), q = 1, \ldots, M+1$ is that they are zero everywhere except in the two elements τ_{q-1} and τ_q involving the node z_q, i.e. $z_q = \bar{\tau}_{q-1} \cap \bar{\tau}_q, q = 2, \ldots, M$. For $q = 1$ and $M+1$ the $\Phi_q^{[1]}$ are, in fact, zero everywhere except, respectively, in the elements τ_1 and τ_M. The $\Phi_q^{[1]}$ has value unity at its own node and value zero at all other nodes; i.e. $\Phi_q^{[1]}(z_j) = \delta_{qj}$. Because $\Phi_q^{[1]}(z_q) = 1$, we see immediately from (3.11) that ${}^N u(z_q) \equiv \mathrm{u}_q = c_q, q \in \mathcal{N}_1$. In what follows we shall make use of notations such as ${}^N u, u_\Delta^{[1]}, {}^N u_\Delta^{[1]}$, to emphasize specific features of the finite element solution.*

Remark 3.5 *The form and the local support property of the hat functions is essential to the computational effectiveness of the finite element method. It ensures that the stiffness matrix $[K]$, and the matrix $[K_{12}]$, are sparse.*

Remark 3.6 *In one dimension there is a* ***natural*** *ordering of the vertices; the one we have used. Use of any other ordering would of course lead to a permutation of the maxtrix $[K]$.*

3.3.2 Implementation: one-dimensional problem with piecewise linear basis functions

We start this section with an example.

Example 3.3 Let us consider problem (2.31) with $B(u,v)$ and $F(v)$ as in (2.17) and (2.18) and with $a=1, c=0, \alpha_2=0, f=1, l=1, g_1=3$ and $G_2=2$; see (2.17) and (2.18). We take a uniform mesh Δ in which $x_q^\Delta = q/(M+1)$, $q=1,\ldots,M+1$ and $h_q^\Delta = h = 1/M$. We shall apply the finite element method with piecewise linear basis functions to this problem and will illustrate Steps 1 and 3 of the numerical process.

Step 1 By suitable computation the unconstrained stiffness matrix $\left[\bar{K}\right]$ as in (3.16) has the form

$$\left[\bar{K}\right]_{(M+1)\times(M+1)} = \frac{1}{h}\begin{bmatrix} 1 & -1 & & & \\ -1 & 2 & -1 & & \\ & & & & \\ & & -1 & 2 & -1 \\ & & & -1 & 1 \end{bmatrix}, \tag{3.23a}$$

and the unconstrained load vector is

$$\bar{\mathbf{f}}_{(M+1)\times 1} = h\begin{bmatrix} \frac{1}{2} \\ 1 \\ \\ \\ 1 \\ 5/2 \end{bmatrix}. \tag{3.23b}$$

Note that we have $N=M+1, N_1=M, N_2=1, \mathcal{N}_1=\{2,3,\ldots,M+1\}, \mathcal{N}_2=\{1.\}$. The augmentation is 3, so that the constrained stiffness matrix $[K]$ as in (3.14a) is

$$[K]_{M\times M} = \frac{1}{h}\begin{bmatrix} 2 & -1 & & & \\ -1 & 2 & -1 & & \\ & & & & \\ & & -1 & 2 & -1 \\ & & & -1 & 1 \end{bmatrix}, \tag{3.24}$$

and the constrained augmented vector $\hat{\mathbf{f}}$ is from (3.13)

$$\hat{\mathbf{f}}_{M\times 1} = h\begin{bmatrix} 4 \\ 1 \\ 1 \\ \\ \\ 1 \\ 5/2 \end{bmatrix}. \tag{3.25}$$

Now, from (3.11) the finite element solution ${}^N u(x)$ is

$$ {}^N u = \sum_{i \in \mathcal{N}_1} c_i \Phi_i^{[1]}(x) + 3\Phi_1^{[1]}(x) \equiv u_\Delta^{[1]}, \tag{3.26} $$

where $\mathbf{c} = \{c_i\}_{i=2}^{N_1}$ is the solution of the linear equation system

$$ \begin{bmatrix} 2 & -1 & & & \\ -1 & 2 & -1 & & \\ & & & & \\ & & -1 & 2 & -1 \\ & & & -1 & 1 \end{bmatrix} \begin{bmatrix} c_2 \\ c_3 \\ \\ \\ c_{M+1} \end{bmatrix} = h^2 \begin{bmatrix} 4 \\ 1 \\ \\ 1 \\ 5/2 \end{bmatrix}. \tag{3.27} $$

Step 2 Involves the solving of (3.27) for $\mathbf{c}$.

Step 3 Requires the construction of ${}^N u$, or other quantities of interest, from the c and the $\Phi_i' s$.

Exercise 3.1 For the case $M = 4$ and the natural ordering of the nodal points, $x_1, \ldots, x_5$, create the constrained and unconstrained matrices, the constrained load vector $\mathbf{f}$ and the augmented vector $\hat{\mathbf{f}}$, and the constrained system of equations (3.13) with the data $a = 1, c = 0, \alpha_2 = 1, f = -2, g_1 = 4, G_2 = 5, l = 2$. Hence, evaluate the approximation ${}^4 u(x)$.

Exercise 3.2 Repeat Exercise 3.1 using the nodal ordering x_1, x_3, x_5, x_4, x_2. Can you see a relation between the unconstrained matrices constructed in Exercises 3.1 and 3.2?

Exercise 3.3 Construct the system of linear equations for the solution of the problem

$$ -u''(x) + 5u(x) = 2, \; 0 < x < 1, $$
$$ u(0) = 3, u(1) = -2, $$

for $M = 4$

We now address briefly the task of constructing efficiently the constrained stiffness matrix $[K]$ and the constrained augmented load vector $\hat{\mathbf{f}}$ as in Step 1 of the algorithm. The approach to this is to work locally element by element and for each element to compute using (3.14a) specific *local* stiffness matrices and load vectors. These local matrices and vectors are then *assembled* to produce the global unconstrained stiffness matrix $[\bar{K}]$ as in (3.16) and the global unconstrained vector $\bar{\mathbf{f}}$. Finally, the constraining process using the boundary condition of the boundary value problem produces (3.13).

As a result of the above, the process has three parts:

(1) Computation of the local stiffness matrices and load vectors using the coordinates of the nodes.
(2) The assembly of $[\bar{K}]$ and $\bar{\mathbf{f}}$.
(3) The constraining process and augmentation of $\mathbf{f}$.

Part (1) We realize that on every element τ_q there are only two non-zero basis functions $\Phi_q^{[1]}$ and $\Phi_{q+1}^{[1]}$, each taking, respectively, the value unity at its own node and zero at the other node of the element, as shown in Fig. 3.2.

We define the local stiffness matrix

$$[\bar{K}^{(q)}]_{2\times 2} \equiv \left[B_{\tau_q}\left(\Phi_i^{[1]}, \Phi_j^{[1]}\right)\ i,j = q, q+1\right], \tag{3.28a}$$

where

$$B_{\tau_q}\left(\Phi_i^{[1]}, \Phi_j^{[1]}\right) \equiv \int_{\tau_q}\left(a\frac{d\Phi_i^{[1]}}{dx}\frac{d\Phi_j^{[1]}}{dx} + c\Phi_i^{[1]}\Phi_j^{[1]}\right)dx,$$

$$q = 1, \ldots, M-1\ \ i,j = q, q+1, \tag{3.28b}$$

and

$$B_{\tau_M}\left(\Phi_i^{[1]}, \Phi_j^{[1]}\right) = \int_{\tau_M}\left(a\frac{d\Phi_i^{[1]}}{dx}\frac{d\Phi_j^{[1]}}{dx} + c\Phi_i^{[1]}\Phi_j^{[1]}\right)dx + \alpha_2\Phi_i^{[1]}(l)\Phi_j^{[1]}(l),$$

$$i,j = M, M+1. \tag{3.28c}$$

Similarly we define the local load vectors

$$\bar{\mathbf{f}}_{0q} = \begin{bmatrix} F_{\tau_q}\left(\Phi_q^{[1]}\right) \\ F_{\tau_q}\left(\Phi_{q+1}^{[1]}\right) \end{bmatrix}, \tag{3.29a}$$

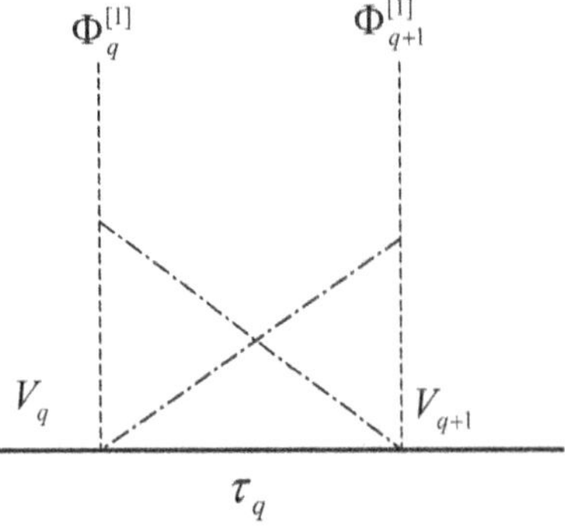

Fig. 3.2 Basis functions $\Phi_q^{[1]}$ and $\Phi_{q+1}^{[1]}$ on the element τ_q.

where

$$F_{\tau q}\left(\Phi_j^{[1]}\right) \equiv \int_{\tau q} f\Phi_j^{[1]} dx,$$
$$q = 1, \ldots, M, \;\; j = q, q+1, \tag{3.29b}$$

and

$$F_{\tau_M}\left(\Phi_{M+1}^{[1]}\right) \equiv \int_{\tau_N} f\Phi_{M+1}^{[1]} dx + G_2\Phi_{M+1}^{[1]}(l). \tag{3.29c}$$

In order to automate the computations we shall use a master line (in the two-dimensional case this will become the master plane), and on the master line we define the *master element* $\tilde{\tau} \equiv (-1, 1)$ in the master variable ξ. This is sometimes called the *standard* element. Consider the element τ_q on the *physical* (x) line. We use a one-to-one mapping of the master element $\tilde{\tau}$ onto the physical element τ_q,

$$x = x_q(\xi), \; \xi \in \tilde{\tau}, \; x \in \tau_q. \tag{3.30}$$

For this we take the linear mapping

$$x_q(\xi) = \frac{x_{q+1} + x_q}{2} + \xi\frac{x_{q+1} - x_q}{2}, \tag{3.31}$$

which obviously depends on the (known) coordinates of the nodes. Clearly $x_q(-1) = x_q$ and $x_q(1) = x_{q+1}$. The points $\xi = -1, 1$ are the *master nodes,* ${}^1\tilde{z}, {}^2\tilde{z}$.

Using this mapping and denoting

$$\tilde{\Phi}_q(\xi) \equiv \Phi^{[1]}(x_q(\xi)), \; \tilde{a}(\xi) \equiv a(x_q(\xi)), \tag{3.32}$$

etc. we see that

$$\tilde{\Phi}_q(\xi) = \frac{1}{2}(1 - \xi) \equiv {}^1\tilde{\Phi}(\xi),$$
$$\tilde{\Phi}_{q+1}(\xi) = \frac{1}{2}(1 + \xi) \equiv {}^2\tilde{\Phi}(\xi),$$

independent of q. The functions ${}^j\tilde{\Phi}, j = 1, 2$ are called *shape functions.*

We recall that for a function $\zeta(x) \in C(x_q, x_{q+1})$ that is mapped onto a function $\zeta(\xi), -1 \leqq \xi \leqq 1$,

$$\int_{x_q}^{x_{q+1}} \zeta(x) dx = \int_{-1}^{1} \tilde{\zeta}(\xi) \, |J| \, d\xi, \tag{3.33}$$

where $|J| = dx/d\xi$ is the determinant of the Jacobian of the mapping. Note that we need here a one-to-one mapping, which of course the linear mapping is.

Thus, we can rewrite (3.28b and c), respectively, as

$$B_{\tau_q}\left(\Phi_i^{[1]}, \Phi_j^{[1]}\right) \equiv \int_{-1}^{1} \tilde{a}(\xi)\frac{d\tilde{\Phi}_i}{d\xi}\frac{d\tilde{\Phi}_j}{d\xi}\frac{1}{\left(\frac{dx_q}{d\xi}\right)^2}\frac{dx_q}{d\xi}d\xi + \int_{-1}^{1} \tilde{c}(\xi)\tilde{\Phi}_i\tilde{\Phi}_j\frac{dx_q}{d\xi}d\xi \qquad (3.34a)$$

$$= \frac{2}{h_q}\int_0^1 \tilde{a}\frac{d\tilde{\Phi}_i}{d\xi}\frac{d\tilde{\Phi}_j}{d\xi}d\xi + \frac{h_q}{2}\int_{-1}^{1} \tilde{c}(\xi)\tilde{\Phi}_i, \tilde{\Phi}_j d\xi,$$

$$i, j = q, q+1, q = 1, \ldots, M-1,$$

$$B_{\tau_M}\left(\Phi_i^{[1]}, \Phi_j^{[1]}\right) \equiv \frac{2}{h_M}\int_{-1}^{1} \hat{a}\frac{d\tilde{\Phi}_i}{d\xi}\frac{d\tilde{\Phi}_j}{d\xi}d\xi + \frac{h_M}{2}\int_{-1}^{1} c\tilde{\Phi}_i, \tilde{\Phi}_j d\xi + \alpha_2\tilde{\Phi}_i(1)\tilde{\Phi}_j(1) \quad (3.34b)$$

$$i, j = M, M+1.$$

The $F_{\tau_q}(\Phi_j^{[i]}), i = 1, 2$, are computed in a similar manner using (3.29b and c).

In Fig. 3.3 we illustrate the one-dimensional master line, the master element and the shape functions ${}^1\tilde{\Phi}, {}^2\tilde{\Phi}$.

The shape functions are defined locally as *polynomials* on the master element $\tilde{\tau}$, and they thus have local numbering, 1, 2, associated with the master nodal points. The corresponding mapped shape functions $\Phi_q(x), \Phi_{q+1}(x)$ are often called *pull-back polynomials.*

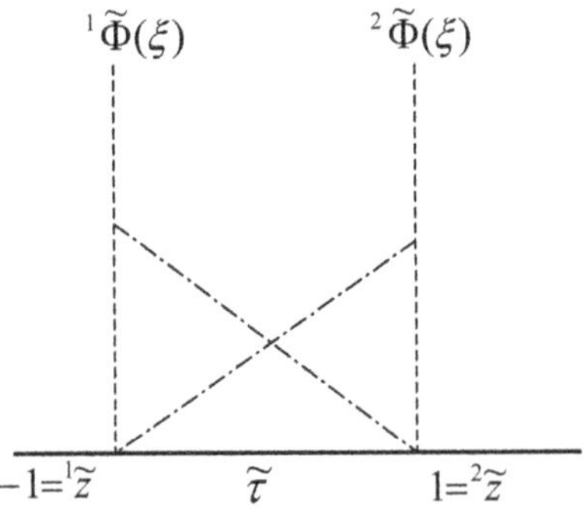

Fig. 3.3 The master element $\tilde{\tau}$, the master nodal points ${}^1\tilde{z}, {}^2\tilde{z}$, and the master shape functions ${}^1\tilde{\Phi}(\xi), {}^2\tilde{\Phi}(\xi)$.

The expressions $B_{\tau_q}\left(\Phi_i^{[1]},\Phi_j^{[1]}\right)$ can usefully be written as the sum of two terms; or three terms in the case of B_{τ_M}. Thus,

$$B_{\tau_q}\left(\Phi_i^{[1]},\Phi_j^{[1]}\right)=\tilde{B}_q^{(1)}\left(\tilde{\Phi}_i,\tilde{\Phi}_j\right)+\tilde{B}_q^{(2)}\left(\tilde{\Phi}_i,\tilde{\Phi}_j\right), \tag{3.35a}$$

$$B_{\tau_M}\left(\Phi_i^{[1]},\Phi_j^{[1]}\right)=\tilde{B}_M^{(1)}\left(\tilde{\Phi}_i,\tilde{\Phi}_j\right)+\tilde{B}_M^{(2)}\left(\tilde{\Phi}_i,\tilde{\Phi}_j\right)+\tilde{B}_M^{(3)}\left(\tilde{\Phi}_i,\tilde{\Phi}_j\right), \tag{3.35b}$$

where

$$\tilde{B}_q^{(1)}\left(\tilde{\Phi}_i,\tilde{\Phi}_j\right)=\frac{2}{h_q}\int_{-1}^{1}\hat{a}(\xi)\frac{d^m\tilde{\Phi}(\xi)}{d\xi}\frac{d^n\tilde{\Phi}(\xi)}{d\xi}d\xi,$$

$$\tilde{B}_q^{(2)}\left(\tilde{\Phi}_i,\tilde{\Phi}_j\right)=\frac{h_q}{2}\int_{-1}^{1}\tilde{c}(\xi)^m\Phi(\xi)^n\Phi(\xi)d\xi \tag{3.35c}$$

$$\tilde{B}_q^{(2)}\left(\tilde{\Phi}_i,\tilde{\Phi}_j\right)=\alpha_2\,{}^m\tilde{\Phi}(1)^n\tilde{\Phi}(1),$$

$$i,j=q,q+1;\ m,n=1,2.$$

Obviously, the m and n have to be related to the i and j; for example (see Figs. 3.2 and 3.3) we have here that $\tilde{\Phi}_q={}^1\Phi$ and $\tilde{\Phi}_{q+1}={}^2\Phi$. From this we see that it is important to assume that both $q,q+1$ and ${}^1\tilde{z},{}^2\tilde{z}$ are numbered in a consistent way; e.g. from left to right.

The index q for the $\tilde{B}'$s is included in order to emphasize the dependence on the mapping $x_q(\xi)$ and the association of the nodes z_q and z_{q+1} with the master nodes ${}^1\tilde{z}$ and ${}^2\tilde{z}$. Hence, referring back to (3.28b and c) we have that

$$\left[B_{\tau_q}\left(\Phi_i^{[1]},\Phi_j^{[1]}\right)\right]\equiv\left[\bar{K}^{(q)}\right]_{2\times 2}\equiv\left[{}^1\bar{K}^{(q)}\right]+\left[{}^2\bar{K}^{(q)}\right]+\left[{}^3\bar{K}^{(q)}\right]$$
$$i,j=q,q+1,\quad q=1,2,\ldots,M, \tag{3.36}$$

where the $[{}^i\bar{K}^{(q)}]$ are 2×2 matrices. Similarly, we may write the local load vectors $\bar{\mathbf{f}}^{(q)}$ from (3.29a and b) as

$$\bar{\mathbf{f}}^{(q)}=[{}^1\bar{f}^{(q)},\ {}^2\bar{f}^{(q)}],q=1,\ldots,M. \tag{3.37}$$

In this way, we can compute all the local stiffness matrices $[\bar{K}^{(q)}]_{2\times 2}$, as in (3.28a) and local load vectors as in (3.29a). We emphasize again that they depend on the positions of the nodes $z_q, q=1,\ldots,M+1$.

Part (2) *Assembly process for unconstrained stiffness matrix* $[\bar{K}]$ *and load vector* $\bar{\mathbf{f}}$.

Although we are currently working in a one-dimensional setting, we shall explain the main ideas of the assembly of the unconstrained stiffness matrix in such a manner that the assembly process is directly applicable to the two-dimensional contexts. The process has three subparts:

(1) identification of the global numbering of the nodes, the elements and the relation between the elements and nodes that belong to $\Gamma_{\rm D}$;
(2) construction of the local stiffness matrices and load vectors;
(3) construction of $[\bar{K}]$ and $\bar{\mathbf{f}}$ by an assembly process.

Subpart 1: Identification

We assume that the vertices are numbered and that the ordering is natural. We first construct a table of the coordinates of all the vertices $z_q = V_q$. This defines the mesh. We also mark the boundary vertices (= nodes). Further, we number the elements globally and with every element τ_q we associate the global numbers of its vertices; i.e. the end points. To τ_q we associate the pair of integers l, s, so that $\tau_q = (z_l, z_s)$. This pair is oriented so that z_l is the left-hand end of τ_q, and z_s is the right-hand end; note that the coordinates of z_l and z_s are already specified. This orientation is essential in order to match the orientation of the master elment nodes ${}^1\tilde{z}, {}^2\tilde{z}$.

Subpart 2: Construction of local stiffness matrices and load vectors

Because the coordinates of the vertices are available, the local matrices $[\bar{K}^{(q)}]$ and load vectors $\bar{\mathbf{f}}^{(q)}$ can be computed.

Subpart 3: Construction of the unconstrained matrix $[\bar{K}]$ and load vector $\bar{\mathbf{f}}$ by the assembly process

The natural orderings of the vertices (= nodes z^q with coordinates x_q^{Δ}) and of the elements τ_q are shown in Fig. 3.1. The unconstrained matrix $[\bar{K}]$ has order $(M+1) \times (M+1)$, and is constructed so that $[\bar{K}] = \sum_{q=1}^{M} [\bar{K}_q]$, where $\bar{K}_q$ are $(M+1) \times (M+1)$ matrices each constructed to incorporate the local stiffness matrix $[\bar{K}] \equiv \begin{bmatrix} k_{11}^q & k_{12}^q \\ k_{21}^q & k_{22}^q \end{bmatrix}$ associated with the element τ_q.

For the element τ_q with nodes z_1 and z_2 we associate z_1 with ${}^1\tilde{z}$ and z_2 with ${}^2\tilde{z}$ and to construct $[\bar{K}_1]$ we utilize the global node numberings 1,2 of z_1, z_2 to position $[\bar{K}_1]$ and to insert zero values in all other positions. Thus,

$$[\bar{K}_1] \equiv \begin{bmatrix} k_{11}^1 & k_{12}^1 & 0 & 0 & & 0 \\ k_{21}^1 & k_{22}^1 & 0 & 0 & & 0 \\ 0 & 0 & 0 & 0 & & 0 \\ 0 & 0 & 0 & 0 & & 0 \\ & & & & & \\ 0 & 0 & 0 & 0 & & 0 \end{bmatrix}_{(M+1)\times(M+1)} . \tag{3.38}$$

In a similar way for element τ_2 with node numberings 2, 3 we insert $[\bar{K}^{(2)}]$ in the (2,2), (2,3), (3,2) and (3,3) positions in $[\bar{K}_2]$ and complete this matrix by inserting zero values in all other positions. Thus,

$$[\bar{K}_2] \equiv \begin{bmatrix} 0 & 0 & 0 & 0 & & 0 \\ 0 & k_{11}^2 & k_{12}^2 & 0 & & \\ 0 & k_{21}^2 & k_{22}^2 & 0 & & 0 \\ 0 & 0 & 0 & 0 & & 0 \\ & & & & & \\ & & & & & \\ 0 & 0 & 0 & 0 & & 0 \end{bmatrix}_{(M+1)\times(M+1)} . \tag{3.39}$$

Proceeding in this way we construct all the matrices $[\bar{K}_q], q = 1, 2, \ldots, M$, and by adding them we obtain the unconstrained stiffness matrix

$$[\bar{K}] \equiv \begin{bmatrix} k_{11}^1 & k_{12}^1 & & & & \\ k_{21}^1 & k_{22}^1 + k_{11}^2 & & & & \\ & k_{12}^2 & k_{22}^2 + k_{11}^3 & k_{12}^3 & & \\ & & k_{21}^3 & k_{22}^3 + k_{11}^4 & & \\ & & & & & \\ & & & & & k_{22}^M \end{bmatrix}_{(M+1)\times M+1} . \tag{3.40a}$$

Referring to Example (3.3) we can see that we have obtained exactly the matrix of (3.23a). We construct the unconstrained load vector $\bar{\mathbf{f}}$ in a similar way by amalgamating the local load vectors $\bar{\mathbf{f}}^{(q)}, q = 1, \ldots, M$, so that

$$\bar{\mathbf{f}}_{(M+1)\times 1} = \begin{bmatrix} {}^1\bar{\mathbf{f}}^{(1)} \\ {}^2\bar{\mathbf{f}}^{(1)} + {}^1\bar{\mathbf{f}}^{(2)} \\ {}^2\bar{\mathbf{f}}^{(2)} + {}^1\bar{\mathbf{f}}^{(3)} \\ \\ {}^2\bar{\mathbf{f}}^{(M)} \end{bmatrix} . \tag{3.40b}$$

Part (3) *The constraining process*

From the unconstrained stiffness matrix and load vector we now wish to construct the respective *constrained* stiffness matrix and load vector. In general, in order to do this we first identify those nodes $\{z_m\}$ that belong to the Dirichlet part Γ_{D} of the boundary Γ. In the unconstrained global stiffness matrix $[\bar{K}]$ for each value of m we delete the mth column and row. Similarly, in the unconstrained load vector $\bar{\mathbf{f}}$ for each time we delete the mth entry. In this way the $N \times N$ matrix $[\bar{K}]$ is, by the deletion of N_2 rows and columns (because there are N_2 nodes on Γ_{D}), constrained to become the $N_1 \times N_1$ matrix $[K]$ as in (3.13). Similarly, $\bar{\mathbf{f}}$ is constrained to become the $N_1 \times 1$ vector $\mathbf{f}$ of (3.13).

This is achieved in the present one-dimensional example by noting that Γ_D consists of the point $x = 0$ at which we have the node $x_{\textcircled{1}}^{\Delta}$. Thus in the unconstrained stiffness matrix $[\bar{K}]$ we delete the first row and column to get

$$[K]_{M\times M} \equiv \begin{bmatrix} k_{22}^1 + k_{11}^2 & k_{12}^2 & & & \\ k_{21}^2 & k_{22}^2 + k_{11}^3 & k_{12}^3 & & \\ & k_{21}^3 & k_{22}^3 + k_{11}^4 & & \\ & & & \ddots & \\ & & & & k_{22}^M \end{bmatrix}_{M\times M} . \tag{3.41}$$

Similarly, in the unconstrained load vector we delete the first entry to get

$$\bar{\mathbf{f}}_{M\times 1} = \begin{bmatrix} {}^2\bar{\mathbf{f}}^{(1)} + {}^1\bar{\mathbf{f}}^{(2)} \\ {}^2\bar{\mathbf{f}}^{(2)} + {}^1\bar{\mathbf{f}}^{(3)} \\ \vdots \\ {}^2\bar{\mathbf{f}}^{(M)} \end{bmatrix}_{M\times 1} , \tag{3.42}$$

and the augmented load vector is obtained by using the Dirichlet condition at $x(\overset{\Delta}{\textcircled{1}})$ as

$$\hat{\mathbf{f}}_{M\times 1} = \begin{bmatrix} {}^2\bar{\mathbf{f}}^{(1)} + {}^1\bar{\mathbf{f}}^{(2)} - k_{21}^1 g_1 \\ {}^2\bar{\mathbf{f}}^{(2)} + {}^1\bar{\mathbf{f}}^{(3)} \\ \vdots \\ {}^2\bar{\mathbf{f}}^{(M)} \end{bmatrix}_{M\times 1} . \tag{3.43}$$

Remark 3.7 *We have shown above the main idea behind the method. During the implementation of the method the matrices are not filled with zeros and the constraint is applied simultaneously to the assembly process. This leads to matrices having compact forms as a result of the zeros being avoided. Many 'tricks' are used in practice in the implementation.*

3.3.3 Complete process for one-dimensional problem

We are now in a position to describe the complete piecewise linear process for problem (2.31) with the discretisation of Section 3.3.1.

In this case, we have that

$$\begin{aligned} \mathcal{N} &\equiv \{1, \ldots, M+1\}, \\ \mathcal{N}_1 &\equiv \{2, \ldots, M+1\}, \\ \mathcal{N}_2 &\equiv \{1\}. \end{aligned} \tag{3.44}$$

As in *subpart (2)* we identify the vertices (nodes) z_k of the partition and create the list $\{k; x_q\}, q = 1, \ldots, M+1\}$, marking the Dirichlet boundary nodal point $\{1; x_1\}$.

With every element $\tau_q, q = 1, \ldots, M$ we associate the nodes z_k, z_{k+1} and we construct the list $\{q; q, q+1\}, q = 1, \ldots, M$. The *identification* is then complete.

For each element we construct the local stiffness matrices

$$[\bar{K}^{(q)}] = \frac{1}{h}\begin{bmatrix} 1 & -1 \\ -1 & 1 \end{bmatrix} = \begin{bmatrix} k_{11}^{(q)} & k_{12}^{(q)} \\ k_{21}^{(q)} & k_{22}^{(q)} \end{bmatrix}, \tag{3.45a}$$

as in (3.35) and the local load vectors,

$$\bar{\mathbf{f}}_q = h\begin{bmatrix} \frac{1}{2} \\ \frac{1}{2} \end{bmatrix} = \begin{bmatrix} {}^{1}\bar{f}^{(q)} \\ {}^{2}\bar{f}^{(q)} \end{bmatrix}, \tag{3.45b}$$

$q = 1, \ldots, M$.

We now construct the unconstrained stiffness matrix $[\bar{K}]$ and unconstrained load vector $\bar{\mathbf{f}}$ by the assembly of the local matrices into $[\bar{K}^{(q)}], q = 1, 2, \ldots, m$ and by inserting the $\bar{\mathbf{f}}_q$ into $\bar{\mathbf{f}}$.

The unconstrained global stiffness matrix and load vector are thus as in (3.23a) and (3.23b).

We finally construct the constrained global stiffness matrix and load vector and the constrained augmented load fector $\hat{\mathbf{f}}$. Because the Dirichlet condition is imposed at $x = 0$ (node 1) we delete the first row and column in (3.23a) to get (3.24), and the first element in $\bar{\mathbf{f}}$ in (3.23b) to get (3.25). In (3.25) we have that

$$\hat{\mathbf{f}} = h\begin{bmatrix} 1 + 3\,k_{1,2} \\ \vdots \\ \vdots \\ \frac{1}{2} \end{bmatrix} = \begin{bmatrix} 4 \\ 1 \\ \vdots \\ \frac{1}{2} \end{bmatrix}, \tag{3.46}$$

where the $3\,k_{1,2}$ is the augmentation arising from the prescribed Dirichlet boundary condition $g_1 = 3$. We have thus constructed the linear system

$$[\bar{K}]\mathbf{c} = \hat{f}, \tag{3.47}$$

which allows us to compute the finite element solution

$$u_{\Delta}^{[1]}(x) = \sum_{i \in \mathcal{N}_1} c_i \Phi_i^{[1]}(\mathbf{x}) + \sum_{i \in \mathcal{N}_2} d_i \Phi_i^{[1]}(\mathbf{x}). \tag{3.48}$$

Exercise 3.4 Consider the problem (2.14), where $I = (0, 4)$ and assume that $a(x) = 3$, $c(x) = 0, g_1 = 5, \alpha_2 = 0, G_2 = 0$ and $f = \sin\frac{5x}{8}$. Take $M = 4$

(1) Find the exact solution $\mathfrak{u}(x)$ with this data.
(2) Determine the stiffness matrix $[K]$ as in (3.41).
(3) Determine the augmented load vector f as in (3.43), using exact (analytic) integration.
(4) Calculate the finite element solution $u_{\Delta}(x)$.
(5) Compare $u_{\Delta}(x)$ with $\mathfrak{u}(x)$.

Exercise 3.5 Consider the problem and data of Exercise 3.4 and repeat (1)–(5), except that in (4) use first Simpson's rule and then trapezoidal quadrature to compute the augmented load vector.

Exercise 3.6 Repeat Exercise 3.4 but take $a(x) = 1 + x$.

3.3.4 The finite element method with piecewise quadratic functions

Again, in the one-dimension setting we use the finite element method to approximate the solution of (2.31), but now with piecewise quadratic functions. In this we proceed completely analogously to what we did in Section 3.1 with the linear shape functions, but now work in the piecewise quadratic setting.

Keeping the same partition as before we let

$$S_{\Delta}^{[2]} \equiv \left\{ v(x), v \in C^0(\bar{I}) \mid v_{\tau q} \in \mathbb{P}^{[2]}(\bar{\tau}_q) \right\} \tag{3.49a}$$

$$S_{\Delta,0}^{[2]} \equiv \{ v(x), v \in S_{\Delta}^{[2]} \mid v(x) = 0,\ x \in \Gamma_{\mathrm{D}} \}. \tag{3.49b}$$

Here, $\mathbb{P}^2(\bar{\tau}_q)$ is the space of all polynomials of degree 2 on τ_q. In $S_{\Delta}^{[2]}$ we define the piecewise quadratic basis functions

$$\Phi_m^{[2]}(x) \equiv \Phi_m^{[1]}(x), m = 1, \ldots, M+1, \tag{3.50a}$$

$$\Phi_m^{[2]}(x) = \begin{cases} \dfrac{4(x - x_q)(x_{q+1} - x)}{h_q^2}, & x \in \tau_q, \\ 0, \ x \notin \tau_q, \end{cases}$$

$$q = 1, \ldots, M,\ m = M+2, \ldots, 2M+1, \tag{3.50b}$$

where the nodes z_m are numbered as in Fig. 3.4, so that $z_m, m = 1, \ldots, M+1$ are the same as in the previous piecewise linear setting and $z_m, m = M+2, \ldots, 2M+1$ are at the points $x_{q+\frac{1}{2}} = (x_q + x_{q+1})/2$, where $m = q + M + 1$.

As in the previous case we have again achieved the property that

$$\Phi_m^{[2]}(x_m) = 1, m = 1, \ldots, M+1, \tag{3.51a}$$

$$\Phi_m^{[2]}(x_{m-M-\frac{1}{2}}) = 1, m = M+2, \ldots, 2M+1, \tag{3.51b}$$

when the additional nodal points are located at the centres of the elements.

Remark 3.8 *We have created the quadratic polynomials in the element by defining the basis functions either as linear functions or as quadratic polynomials that are zero at the nodes $z_m, m = 1, \ldots, m+1$. Such polynomials are called bubble functions. We now no longer have that $u(z_m) = c_m$. This condition could, however, have been achieved if we had taken the $\Phi_j^{[2]}, j = 1, \ldots, M+1$ as quadratic polynomials that took zero value at the midpoint of the element.*

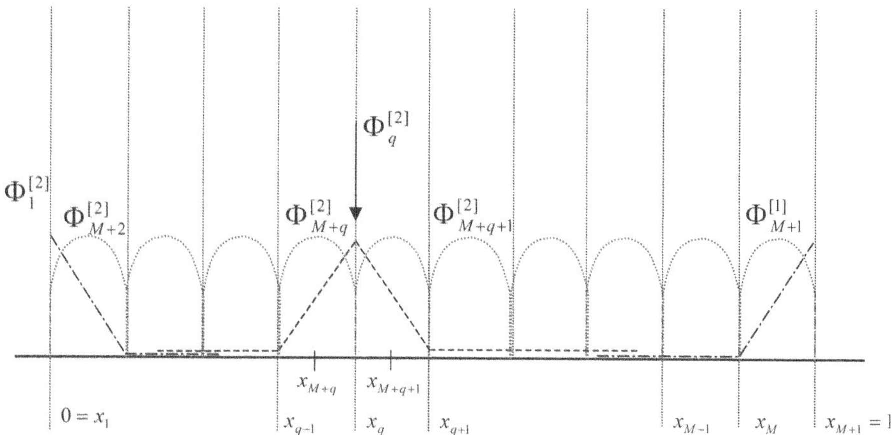

Fig. 3.4 The partition with the basis function $\Phi_q^{[2]}, q = M + 2, \ldots, 2M + 1$.

*Our choice of basis functions has the advantage that when proceeding from the linear to the quadratic case we merely add extra unknowns and do not need to recompute the parts of the matrix arising from the linear case. It also has the added advantage from the computational point of view that it produces more sparse matrices than those arising from other strategies. Our approach is sometimes called **hierarchical**. This approach, when used in 2D, has analogously side shape functions and internal shape functions and these produce many advantages computationally.*

As in the piecewise linear case we use the master element $\tilde{\tau} \equiv (-1, 1)$ on the master ξ line. We again map the elements $\tau_k, k = 1, \ldots, M$ in turn onto this master element using the *linear* mapping functions $x_k(\xi)$ as in (3.31). The master elements, the shape functions and the local numbering are shown in Fig. 3.5, where we see that the quadratic master shape function has the form

$$ {}^3\tilde{\Phi}(\xi) = (1 - \xi^2). \tag{3.52} $$

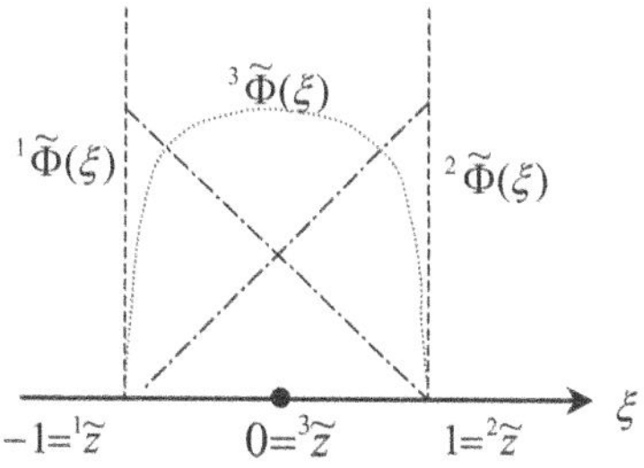

Fig. 3.5 The shape functions $\Phi(\xi)$ and their local numberings on the master element.

Example 3.4 We now proceed to construct the global stiffness matrix and load vector for problem (2.31), with the same specific parameter values as in Section 3.3.1; (i.e. $a = 1, c = 0, \alpha_2 = 0, f = 1, l = 1, g_1 = 3, G_2 = 2$). We use the mesh partition and shape functions as in Fig. 3.4.

Corresponding to the local stiffness matrices of (3.45) we now have

$$[\bar{K}^{(q)}] = \frac{1}{h}\begin{bmatrix} 1 & -1 & 0 \\ -1 & 1 & 0 \\ 0 & 0 & 4/3 \end{bmatrix},$$

$$\bar{\mathbf{f}}_q = h\begin{bmatrix} 1/2 \\ 1/2 \\ 2/3 \end{bmatrix}. \tag{3.53}$$

Corresponding, respectively, to (3.40a and b) we now have the unconstrained global stiffness matrix and vector

$$[\bar{K}]_{(2M+1)\times(2M+1)} = \frac{1}{h}\begin{bmatrix} 1 & -1 & & & & & & & \\ -1 & 2 & -1 & & & & & & \\ & & & & & & & & \\ & & -1 & 2 & -1 & & & & \\ & & & -1 & 1 & & & & \\ & & & & & 4/3 & & & \\ & & & & & & 4/3 & & \\ & & & & & & & \ddots & \\ & & & & & & & & 4/3 \end{bmatrix}, \tag{3.54a}$$

$$\bar{\mathbf{f}}_{(2M+1)\times 1} = h\begin{bmatrix} 1/2 \\ 1 \\ \vdots \\ 1 \\ 1/2 \\ 2/3 \\ \vdots \\ 2/3 \end{bmatrix}. \tag{3.54b}$$

We still have the single point $x = 0$ as the Dirichlet boundary Γ_D, so that we delete the first row and first column of $[\bar{K}]$ and the first row of $\bar{\mathbf{f}}$, corresponding to the first node z_1, to get

$$[K]_{2M\times 2M} = \frac{1}{h}\begin{bmatrix} 2 & -1 & & & & & & \\ -1 & 2 & -1 & & & & & \\ & & & & & & & \\ & & -1 & 2 & -1 & & & \\ & & & -1 & 1 & & & \\ & & & & & 4/3 & & \\ & & & & & & \ddots & \\ & & & & & & & \ddots \\ & & & & & & & & 4/3 \end{bmatrix}, \tag{3.55a}$$

$$\bar{\mathbf{f}}_{2M\times 1} = h\begin{bmatrix} 1 \\ 1 \\ \vdots \\ 1 \\ 1/2 \\ 2/3 \\ \vdots \\ 2/3 \end{bmatrix}. \tag{3.55b}$$

Finally, we compute the aumentation that is

$$\begin{bmatrix} -3 \\ 0 \\ 0 \\ \vdots \\ \vdots \\ 0 \end{bmatrix}_{2M\times 1}, \tag{3.56}$$

and obtain the constrained augmented system of equations for the piecewise quadratic finite element solution, as

$$\begin{bmatrix} 2 & -1 & & & & & & \\ -1 & 2 & -1 & & & & & \\ & & & & & & & \\ & & -1 & 2 & -1 & & & \\ & & & -1 & 1 & & & \\ & & & & & 4/3 & & \\ & & & & & & \ddots & \\ & & & & & & & \ddots \\ & & & & & & & & 4/3 \end{bmatrix} \begin{bmatrix} c_2 \\ \\ \\ c_{M+1} \\ c_{M+2} \\ \\ \\ c_{2M+1} \end{bmatrix} = h^2 \begin{bmatrix} -2 \\ 1 \\ \\ 1/2 \\ 2/3 \\ \\ \\ 2/3 \end{bmatrix}, \quad (3.57)$$

which are solved to give $\mathbf{c}$, and hence from (3.11)

$${}^{2M+1}u(x) = \Sigma_{i=2}^{2M+1} c_i \Phi_i^{[2]}(x) + 3\Phi_1(x). \quad (3.58)$$

For this simple case we have seen that the augmentation is straightforward. In the more general case, where $\mathbf{c}$ is non-zero, the augmentation is more complicated.

Note that N, which denotes the number of degrees of freedom, is the order of the maxtrix $[K]$. In the piecewise linear case $N = M$, whilst in the piecewise quadratic case $N = 2M + 1$.

Exercise 3.7 Consider the problem of type (2.14) in which $\mathrm{u}(x)$ satisfies

$$\frac{-d}{dx}\left((1+x)\frac{d\mathrm{u}}{dx}(x)\right) + \mathrm{u}(x) = 0, \quad x \in I \equiv (0,5),$$

$$g_1 = \mathrm{u}(0) = 2,$$

$$(1+5)\frac{d\mathrm{u}}{dx}(5) = 0.$$

Derive the weak form of this problem.

Exercise 3.8 Consider the problem of Exercise 3.7. Using the weak form that you have just derived, a finite element solution will be obtained using quadratic elements on a uniform mesh with $M = 3, (N = 7)$. Compute the local stiffness matrices for these quadratic elements, employing the basis functions of (3.49), and the local load vectors, employing the node ordering of Fig. 3.6.

3.3.5 Engineering application: one-dimensional heat-transfer problem

In Section 2.1.5 we introduced a simple one-dimensional problem of engineering character involving heat transfer in a bimaterial pipe consisting of two types of steel. This problem will be used throughout this book to illustrate various aspects of the finite element method. In Section 2.1.5 we also analyzed the influence of the input

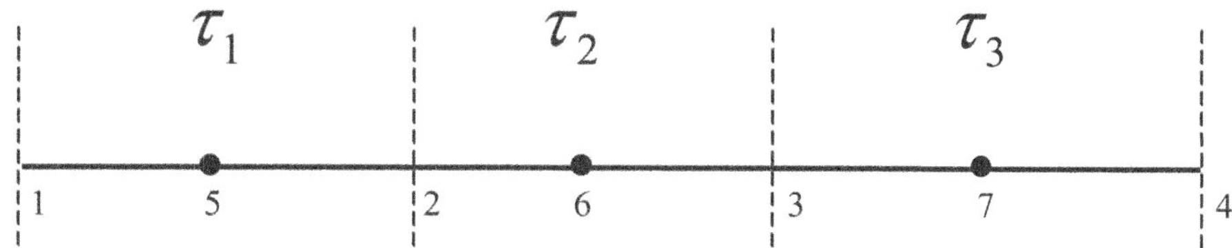

Fig. 3.6 Specific nodal ordering.

data on the exact output values of the quantities of interest $Q_i, i = 1, 2, 3$; this was related to validation.

We now apply the finite element method to the 1D Eng Problems 1 and 2 of Section 2.1.5, assuming that the values of γ_{ss}, γ_{cs} and β as in (2.53 a–c) have been validated and contain no uncertainties. Hence, problem (2.49) is defined and has exactly one solution. We are interested in the accuracy of the finite element method, that is in *verification*. As in Section 2.1.5 we shall assess this in terms of the quantities of interest $Q_i, i = 1, 2, 3$, the exact values of which can for 1D Eng Problems 1 and 2 be expressed analytically and are given in Section 2.1.5. For each problem we thus now

(1) calculate approximations Q_1^Δ to $Q_1 \equiv u(6.5)$;
(2) calculate approximations Q_2^Δ to $Q_2 \equiv a_2 \dfrac{du}{dx}(6.5)$;
(3) evaluate the total energy of the approximation $\equiv Q_3^\Delta$.

The calculations for Q_i^Δ and all the finite element computations are performed in double precision. **We note that the values of Q_i and Q_i^Δ are all reported with five digits after the decimal point. This level of accuracy is of course not needed in the engineering context; rather it is needed for assessing the errors in the finite element values**; in fact still more digits are needed for the assessment, although these are not reported. In order to ensure that the finite element solutions are not influenced by inaccurate coefficients in the stiffness matrices and load vectors, we compute these analytically (rather than with numerical integration). Similarly, the solving of the systems of linear equations and all the post-processing computations are undertaken in such a way that the errors of the finite element method are not influenced by any of these factors; this is necessary because in the computed results we present relative errors of order 10^{-12}. Further, we shall develop in later sections theoretical results that we wish to compare in a meaningful way with the computed results. We note that the theoretical results are developed for an ideal finite element method, where no 'crimes' are committed.

We now apply the finite element method to 1D Eng Problems 1 and 2 of Section 2.1.5 using a uniform mesh over $I \equiv (3, 6.5)$ with the number of elements $M = 1, 2, \ldots,$ with corresponding mesh lengths h and for elements of degree $p = 1, 2$, see Fig. 3.7. For Problem 1 the interface between the two types of steel is located at a nodal point when $M = 7, 14, 21, \ldots,$ with $h = 3.5 \,/\, k, k \geqq 1$ an integer. Contrastingly, for values of M that are not multiples of 7 the interface will be located at an interior point of an element with the same relative position of the element. In Problem 2, where the

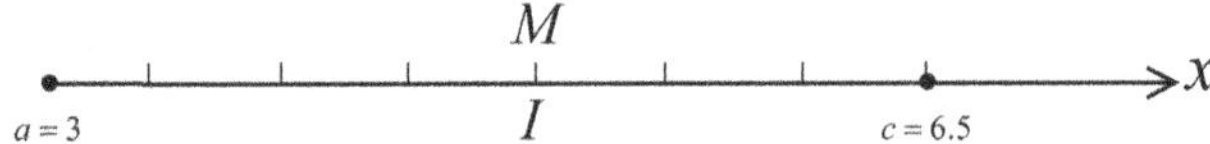

Fig. 3.7 Mesh for one-dimensional problem, see (2.49a–c).

interface is always located at an irrational point, the interface is always located at an element interior point and the relative position is virtually random. We shall see that these contrasting features lead to very differing results. The results of the numerical experiments are given here without explanation; which will be given in later sections. The purpose here is to illustrate the use of the finite element method to highlight interesting features, and to create curiosity about these.

Tables 3.1–3.4 give the values of $Q_i^\Delta \equiv Q_{i,p}^\Delta \equiv Q_i\left(u_\Delta^{[p]}\right)$ $i = 1, 2, 3$ with $p = 1, 2$ for Problems 1 and 2 taking $M = 1, \ldots, 25$.

We recall that $Q_1^\Delta \equiv u_\Delta^{[p]}(c)$ is the calculated value of the temperature at the outer surface of the pipe, whilst Q_2^Δ is the heat transferred from the pipe outer surface to the surrounding environment. We shall compute Q_2^Δ in two different ways; i.e. with two different *methods of extraction.*

(1) The first (direct) way. The flux F at the pipe surface $x = c$, see Section 2.1.5 is

$$F(c) = a(c)\frac{du}{dx}(c) = c\gamma_{cs}\frac{du}{dx}(c),$$

so that the heat transferred to the environment is

$$Q_2 = -2\pi c\, a(c)\frac{du}{dx}(c),$$

and we define

$$Q_2^{\Delta,1} \equiv -2\pi c\, a(c)\frac{du_\Delta^{[p]}}{dx}(c). \tag{3.59}$$

(2) The second (indirect) way. The second way is to use the boundary condition (2.49c), where

$$\frac{du}{dx} = -\alpha_2\left(\mathrm{u}(c) - T_1\right),$$

and

$$Q_2 = 2\pi c\alpha_2\left(\mathrm{u}(c) - T_1\right).$$

We define

$$Q_2^{\Delta,2} = 2\pi c\alpha_2\left(u_\Delta^{[p]}(c) - T_1\right), \tag{3.60}$$

and note that $Q_2^{\Delta,1} \neq Q_2^{\Delta,2}$, although both methods (1) and (2) give the same value for Q_2.

Table 3.1 Results for Problem 1 using linear elements ($p = 1$). The exact values are: $Q_1 = 200.03412$, $Q_2 = 4043.99611$, $Q_3 = 153\ 754.27310$.

M	h	Q_1^Δ	$Q_2^{\Delta,1}$	$Q_2^{\Delta,2}$	Q_3^Δ
1	3.50000	226.13305	6710.95148	4630.24046	176 147.15780
2	1.75000	218.31224	5147.49898	4454.56642	169 436.90380
3	1.16667	214.62058	4802.64983	4371.64279	166 269.45434
4	0.87500	211.60759	4614.55918	4303.96385	163 684.30925
5	0.70000	208.49699	4475.05704	4234.09243	161 015.41770
6	0.58333	204.85066	4347.25632	4152.18713	157 886.86543
7	0.50000	200.18502	4209.281113	4047.38570	153 883.74584
8	0.43750	201.76968	4225.17435	4082.98099	155 243.38557
9	0.38889	202.78404	4232.37552	4105.76599	156 113.70865
10	0.35000	203.31072	4231.52201	4117.59642	156 565.59777
11	0.31818	203.37390	4222.35997	4119.01549	156 619.80228
12	0.29167	202.94220	4203.63077	4109.31854	156 249.40605
13	0.26923	201.91379	4172.63360	4086.21812	155 367.03481
14	0.25000	200.07201	4124.15789	4044.84716	153 786.78088
15	0.23333	200.89218	4137.53362	4063.227019	154 490.48874
16	0.21875	201.48852	4146.43716	4076.66538	155 002.14724
17	0.20588	201.85467	4150.62402	4084.89016	155 316.31057
18	0.19444	201.96707	4149.47978	4087.41491	155 412.74885
19	0.18421	201.77700	4141.83547	4083.14549	155 249.66915
20	0.17500	201.19417	4125.59037	4070.05358	154 749.59492
21	0.16667	200.05097	4096.89901	4044.37466	153 768.73266
22	0.15909	200.60209	4107.01475	4056.75408	154 241.59169
23	0.15217	201.02066	4114.31712	4066.15622	154 600.76275
24	0.14583	201.29252	4118.46343	4072.26272	154 833.97789
25	0.14000	201.39234	4118.86199	4074.50501	154 919.62723

The final column in these tables reports $Q_3^\Delta \equiv \frac{1}{2}B\left(u_\Delta^{[p]}, u_\Delta^{[p]}\right) - F\left(u_\Delta^{[p]}\right)$, see (2.50). Because the Dirichlet condition at $x = a$ is non-homogeneous (see 2.49c) with $u(a) = 500$, the total energy $\Pi(u) > 0$.

From Tables 3.1–3.4 we see that, whilst the values of $Q_i^\Delta, i = 1, 2, 3$ seem to converge to the exact values Q with decreasing h, this convergence is non-monotone.

Table 3.2 Results for Problem 1 using quadratic elements ($p = 2$). The exact values are: $Q_1 = 200.03412$, $Q_2 = 4043.99611$, $Q_3 = 153\ 754.27310$.

M	h	Q_1^{Δ}	$Q_2^{\Delta,1}$	$Q_2^{\Delta,2}$	Q_3^{Δ}
1	3.50000	213.15952	2928.99934	4338.82398	165 015.86809
2	1.75000	207.54111	4144.11186	4212.62113	160 195.27447
3	1.16667	203.93471	1404.75196	4131.61265	157 100.97840
4	0.87500	202.12349	4076.70070	4090.92850	155 546.95831
5	0.70000	201.65895	4071.67366	4080.49382	155 148.38267
6	0.58333	201.58726	4072.87730	4078.88341	155 086.86977
7	0.50000	200.03431	4039.68447	4044.00037	153 754.43562
8	0.43750	201.21616	4067.25492	4070.54754	154 768.46278
9	0.38889	201.38633	4071.78623	4074.36999	154 914.46969
10	0.35000	201.06027	4064.96979	4067.04592	154 634.71069
11	0.31818	200.77491	4058.93157	4060.63601	154 389.87027
12	0.29167	200.74255	4058.48326	4059.90919	154 362.10796
13	0.26923	200.77757	4059.48500	4060.69592	154 392.15874
14	0.25000	200.03413	4042.95973	4043.99638	153 754.28340
15	0.23333	200.65537	4057.04718	4057.95096	154 287.30881
16	0.21875	200.77897	4059.93416	4060.72722	154 393.35426
17	0.20588	200.62638	4056.59935	4057.29983	154 262.43780
18	0.19444	200.48499	4053.50061	4054.12382	154 141.12285
19	0.18421	200.48582	4053.58395	4054.14239	154 141.83241
20	0.17500	200.52132	4054.43643	4054.93979	154 172.29089
21	0.16667	200.03412	4043.54143	4043.99617	153 754.27514
22	0.15909	200.45562	4053.04926	4053.46408	154 115.92271
23	0.15217	200.54824	4055.16527	4055.54458	154 195.39209
24	0.14583	200.45049	4053.00092	4053.34875	154 111.51762
25	0.14000	200.35821	4050.95592	4051.27603	154 032.34539

The accuracy of the computed values for 1D Eng Problem 1 is clearly superior when $M = 7, 14, 21$, for which the interface coincides with a node of the mesh, compared with other values of M when it does not. Also, $Q_2^{\Delta,1} \neq Q_2^{\Delta,2}$ and it is not possible to see from the tables which is superior. Contrastingly, see Tables 3.3 and 3.4, for 1D Eng Problem 2 there are not specific values of M where the accuracy is superior, and the values seem to converge in a 'chaotic' way.

Table 3.3 Results for Problem 2 using linear elements ($p = 1$). The exact values are: $Q_1 = 197.58989$, $Q_2 = 3898.093$, $Q_3 = 151\,657.12854$.

M	h	Q_1^Δ	$Q_2^{\Delta,1}$	$Q_2^{\Delta,2}$	Q_3^Δ
1	3.50000	225.26855	6732.13558	4610.82170	175 405.41571
2	1.75000	216.85993	5109.80195	4421.94400	168 190.81886
3	1.16667	212.58070	4752.31185	4325.82233	164 519.24158
4	0.87500	208.84068	4547.92261	4241.81244	161 310.30011
5	0.70000	204.72149	4385.42382	4149.28561	157 776.03563
6	0.58333	199.55027	4222.60365	4033.12784	153 339.13582
7	0.50000	199.18727	4185.97276	4024.97381	153 027.67457
8	0.43750	200.49402	4195.52199	4054.32654	154 148.86643
9	0.38889	201.14353	4194.38940	4068.91621	154 706.15076
10	0.35000	201.17162	4182.14335	4069.54719	154 730.25216
11	0.31818	200.52177	4156.68707	4054.94997	154 172.67976
12	0.29167	199.01256	4113.33577	4021.04939	152 877.77293
13	0.26923	198.09126	4084.95453	4000.35488	152 087.30072
14	0.25000	198.92646	4097.92156	4019.11537	152 803.89889
15	0.23333	199.46170	4104.81434	4031.13818	153 263.13631
16	0.21875	199.67055	4104.90234	4035.82946	153 442.32990
17	0.20588	199.48952	4096.64215	4031.76321	153 287.01044
18	0.19444	198.79172	4077.07073	4016.08891	152 688.29629
19	0.18421	197.71910	4049.37520	3991.99539	151 767.99169
20	0.17500	198.34169	4060.64262	4005.98012	152 302.16921
21	0.16667	198.78443	4068.07998	4015.92511	152 682.03974
22	0.15909	199.02087	4071.05676	4021.23614	152 884.90616
23	0.15217	199.00235	4068.44399	4020.82006	152 869.01312
24	0.14583	198.64079	4058.22349	4012.69855	152 558.79423
25	0.14000	197.77033	4036.61733	3993.14607	151 811.94450

In order to understand these tables better, we now compute the errors (this is possible because we know the exact values) and display them in graphical form.

In Figs. 3.8–3.15 we give on $\ln - \ln$ scales the relative errors in $Q_i^\Delta, i = 1, 2, 3$, where $\mathrm{rel} \equiv \mathrm{rel}(Q_i^\Delta) \equiv \left|Q_i - Q_i^\Delta\right| \,/\, Q_i$ for elements of order $p = 1, 2$, $M = 1, \ldots, 315$ and $h_{\min} = 0.01111$ for 1D Eng Problems 1 and 2. The reason for the selection of the $\ln - \ln$ scale is that when $\mathrm{rel} = \mathcal{C}h^\alpha$ then $\ln \mathrm{rel} = \ln \mathcal{C} - \alpha \ln h$ and the resulting plot is a straight line with slope α and position given by $\ln \mathcal{C}$, which we call a factor.

Table 3.4 Results for Problem 2 using quadratic elements ($p = 2$). The exact values are: $Q_1 = 197.58989$, $Q_2 = 3989.093$, $Q_3 = 151\ 657.12854$.

M	h	Q_1^Δ	$Q_2^{\Delta,1}$	$Q_2^{\Delta,2}$	Q_3^Δ
1	3.50000	211.25331	2782.05524	4296.00593	163 380.33909
2	1.75000	204.74981	4082.43225	4149.92185	157 800.33832
3	1.16667	200.93752	4037.86572	4064.28872	154 529.39341
4	0.87500	199.46086	4017.09959	4031.11938	153 262.41819
5	0.70000	199.29359	4018.65668	4027.36199	153 118.89650
6	0.58333	198.55662	4004.90212	4010.80799	152 486.58022
7	0.50000	198.69968	4009.73746	4014.02137	152 609.32217
8	0.43750	199.05354	4018.71673	4021.97006	152 912.93967
9	0.38889	198.67862	4011.00317	4013.54835	152 591.25429
10	0.35000	198.35943	4004.33341	4006.37857	152 317.38876
11	0.31818	198.36287	4004.77414	4006.45583	152 320.34009
12	0.29167	198.28937	4003.39839	4004.80496	152 257.28150
13	0.26923	197.98327	3996.73689	3997.92909	151 994.64227
14	0.25000	198.38591	4005.94622	4006.97339	152 340.10909
15	0.23333	198.28132	4003.73227	4004.62417	152 250.37581
16	0.21875	198.08743	3999.48769	4000.26894	152 084.01806
17	0.20588	198.06746	3999.12973	3999.82029	152 066.88111
18	0.19444	198.10517	4000.05228	4000.66728	152 099.23342
19	0.18421	197.69140	3990.82325	3991.37304	151 744.21969
20	0.17500	198.09976	4000.04924	4000.54585	152 094.59545
21	0.16667	198.11002	4000.32630	4000.77618	152 103.39340
22	0.15909	197.97323	3997.29451	3997.70362	151 986.02998
23	0.15217	197.92876	3996.33095	3996.70476	151 947.87641
24	0.14583	197.97759	3997.45850	3997.80156	151 989.77121
25	0.14000	197.72644	3991.84480	3992.16023	151 774.28823

From Fig. 3.8 for Problem 1 we note the following interesting features:

(1) The error oscillates and for $h = 0.5/k, k = 1, 2, \ldots$, it has a minimal value. For these particular values of h the relative error is such that rel $\approx Ch^\alpha$, with $\alpha = 2$ for $p = 1$ and $\alpha = 4$ for $p = 2$. The slopes associated with $\alpha = 2$ and $\alpha = 4$ are indicated in Figs. 3.8 and 3.9. Note that these particular values of h are exactly those for which the interface $x = b$ coincides with a nodal point. In Fig. 3.8 in

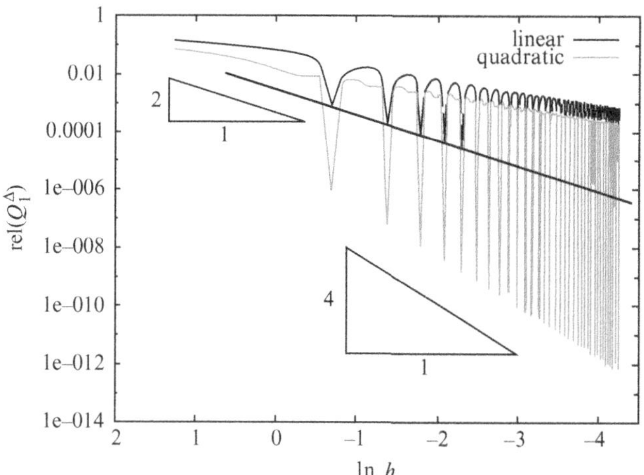

Fig. 3.8 Plots of rel(Q_1^Δ) v $\ln h$ for Problem 1 with $p = 1$ and 2.

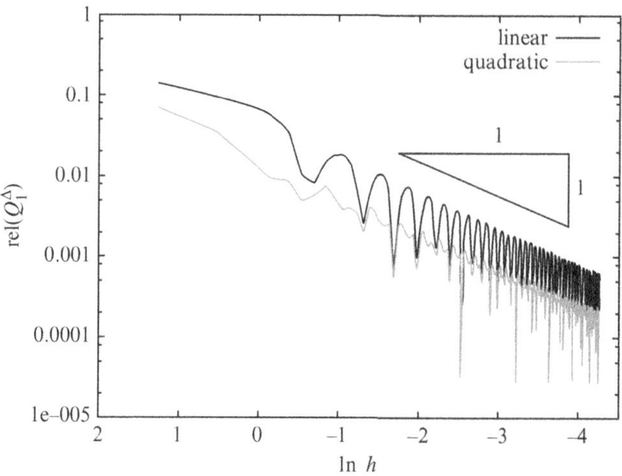

Fig. 3.9 Plots of rel(Q_1^Δ) v ln$h$ for Problem 2 with $p = 1$ and 2.

order to make more visible the envelope of the points of minimum value for $\alpha = 2$ a heavy line has been drawn through them.

(2) The upper envelope of the relative errors lies on the line with slope $\alpha = 1$ for both $p = 1$ and 2. For $h_k = 3.5 \ / \ (7k + s), k = 1, 2, 3$, with $1 \leqq 5 \leqq 6$ fixed, the interface lies each time in the same relative position in an element. Considering only these k and the associated $M = 7k + s$, the corresponding ln rel lie on a straight line with slope $\alpha = 1$ independent of s.

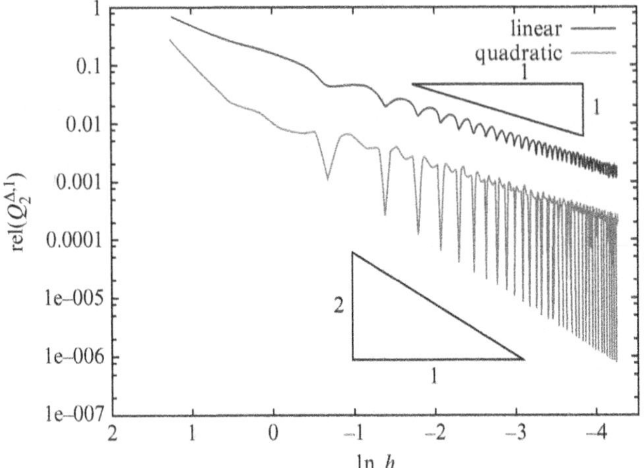

Fig. 3.10 Plots of rel($Q_2^{\Delta,1}$) v ln$h$ for Problem 1 with $p = 1$ and 2.

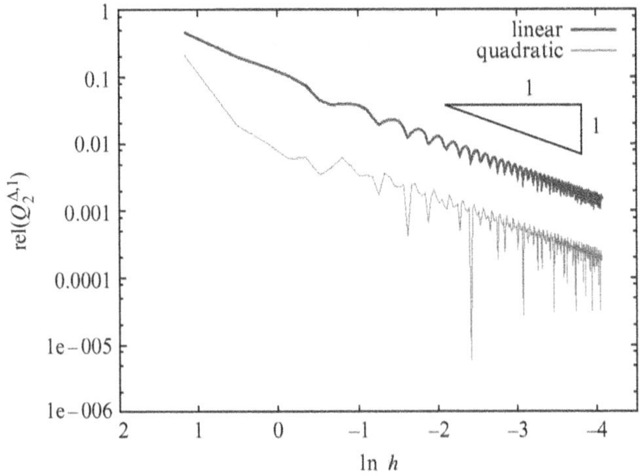

Fig. 3.11 Plots of rel($Q_2^{\Delta,1}$) v ln$h$ for Problem 2 with $p = 1$ and 2.

(3) As h approaches the value 3.5 / 7k, $M = 7k$, the error decreases rapidly.

(4) For arbitrary values of h we can only say that

$$C_1 h^2 \leqq \text{rel} \leqq C_2 h \text{ for } p = 1,$$
$$C_1 h^4 \leqq \text{rel} \leqq C_2 h \text{ for } p = 2,$$

i.e. the errors lie in a zone bounded by two straight lines.

(5) If a node coincides with the interface then $p = 2$ gives the best results. If not, then $p = 1$ is preferable because less computational effort is required to achieve similar accuracy.

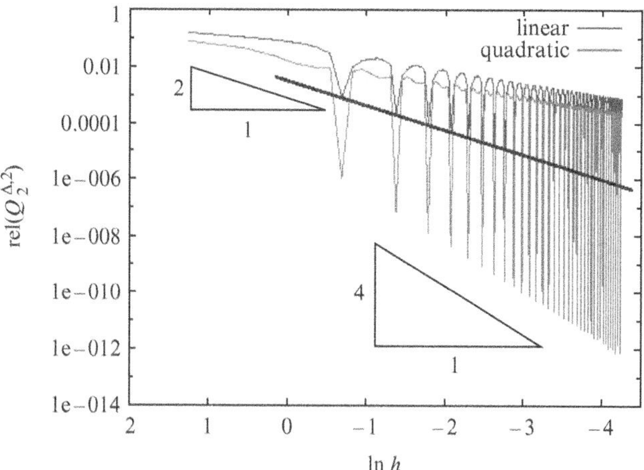

Fig. 3.12 Plots of $\text{rel}(Q_2^{\Delta,2})$ v $\ln h$ for Problem 1 with $p = 1$ and 2.

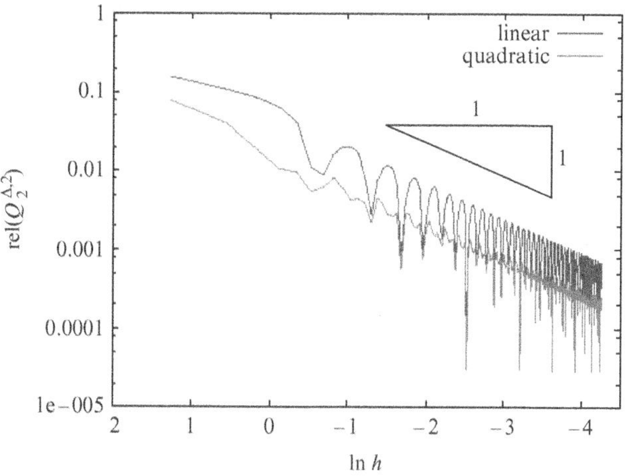

Fig. 3.13 Plots of $\text{rel}(Q_2^{\Delta,2})$ v $\ln h$ for Problem 2 with $p = 1$ and 2.

Analogous results are given in Fig. 3.9 for Problem 2, where of course no nodal point ever coincides with the interface. We note the following features:

(1) The upper, and especially the lower, envelopes are not well represented by straight lines, due to the 'chaotic' behaviour of the error.

(2) We can only say that

$$\mathcal{C}_1 h \leqq \text{rel}(Q_1^{\Delta}) \leqq \mathcal{C}_2 h,$$

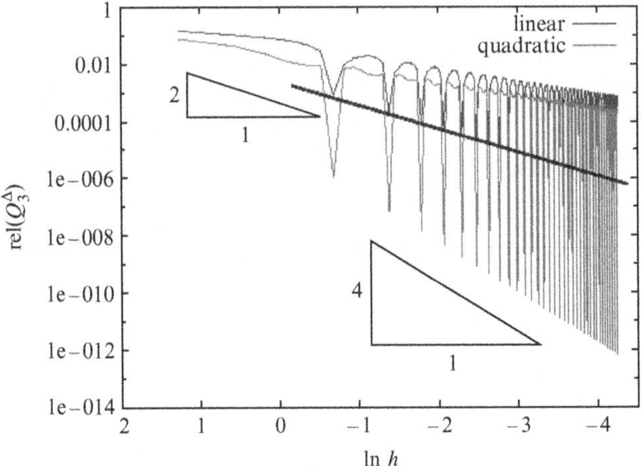

Fig. 3.14 Plots of rel(Q_3^Δ) v $\ln h$ for Problem 1 with $p = 1$ and 2.

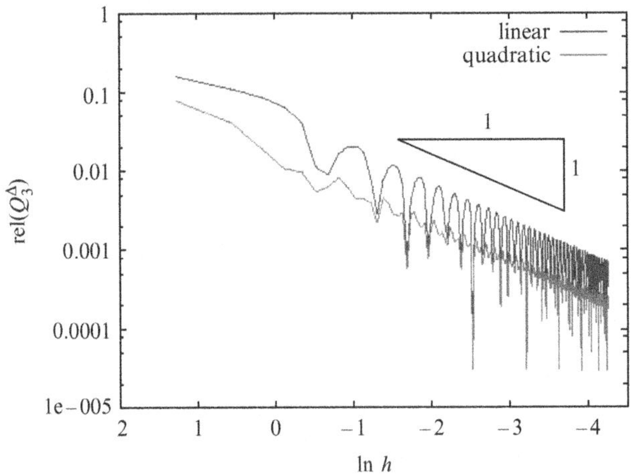

Fig. 3.15 Plots of rel(Q_3^Δ) v $\ln h$ for Problem 2 with $p = 1$ and 2.

for both $p = 1$ and 2; i.e. that the errors lie in a zone bounded above and below by two straight lines with the same slope. There are no specific mesh sizes that produce higher accuracy.

Plots of ln rel($Q_2^{\Delta,1}$) v ln h for Problems 1 and 2 with $p = 1$ and 2 are given, respectively, in Figs. 3.10 and 3.11. We again see the following features:

(1) For both problems there is again oscillating behaviour and we now have in each case that $\ln \text{rel} \leq Ch$, for $p = 1$ and 2.

(2) In Fig. 3.10 for Problem 1 when a node coincides with the interface we have for $p = 1$ that rel $\leqq Ch$ and hence the accuracy has the same form for general h; i.e. there are no special meshes that produce higher accuracy. The situation is different for $p = 2$, where we have $\text{rel}(Q_2^{\Delta,1}) \leqq Ch^2$ and there is improvement when a node coincides with the interface.

(3) In Fig. 3.11 for Problem 2 the behaviour is ever more 'chaotic' with $Ch \leqq \text{rel}(Q_2^{\Delta,1}) \leqq Ch$, so that the errors lie in a strip bounded above and below with lines of slope $\alpha = 1$.

The accuracy of $Q_2^{\Delta,2}$ is now considered.

(1) Figures 3.12 and 3.13 contain plots of $(\text{rel}(Q_2^{\Delta,2}))$ v $\ln h$, respectively, for Problems 1 and 2 for both $p = 1$ and 2. We have $Ch^{2p} \leqq \text{rel}(Q_2^{\Delta,2}) \leqq Ch$.

(2) For Problem 1 and meshes for which a nodal point coincides with the interface we have

$$\text{rel}(Q_2^{\Delta,2}) \approx \begin{cases} Ch^2, & p = 1, \\ Ch^4, & p = 2, \end{cases}$$

so that in the case $Q_2^{\Delta,1}$ and $Q_2^{\Delta,2}$ lead to very different accuracies, although they are both extracted from the same finite element solutions.

(3) For Problem 2 the behaviour of $Q_2^{\Delta,2}$ is again 'chaotic' and both $Q_2^{\Delta,2}$ and $Q_2^{\Delta,1}$ have the same accuracy.

Finally, Figs. 3.14 and 3.15 show results for Q_3^{Δ}, respectively, for Problems 1 and 2, and we see the following:

(1) In Problem 1 for meshes where a node coincides with the interface

$$\text{rel}(Q_3^{\Delta}) \approx \begin{cases} Ch^2, & p = 1, \\ Ch^4, & p = 2, \end{cases}$$

whereas for general meshes we have that

$$Ch^2 \leqq \text{rel}(Q_3^{\Delta}) \leqq Ch, \ p = 1,$$
$$Ch^4 \leqq \text{rel}(Q_3^{\Delta}) \leqq Ch, \ p = 2.$$

We may summarize all the above results as follows:

(1) When the different quantities of interest $Q_i^{\Delta}, i = 1, 2, 3$ are computed using the finite element method, they have differing accuracies.

(2) The accuracy in each case depends on both the mesh and the sizes of the elements. When a mesh is such that a node coincides with the interface better accuracy is obtained, with the exception of Q_2^{Δ} for $p = 1$ in Problem 1.

(3) There can be largely differing values in the accuracy of the computed quantities of interest if different methods of extraction from the same finite element solution are used.

Again, heavy lines have been drawn on Figs. 3.12 and 3.14 to improve visibility.

3.4 Two-dimensional finite element method

We now move to the two-dimensional setting as considered in Section 2.2, and consider problems of the type (2.63) for which the weak form is (2.72). For the two-dimensional finite element method we proceed in a very similar manner to that of the one-dimensional case in Section 3.3.

We recall that in two dimensions $\Omega \subset \mathbb{R}^2$ is a bounded polygonal domain with boundary Γ and that a Dirichlet boundary condition is imposed on the part Γ_{D} of Γ, where $\Gamma = \Gamma_{\mathrm{D}} \cup \Gamma_N$; see Fig. 3.16.

The finite element method is again a specific form of the Galerkin method for constructing ${}^N S$ and ${}^N S_0$, but now in the two-dimensional setting, and is based on piecewise polynomial functions of $\mathbf{x} \equiv (x_1, x_2)$ defined over partitions of Ω. As before, we consider separately the method using piecewise linear and piecewise quadratic functions.

3.4.1 The finite element method with piecewise linear functions

On $\bar{\Omega} \equiv \Omega \cup \Gamma$ we define a mesh Δ of *triangular* elements τ_q, see for example Fig. 3.16. Note that now the τ_q have vertices and sides in common, but they do not overlap. We keep the same notation as before, so that the vertices of the mesh are V_q, the nodal points will be $z_q \equiv V_q$. In addition, we shall call the sides of the triangles the edges

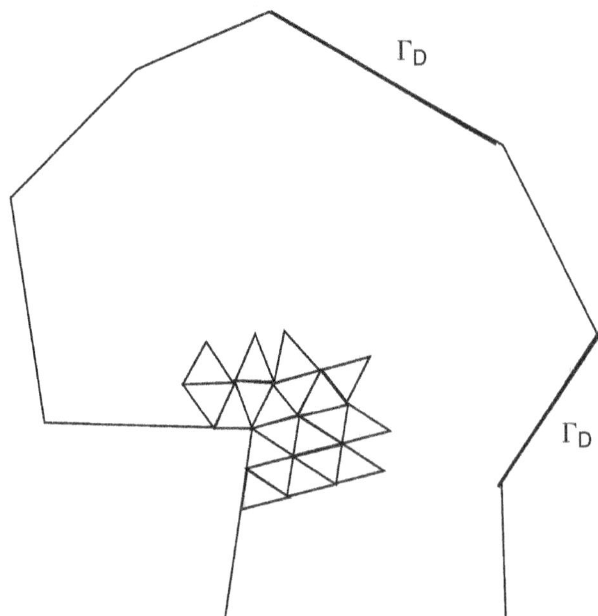

Fig. 3.16 Polygonal domain Ω with boundary Γ and with Dirichlet boundary Γ_{D} and an illustrative mesh of triangular elements.

and denote these by e_j; note that each edge belongs to exactly two triangles. We shall write $\Delta = \{\tau_q\}, E = \{e_j\}$.

For problem (2.64) we now define the spaces

$$S_\Delta^{[1]} \equiv \left\{v(\mathbf{x}), v \in C^0(\bar{\Omega}) \mid v\,|_{\tau q} \in \mathbb{P}^{[1]}(\bar{\tau}_q)\right\}, \tag{3.61a}$$

$$S_{\Delta,0}^{[1]} \equiv \left\{v(\mathbf{x}), v \in S_\Delta^{[1]} \mid v(\mathbf{x}) = 0,\ \mathbf{x} \in \Gamma_D\right\}, \tag{3.61b}$$

$$S_{\Delta,g}^{[1]} \equiv \left\{v(\mathbf{x}), v \in S_\Delta^{[1]} \mid v(\mathbf{x}) = g(\mathbf{x}),\ \mathbf{x} \in \Gamma_D\right\}, \tag{3.61c}$$

again denoting by $\mathbb{P}^{[1]}(\bar{\tau}_q)$ the space of polynomials of degree 1 on τ_q, and assuming that g is the trace of $S_{\Delta,}^{[1]}$ on Γ_D . Because any linear function in x_1 and x_2 on τ_q is completely determined by its values at three (non-collinear) nodal points, the vertices of the triangles will also be the nodal points; i.e. $z_q \equiv V_q$ in this piecewise linear case.

The basis functions $\Phi_q^{[1]}(\mathbf{x})$ will be *pyramids*; that is they are piecewise linear functions that are linear in each element, satisfying (see (3.21))

$$\Phi_q^{[1]}(z_j) = \begin{cases} 1 \; j = q, \\ 0 \; j \neq q. \end{cases} \tag{3.62}$$

By virtue of the location of the nodes z_q at the vertices of the elements, it also follows that $\Phi_q^{[1]}(\mathbf{x}) \in C^0(\bar{\Omega})$.

With each node z_q we now associate a *patch* σ_q of elements that forms the *support* for the pyramid $\Phi_q^{[1]}(\mathbf{x})$, see Fig. 3.17, each of which has z_q as a node. Condition (3.62) means that

$$\Phi_q^{[1]}(\mathbf{x}) = 0,\ \mathbf{x} \notin \sigma_q. \tag{3.63}$$

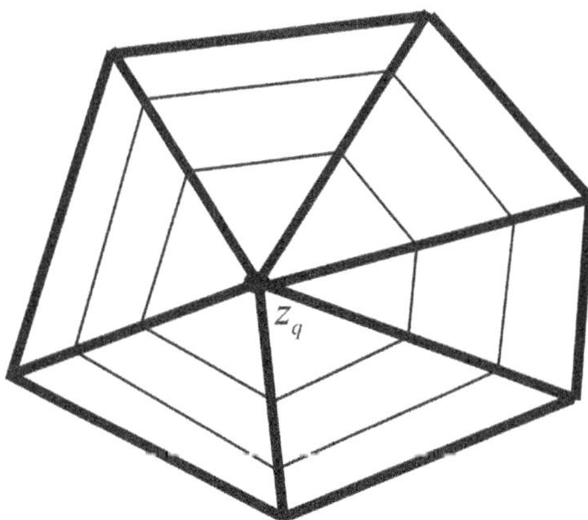

Fig. 3.17 The patch σ_q associated with the node z_q, showing the level lines of the basis function $\Phi_q^{[1]}$

If the number of the nodal points of the mesh is N, then the dimension of $S_{\Delta}^{[1]}$ is N and as before we take $\mathcal{N} \equiv \{1, \ldots, N\}$.

Some of the nodal points are located on Γ_{D}, so again as before the indices of these points lie in the set $\mathcal{N}_2$. Thus, $\mathcal{N}_1 = \mathcal{N} - \mathcal{N}_2$ and any ${}^N u(\mathbf{x}) \in S_{\Delta}^{[1]}$ can be written in the form (see 3.11)

$$u_{\Delta}^{[p]}(\mathbf{x}) \equiv^N u(\mathbf{x}) = \sum_{q \in \mathcal{N}_1} c_q \Phi_q^{[1]}(\mathbf{x}) + \sum_{q \in \mathcal{N}_2} d_q \Phi_q^{[1]}(\mathbf{x}), \tag{3.64}$$

where $c_q = u(z_q), q \in \mathcal{N}_1$ and $d_q = u(z_q), q \in \mathcal{N}_2$.

In order to set up the equation system (3.13) for the two-dimensional piecewise linear case, we need to construct the local stiffness matrix and load vector for each element τ_q. This will again be done using the master element technique.

Example 3.5 We consider the problem of type (2.55) in which $\mathfrak{u}(\mathbf{x})$ satisfies

$$-\nabla . ([\mathcal{A}(\mathbf{x})] \nabla \mathfrak{u}(\mathbf{x})) = 1, \ \mathbf{x} \in \Omega,$$

with $\mathcal{A}(\mathbf{x}) = \begin{bmatrix} 2 & 0 \\ 0 & 2 \end{bmatrix}$, e.g. the a_{ij} are constants, Ω is the unit square, see Fig. 3.18, and the boundary conditions are

$$\mathfrak{u}(\mathbf{x}) = 3, \ \mathbf{x} \in \Gamma_D \equiv \overline{A_2 \, A_3 \, A_4},$$

$$\frac{\partial u}{\partial n_c}(\mathbf{x}) = 0, \ \mathbf{x} \in \Gamma_N \equiv \overline{A_1 A_2} \cup \overline{A_4 \, A_1}.$$

We note from (2.54) that in this case $\dfrac{\partial u}{\partial n_c}(x)$ is $2\dfrac{\partial u}{\partial n}(x)$.

The associated weak form of the problem is (2.64) with

$$B(u, v) \equiv \int_{\Omega} \sum_{i,j=1}^{2} a_{ij} \frac{\partial u}{\partial x_i} \frac{\partial u}{\partial x_j} dx, \tag{3.65a}$$

$$F(v) \equiv \int_{\Omega} v dx. \tag{3.65b}$$

The mesh Δ and the numbering of the nodes z_q (which are identical with the triangle vertices) and of the elements τ_q are shown in Fig. 3.18. There are 16 nodal points and 18 elements.

As in the one-dimensional case we produce tables indicating the node numbers and coordinates and also associating with every element τ_q the numbers of the nodes (ordered clockwise).

Example 3.6 For the problem of Example 3.5 and the region, mesh and orderings of Fig. 3.18 the element/node table is given in Table 3.5.

Note that the sequence of node numbers for any element can start with any node, as long as the numbering is counterclockwise.

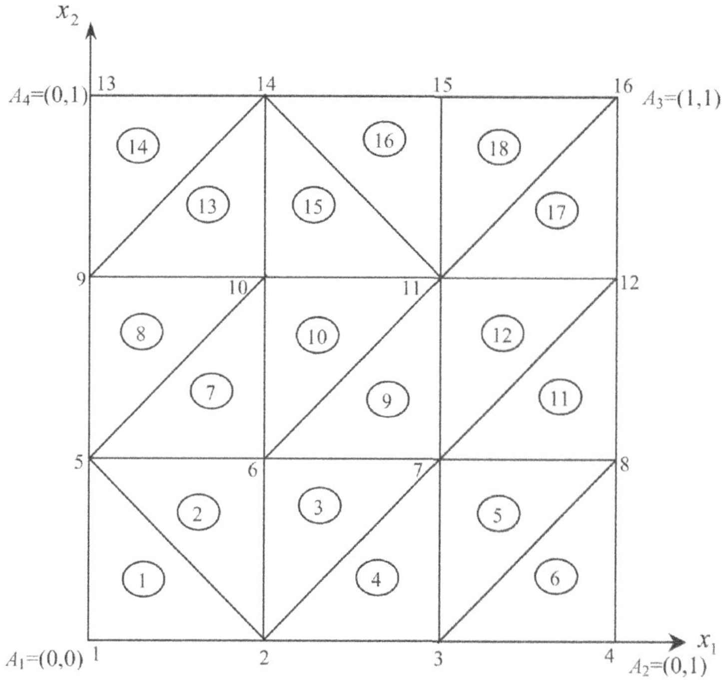

Fig. 3.18 The square domain Ω with boundary $\overline{A_1A_2A_3A_4A_1}$ and uniform triangular mesh with node and element numbering.

Table 3.5 Element/node table.

q		q		q	
1	$(1,2,5)$	7	$(5,6,10)$	13	$(10,14,9)$
2	$(6,5,2)$	8	$(5,10,9)$	14	$(9,14,13)$
3	$(2,7,5)$	9	$(7,11,6)$	15	$(10,11,14)$
4	$(2,3,7)$	10	$(6,11,10)$	16	$(15,14,11)$
5	$(7,3,8)$	11	$(7,8,12)$	17	$(16,11,16)$
6	$(8,3,4)$	12	$(7,12,11)$	18	$(11,16,15)$

We now address the task of constructing the local stiffness matrix and local load vector for an element τ_q of the mesh Δ as in Fig. 3.18 for problem (2.64). For this we use the master triangle $\tilde{\tau}$ in the (ξ_1, ξ_2) plane, as shown in Fig. 3.19, which has the master nodal points ${}^1\tilde{z}, {}^2\tilde{z}, {}^3\tilde{z}$, shape functions ${}^1\tilde{\Phi}(\ (\xi_1, \xi_2)$, ${}^2\tilde{\Phi}(\ (\xi_1, \xi_2)$, ${}^3\tilde{\Phi}(\ (\xi_1, \xi_2)$, and edges ${}^1\tilde{e}, {}^2\tilde{e}, {}^3\tilde{e}$, numbered counterclockwise.

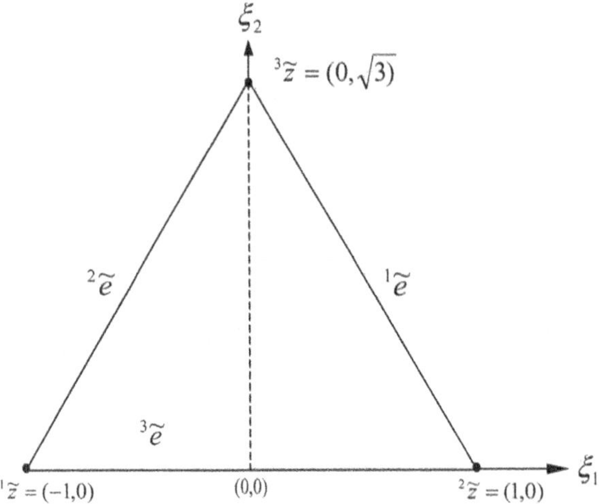

Fig. 3.19 The master triangle $\tilde{\tau}$.

$$ {}^1\tilde{\Phi}(\xi_1, \xi_2) = \frac{1}{2}\left(1 - \xi_1 - \frac{\xi_2}{\sqrt{3}}\right), \tag{3.66a} $$

$$ {}^2\tilde{\Phi}(\xi_1, \xi_2) = \frac{1}{2}\left(1 + \xi_1 - \frac{\xi_2}{\sqrt{3}}\right), \tag{3.66b} $$

$$ {}^3\tilde{\Phi}(\xi_1, \xi_2) = \frac{\xi_2}{\sqrt{3}}. \tag{3.66c} $$

Given an element τ_q in the physical region Ω with nodes $z_1(q), z_2(q), z_3(q)$, having coordinates $(x_{i,1}(q), x_{i,2}(q)), q = 1, 2, 3$, following the practice used in the one-dimensional case, we define the linear mapping

$$ {}^q x(\boldsymbol{\xi}) \equiv \left({}^q x_1(\xi_1, \xi_2), {}^q x_2(\xi_1, \xi_2)\right), $$

which maps the master triangle $\tilde{\tau}$ onto τ_q. This mapping has the form

$$ {}^q x_1(\xi_1, \xi_2) = \sum_{i=1}^{3} x_{i,1}(q)\,{}^i\tilde{\Phi}(\xi_1, \xi_2), \tag{3.67a} $$

$$ {}^q x_2(\xi_1, \xi_2) = \sum_{i=1}^{3} x_{i,2}(q)\,{}^i\tilde{\Phi}(\xi_1, \xi_2). \tag{3.67b} $$

We always order the nodal points in $\tilde{\tau}$ and τ_q counterclockwise. The coordinates of the nodes $x_{i,j}(q), i = 1, 2, 3$ for the element τ_q define the mapping ${}^q x_j(\xi_1, \xi_2), j = 1, 2$ of (3.66). The inverse linear mapping $\xi(x)$ exists because the vertices z_i of the element τ_q are not collinear.

Following the steps as for the one-dimensional case, for problem (2.64) we can now construct the local stiffness matrix $[\bar{K}^{(q)}]_{3\times 3}$ for τ_q as the sum, see (3.35) and (3.36).

$$[\bar{K}^{(q)}]_{3\times 3} = [^1\bar{K}^{(q)}] + [^2\bar{K}^{(q)}] + [^3\bar{K}^{(q)}],\ q = 1, 2, \ldots, M. \tag{3.68}$$

In order to do this we denote by J_q the Jacobian of the mapping (3.67) of $\tilde{\tau}$ onto τ_q

$$[J_q] = \begin{bmatrix} {}^qJ_{11} & {}^qJ_{12} \\ {}^qJ_{21} & {}^qJ_{22} \end{bmatrix} = \begin{bmatrix} \frac{\partial^q x_1}{\partial \xi_1} & \frac{\partial^q x_1}{\partial \xi_2} \\ \frac{\partial^q x_2}{\partial \xi_1} & \frac{\partial^q x_2}{\partial \xi_2} \end{bmatrix}. \tag{3.69}$$

Then, we have

$$\begin{aligned} \frac{\partial}{\partial \xi_1} &= J_{11}\frac{\partial}{\partial x_1} + J_{12}\frac{\partial}{\partial x_2}, \\ \frac{\partial}{\partial \xi_2} &= J_{21}\frac{\partial}{\partial x_2} + J_{22}\frac{\partial}{\partial x_2}, \end{aligned} \tag{3.70}$$

$$\begin{aligned} \frac{\partial}{\partial x_1} &= J^*_{11}\frac{\partial}{\partial \xi_1} + J^*_{12}\frac{\partial}{\partial \xi_2}, \\ \frac{\partial}{\partial x_2} &= J^*_{21}\frac{\partial}{\partial \xi_1} + J^*_{22}\frac{\partial}{\partial \xi_2}, \end{aligned} \tag{3.71}$$

where

$$[J^*] \equiv \begin{bmatrix} J^*_{11} & J^*_{12} \\ J^*_{21} & J^*_{22} \end{bmatrix} = \begin{bmatrix} J_{11} & J_{12} \\ J_{21} & J_{22} \end{bmatrix}^{-1} = [J]^{-1}.$$

This can now be used in the construction of the items of (2.66) so that we have first

$$\begin{aligned} \left[\bar{K}^{(q)}\right] &\equiv {}^1B_{\tau_q}\left(\Phi_i^{[1]}, \Phi_j^{[1]}\right) \equiv \int_{\tau_q} \begin{bmatrix} \frac{\partial \Phi_i^{[1]}}{\partial x_1} \\ \frac{\partial \Phi_i^{[1]}}{\partial x_2} \end{bmatrix}^T [\mathcal{A}] \begin{bmatrix} \frac{\partial \Phi_j^{[1]}}{\partial x_1} \\ \frac{\partial \Phi_j^{[1]}}{\partial x_2} \end{bmatrix} d\mathbf{x} \\ &= \int_{\tilde{\tau}} \begin{bmatrix} \frac{\partial^m \tilde{\Phi}}{\partial \xi_1} \\ \frac{\partial^m \tilde{\Phi}}{\partial \xi_2} \end{bmatrix}^T [J^*]^T \left[\tilde{A}\right] [J^*] \begin{bmatrix} \frac{\partial^n \tilde{\Phi}}{\partial \xi_1} \\ \frac{\partial^n \tilde{\Phi}}{\partial \xi_2} \end{bmatrix} |J|\, d\boldsymbol{\xi}, \end{aligned} \tag{3.72}$$

where $|J|$ is the determinant of J; because of the orientation this is always positive. Equation (3.72) produces $\left[^1\bar{K}^{(q)}\right]$. Similarly,

$$\left[^2\bar{K}^{(q)}\right] \equiv {}^2B_{\tau_q}\left(\Phi_i^{[1]}, \Phi_j^{[1]},\right) \equiv \int_{\tau} c\Phi_i^{[1]}, \Phi_j^{[1]} d\mathbf{x} = \int_{\tilde{\tau}} \tilde{c}^m\tilde{\Phi}^n\tilde{\Phi}\, |J|\, d\boldsymbol{\xi}, \tag{3.73}$$

which produces $\left[^2\bar{K}^{(q)}\right]$.

The constructing of $\left[^3\bar{K}^{(q)}\right]$ from the integral in (2.67) is slightly more complicated, as we have to integrate along element edges that coincide with Γ_{N}. It is usual for the α in (2.67) to be defined on a side $\Gamma_i \equiv \overline{A_iA_{i+1}} \subset \Gamma_{\mathrm{N}}$ in terms of a parameter s_i, starting at A_i. Then, $\Gamma_i = \{x_i(s_i), x_2(s_i)\}$. Let the edges of the master element be ${}^1\tilde{e}, {}^2\tilde{e}, {}^3\tilde{e}$, as in Fig. 3.18. In the finite element calculation for an element τ_q having an edge on $\Gamma_{\rm N}$, we always choose the mapping such that the edge ${}^3\tilde{e}$, of the master element maps onto the edge of τ_q that coincides with $\Gamma_{\rm N}$, say $[z_i, z_j]$. With this convention we have

$$\begin{aligned}\left[{}^3\bar{K}^{(q)}\right] &\equiv {}^3B_{\tau_q}\left(\Phi_i^{[1]}, \Phi_j^{[1]},\right) \\ &= \int_{z_i}^{z_j} \alpha(s_i)\Phi_i^{[1]}(s_i)\Phi_j^{[1]}(s_i)ds_i \\ &= \int_{{}^1\tilde{z}}^{{}^2\tilde{z}} \tilde{\alpha}(\xi_i)^m\tilde{\Phi}(\xi_i)^n\tilde{\Phi}(\xi_i)\,|J_s|\,d\tilde{s} \\ &= \int_{-1}^{1} \tilde{\alpha}(\xi_i)^m\tilde{\Phi}(\xi_1)^n\tilde{\Phi}(\xi_1)\,|J_s|\,d\xi_1, \end{aligned} \tag{3.74}$$

where J_s is the Jacobian of the transformation of τ_q onto the master element in terms of s, and we note that on ${}^3\tilde{e}$ the parameter $\tilde{s}$ becomes ξ_q. By summing (3.71)–(3.73) we obtain the local stiffness matrix $\left[\bar{K}^{(q)}\right]$ for the element τ_q.

The local load vector $\bar{\mathbf{f}}^{(q)}$ for the element τ_q is constructed analogously.

Exercise 3.9 For the problem given in Example 3.5 compute the local stiffness matrices for elements τ_1, τ_2, τ_5, and τ_6. Show that the local stiffness matrices for all the elements of the mesh in Fig. 3.18 can be obtained from these four local stiffness matrices, with possible permutation of rows and columns.

Exercise 3.10 Consider again the problem of Example 3.5, with the mesh and numbering of Fig. 3.18. Using the local stiffness matrices that you have computed in Exercise 3.9, construct the unconstrained stiffness matrix for this problem.

3.4.2 The finite element method with piecewise quadratic functions

Just as in the one-dimensional context we considered the extension of the finite element method from piecewise linear to piecewise quadratic functions, we now do exactly the same in the two-dimensional context. Keeping the same partition of Ω into triangles as in Fig. 3.16 we now define the spaces

$$S_\Delta^{[2]} \equiv \left\{v(\mathbf{x}),\ v \in C^0(\bar{\Omega}) \mid v|_{\tau_q} \in \mathbb{P}^{[2]}(\tau_q)\right\}, \tag{3.75a}$$

$$S_{\Delta,0}^{[2]} \equiv \left\{v(\mathbf{x}),\ v \in S_\Delta^{[2]} \mid v(\mathbf{x}) = 0,\ \mathbf{x} \in \Gamma_D\right\}, \tag{3.75b}$$

$$S_{\Delta,g}^{[2]} \equiv \left\{v(\mathbf{x}),\ v \in S_\Delta^{[2]} \mid v(\mathbf{x}) = g(\mathbf{x}),\ \mathbf{x} \in \Gamma_D\right\}, \tag{3.75c}$$

where $\mathbb{P}^{[2]}(\tau_q)$ is now the space of all polynomials in x_1 and x_2 of degree 2 on τ_q.

In every (physical) triangle τ_q, and the master triangle, we now use quadratic polynomials that have six coefficients, so that correspondingly locally we need six

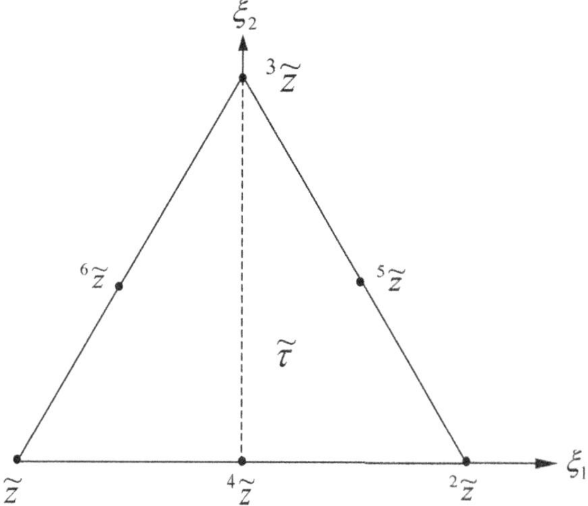

Fig. 3.20 Master triangle for piecewise quadratics, with vertices $(-1,0)(1,0)(0,\sqrt{3})$, and master nodes as shown.

nodes per triangle. For this, we first define the six-node master triangle in the (ξ_1, ξ_2)-plane as in Fig. 3.20, with nodes and associated shape functions ${}^i\tilde{\Phi}(\xi) i = 1, \ldots, 6$. These have the form

$$ {}^1\tilde{\Phi}(\xi_1,\xi_2) = \frac{1}{2}\left(1 - \xi_1 - \frac{\xi_2}{\sqrt{3}}\right), \tag{3.76a}$$

$$ {}^2\tilde{\Phi}(\xi_1,\xi_2) = \frac{1}{2}\left(1 + \xi_1 - \frac{\xi_2}{\sqrt{3}}\right), \tag{3.76b}$$

$$ {}^3\tilde{\Phi}(\xi_1,\xi_2) = \frac{\xi_2}{\sqrt{3}}, \tag{3.76c}$$

$$ {}^4\tilde{\Phi}(\xi_1,\xi_2) = \left(1 - \xi_1 - \frac{\xi_2}{\sqrt{3}}\right)\left(1 + \xi_1 - \frac{\xi_2}{\sqrt{3}}\right), \tag{3.76d}$$

$$ {}^5\tilde{\Phi}(\xi_1,\xi_2) = \frac{2}{\sqrt{3}}\xi_2\left(1 + \xi_1 - \frac{\xi_2}{\sqrt{3}}\right), \tag{3.76e}$$

$$ {}^6\tilde{\Phi}(\xi_1,\xi_2) = \frac{2}{\sqrt{3}}\xi_2\left(1 - \xi_1 - \frac{\xi_2}{\sqrt{3}}\right). \tag{3.76f}$$

Note that at each vertex of the master triangle with (3.76) we have again the same shape functions as we had in the linear case, and that the additional nodes are located at the side midpoints. Also, note that the functions ${}^i\tilde{\Phi}(\xi_1, \xi_2)$ are linearly independent and that a quadratic polynomial can be expressed as a linear combination of these functions.

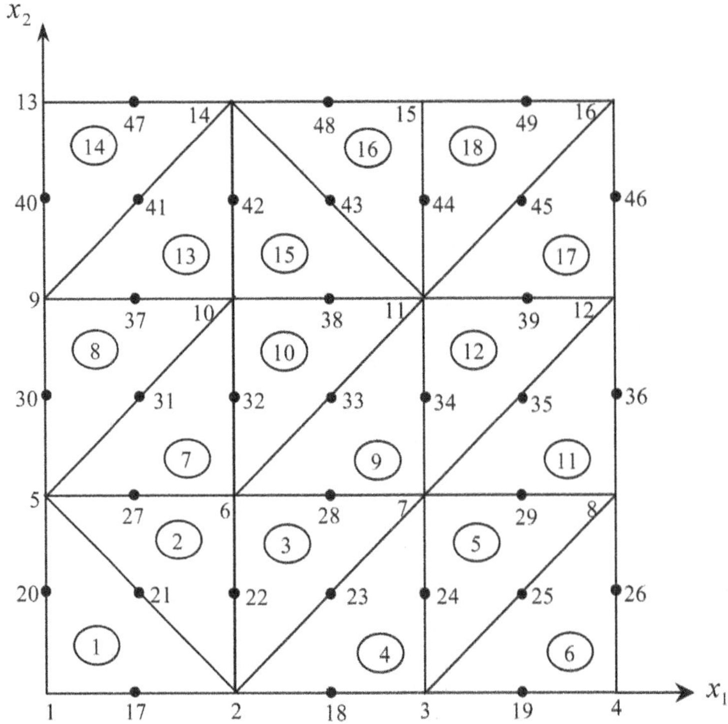

Fig. 3.21 Mesh partition with node and element numberings.

Considering again the problem given in Example 3.5 and the mesh of Fig. 3.18 we add new nodal points at the midpoints of all the element edges. This produces a total of 49 nodes, so that $N = 49$. The global node numbering is shown in Fig. 3.21.

The nodal points at the vertices will be called *vertex* nodes, whilst those at the side midpoints will be called *edge* nodes. Because the edge nodes are located at the edge midpoints, the linear mapping (3.67) based on the vertex nodes maps the master triangle onto the element τ_q of the mesh, and the master edge nodes onto the physical edge nodes (i.e. at the edge midpoints of τ_q).

In order to construct the local stiffness and load vector for the element τ_q for the piecewise quadratic case we proceed exactly as for the linear case, except that with τ_q we associate six nodal points and six shape functions arising from the six shape functions (3.76) of the master triangle. Once more, we respect the counterclockwise node ordering in the element, first with the vertex nodes and then with the edge nodes. The local stiffness matrix will now be of order 6×6.

Remark 3.9 *For the finite element method as set out in this chapter it has continually been necessary to evaluate integrals in order to compute stiffness matrices, see, e.g., (3.14) and (3.29) and load vectors, see, e.g., (3.15) and (3.29). In practice, these integrals will invariably have to be evaluated numerically using a quadrature rule.*

If a finite element approximation is a piecewise pth-order polynomial, and the problem has constant coefficients, the corresponding local stiffness matrices will involve integrals of polynomials of order $2(p-1)$. The choice of quadrature rule must be such that the integrals of these $2(p-1)$th-order polynomials are evaluated exactly. When non-constant coefficients are present in the problem the quadrature rules, even when following the above rule, are not exact. However, when the rule is observed the error in this case is not significant.

The situation for the integrals in the load vectors is similar to the above, but is not so dangerous as for the stiffness matrices. In practice, the quadrature rule used for the load vectors is taken to be the same as that for the stiffness matrices.

From the above it should be clear that we counsel great care in the choice of quadrature rules in this context!

3.4.3 Two benchmark problems

Having so far in Section 3.4 introduced the finite element method for the two-dimensional context, we now wish to demonstrate the performance of the method. This we do in the context of two *benchmark problems*.

A benchmark problem is one that is, as far as possible, representative of a class of problems that is under consideration and for which the exact solution is known. Benchmark problems are used both for verification purposes and for gaining understanding of a numerical technique. Often, a benchmark problem is constructed by first selecting a solution and then creating the input data for the problem from this; such a problem is often referred to as a *manufactured problem*. Alternatively, the solution of a benchmark problem may sometimes be obtained by using 'numerical overkill', which means by using expensive numerical techniques leading to an accurate solution that can be identified with the exact solution of the problem. We note that our one-dimensional engineering problem of Section 2.1.5 can be thought of as a benchmark problem.

We now consider two benchmark problems, chosen deliberately so that one has a smooth solution whilst the other has a non-smooth solution due to the presence of a re-entrant (concave) corner in the domain of the problem, and solve them with finite element methods.

Benchmark Problem 1

Let $\Omega \equiv (0,1) \times (0,1)$ be the unit square domain with boundary $\Gamma \equiv \overline{A_1 A_2 A_3 A_4 A_1}$ as in Fig. 3.18. In Benchmark Problem 1 we consider the problem in which $u(\mathbf{x})$ satisfies

$$-\Delta u(\mathbf{x}) = f(\mathbf{x}),\ \mathbf{x} \in \Omega, \tag{3.77a}$$

$$u(\mathbf{x}) = 0,\ \mathbf{x} \in \overline{A_4 A_1 A_2}, \tag{3.77b}$$

$$\frac{\partial u}{\partial n}(x) = 0,\ \mathbf{x} \in \overline{A_2 A_3 A_4}, \tag{3.77c}$$

with

$$f(\mathbf{x}) = \left(\sin\frac{\pi x_1}{2}\right)\left(\sin\frac{\pi x_2}{2}\right),$$

so that $\mathfrak{u}(\mathbf{x})$ has the form

$$\mathfrak{u}(\mathbf{x}) = \left(\sin\frac{\pi x_1}{2}\right)\left(\sin\frac{\pi x_2}{2}\right) \Big/ \left(\left(\frac{\pi}{2}\right)^2 + \left(\frac{\pi}{2}\right)^2\right).$$

This is a typical problem with a manufactured solution.

Analogously as before we are interested in the total energy $Q_3 \equiv \Pi(\mathfrak{u})$ and its approximation $Q_3^{\Delta} \equiv \Pi\left(u_{\Delta}^{[p]}\right)$. For this problem we have that $Q_3 = -0.0253303$. It has also been shown, see Exercise 2.10, that

$$\|\mathfrak{u}\|_{\mathcal{U}}^2 = |2Q_3|,$$

so that here $\|\mathfrak{u}\|_{\mathcal{U}} = 0.2250708$. Further,

$$2\left\|e_{\Delta}^{[p]}\right\|_{\mathcal{U}}^2 \equiv 2\left\|\mathfrak{u} - u_{\Delta}^{[p]}\right\|_{\mathcal{U}}^2 = Q_3^{\Delta} - Q_3,$$

so that

$$Q_3^{\Delta} = Q_3 + 2\left\|e_{\Delta}^{[p]}\right\|_{\mathcal{U}}^2.$$

As a result of this, instead of computing Q_3^{Δ} we shall instead compute $\left\|u_{\Delta}^{[p]}\right\|_{\mathcal{U}}$, $\left\|e_{\Delta}^{[p]}\right\|_{\mathcal{U}}$ and

$$rel \equiv rel\left(\left\|u_{\Delta}^{[p]}\right\|\right) \equiv \left\|e_{\Delta}^{[p]}\right\|_{\mathcal{U}} \Big/ \left\|u_{\Delta}^{[p]}\right\|_{\mathcal{U}}. \tag{3.78}$$

In the computations we use four types of uniform mesh, regular, chevron, union jack and criss-cross, with different mesh sizes; these patterns are illustrated in Fig. 3.22 for $h = 0.125$.

Again, as in the one-dimensional case, the computations are made in double precision, and the coefficients in the stiffness matrices and load vectors are computed so that all digits given in Tables 3.6 and 3.7 are free of any 'crimes'.

Tables 3.6 and 3.7 show, respectively, the values of $\left\|u_{\Delta}^{[p]}\right\|_{\mathcal{U}}$ and the corresponding relative errors, rel for $p = 1, 2$.

In Fig. 3.23 we show plots of rel against $\ln h$ for $p = 1$ and 2 and see the following:

(1) the accuracy is
$$\text{rel} \approx Ch \quad \text{for } p = 1,$$
$$\text{rel} \approx Ch^2 \ \text{for } p = 2,$$
where the type of mesh influences only the constant C and not the decay of error, which is characterized by the power of h.

(2) The criss-cross pattern is related to the union jack pattern in the sense that rotating the criss-cross pattern by 45° produces a union jack pattern, but of course for any h the criss-cross pattern mesh has twice as many degrees of freedom as the other patterns. Because the computational effort is related to the degrees of freedom in the mesh we have that $hcris = hreg \,/\, \sqrt{2}$.

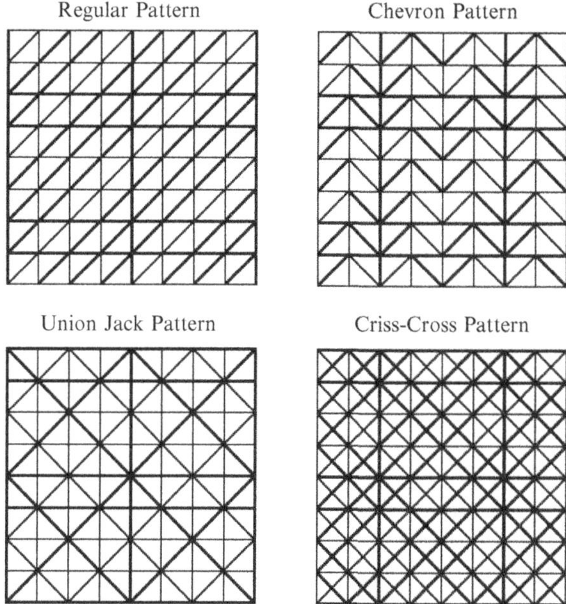

Fig. 3.22 Regular, chevron, union jack and criss-cross pattern meshes.

Table 3.6 Energy norm and relative error for $p = 1$.

h	Regular $\left\|u_\Delta^{[1]}\right\|_\mathcal{U}$	Pattern rel	Chevron $\left\|u_\Delta^{[1]}\right\|_\mathcal{U}$	Pattern rel	Union Jack $\left\|u_\Delta^{[1]}\right\|_\mathcal{U}$	Pattern rel	Criss-Cross $\left\|u_\Delta^{[1]}\right\|_\mathcal{U}$	Pattern rel
0.500000	0.214517	0.302737	0.209484	0.365752	0.210071	0.359045	0.220220	0.206660
0.250000	0.222079	0.162715	0.220958	0.190485	0.221249	0.183700	0.223872	0.103446
0.125000	0.224300	0.083135	0.224020	0.096910	0.224117	0.092382	0.224778	0.051738
0.062500	0.224882	0.041831	0.224811	0.048806	0.224838	0.046258	0.225004	0.025871
0.031250	0.225030	0.020953	0.225012	0.024478	0.225019	0.023137	0.225060	0.012936
0.015625	0.225067	0.010482	0.225062	0.012256	0.225064	0.011570	0.225074	0.006468

(3) The relative performances with the different meshes are dependent on p. For example, the solution with the regular pattern is better than those with the chevron and union jack patterns for $p = 1$, whilst the opposite is the case for $p = 2$. After rescaling, the criss-cross pattern produces the best solution.

We merely present the numerical results here for purposes of illustration and leave the explanation as to why they are so to later sections.

Table 3.7 Energy norm and relative error for $p = 2$.

h	Regular $\left\|u_\Delta^{[1]}\right\|_{\mathcal{U}}$	Pattern rel	Chevron $\left\|u_\Delta^{[1]}\right\|_{\mathcal{U}}$	Pattern rel	Union Jack $\left\|u_\Delta^{[1]}\right\|_{\mathcal{U}}$	Pattern rel	Criss-Cross $\left\|u_\Delta^{[1]}\right\|_{\mathcal{U}}$	Pattern rel
0.500000	0.224562	0.067719	0.224723	0.056224	0.224731	0.55604	0.225030	0.020804
0.250000	0.225045	0.017342	0.225055	0.014519	0.225056	0.014370	0.225076	0.005276
0.125000	0.225077	0.004379	0.225078	0.003662	0.225078	0.003624	0.225079	0.001323
0.062500	0.225079	0.001099	0.225079	0.000918	0.225079	0.000908	0.225079	0.000331
0.031250	0.225079	0.000275	0.225079	0.000230	0.225079	0.000227	0.225079	0.000088
0.015625	0.225079	0.000069	0.225079	0.000057	0.225079	0.000057	0.225079	0.000021

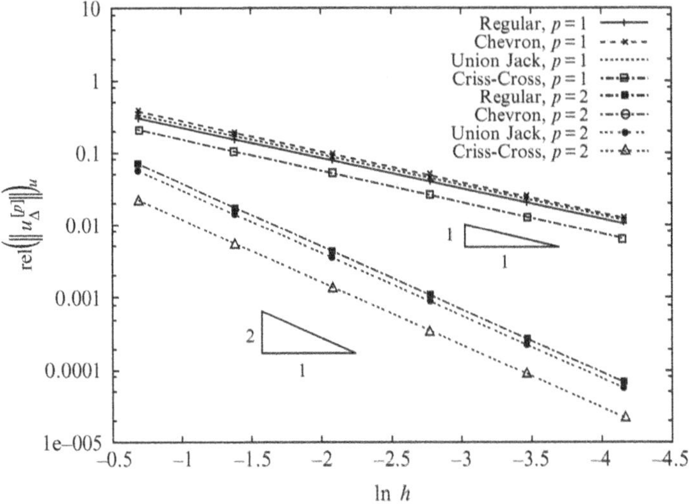

Fig. 3.23 Plots of rel $\left(\left\|u_\Delta^{[p]}\right\|_{\mathcal{U}}\right)$ v ln h for $p = 1, 2$ and the four mesh patterns.

Benchmark Problem 2

For Benchmark Problem 2 we consider once again a problem in an L-shaped domain with a re-entrant corner at the origin. Let $\Omega \equiv (-1,0) \times (-1,1) \cup (0,1) \times (0,1)$ with boundary $\Gamma \equiv \overline{A_1A_2 \ldots A_8A_1}$ as shown in Fig. 3.24, and let $\mathrm{u}(\mathbf{x})$ satisfy

$$-\Delta \mathrm{u}(\mathbf{x}) = 0,\ \mathbf{x} \in \Omega, \tag{3.79a}$$

$$\mathrm{u}(\mathbf{x}) = 0,\ \mathbf{x} \in \overline{A_1A_2}, \tag{3.79b}$$

$$\frac{\partial \mathrm{u}}{\partial n}(\mathbf{x}) = g(\mathbf{x}),\ \mathbf{x} \in \overline{A_2A_3A_4A_5A_6A_7A_1}, \tag{3.79c}$$

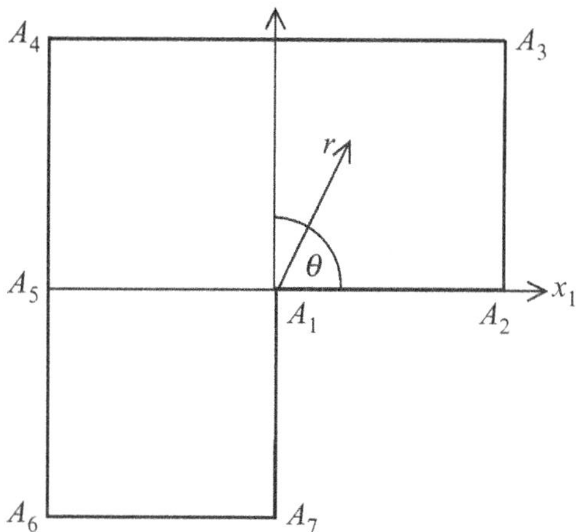

Fig. 3.24 L-shaped domain Ω.

with

$$u(\mathbf{x}) = r^{1/3}\sin(\theta/3), \tag{3.80}$$

(r, θ) are polar coordinates centred on the origin A as shown, and the $g(\mathbf{x})$ results from $u(\mathbf{x})$. This is again a typical problem with a manufactured solution.

Whilst the $g(\mathbf{x})$ is smooth on each segment of the boundary Γ, the solution $u(\mathbf{x})$ has a singularity in the neighbourhood of A; see again Chapter 2. The purpose of this benchmark problem is thus to target the effect of this singularity on the finite element solution.

We compute finite element solutions for Benchmark Problem 2 with regular meshes having regular and criss-cross patterns for $h = 0.125$ as in Fig. 3.25. We restrict ourselves here to the two meshes because results with the regular, chevron and union jack patterns are very similar with respect to the number of degrees of freedom.

Values of rel for $p = 1, 2$ are given in Table 3.8 for both meshes with different mesh sizes h, whilst in Fig. 3.26 we show plots of rel against $\ln h$ for $p = 1, 2$. We see the following:

(1) For $p = 1, 2$ we have that rel $\approx Ch^{1/3}$, and hence the singular behaviour of the solution influences the performance of the finite element method. By contrast to Benchmark Problem 1 the shape of the curves is the same for $p = 1, 2$ and depends on the singularity in $u(\mathbf{x})$. Nevertheless, the factor C, as above, is preferable in the case $p = 2$ (after rescaling as necessary to the number of degrees of freedom).

(2) The criss-cross pattern produces the better solution, again after rescaling. In fact, the relative performance is better in this case than for Benchmark Problem 1. Again, we leave the explanation of the results until later.

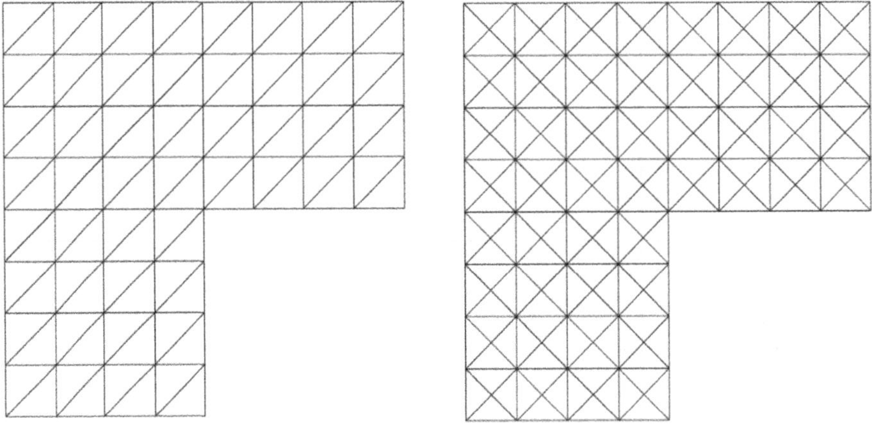

Fig. 3.25 Regular and criss-cross mesh patterns for $h = 0.125$.

Table 3.8 Relative errors $\text{rel} \equiv \text{rel}\left(\left\|u_{\Delta}^{[p]}\right\|_{\mathcal{U}}\right)$ for $p = 1$ and 2 for regular and criss-cross patterns.

	$p = 1$		$p = 2$	
	Regular	Criss-Cross	Regular	Criss-Cross
h	*rel*	*rel*	*rel*	*rel*
0.500000	0.559058	0.509494	0.404861	0.363912
0.250000	0.462103	0.417131	0.326513	0.292549
0.125000	0.375693	0.337461	0.261820	0.234106
0.062500	0.302657	0.271031	0.209172	0.186784
0.031250	0.242461	0.216709	0.166713	0.148744

3.4.4 Engineering application: two-dimensional heat transfer-problems

In Section 2.2.3 we defined the two-dimensional heat-transfer problem, 2D Eng Problem, defined in the cooling pipe with fins of Fig. 2.13, which will be used throughout this book to illustrate various aspects of the finite element method. As explained in Section 2.2.3, due to symmetry we need to consider only the part of the problem in the sector of Fig. 2.14. This in turn is transformed from radial into Cartesian coordinates so that we treat the problem (2.91) in the polygonal domain of Fig. 2.15. We shall compute approximations $Q_1^{\Delta}(A)$, $Q_1^{\Delta}(B)$, $Q_2^{\Delta,1}$, $Q_2^{\Delta,2}$, Q_3^{Δ}, respectively to the quantities of interest $Q_1(A)$, $Q_1(B), Q_2$ and Q_3, as defined in Section 2.2.3, using again the values $T_0 = 500, T_1 = 20,\ \gamma_{\text{ss}} = 0.90,\ \gamma_{\text{cs}} = 2.10$ and $\beta = 0.55$. The non-

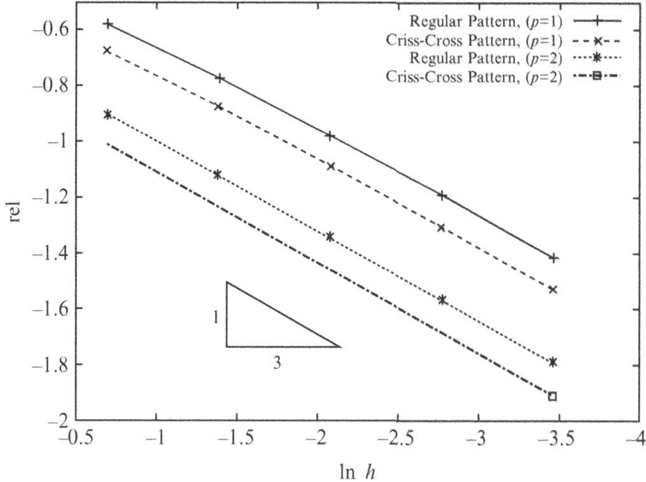

Fig. 3.26 rel v ln h for $p = 1$ and 2.

homogenous Dirichlet boundary condition $T_0 = 500$ is implemented using $\bar{u} = 500$; see (2.32) and (2.72). When applying the finite element method in the region of Fig. 2.15 we shall use the four basic mesh patterns of Fig. 3.27 (respectively, (a) regular, (b) chevron, (c) union jack and (d) criss-cross, with the nomenclature as used earlier) and will create a sequence of refinements of these by successive halving of the elements in each case. We note that, due to the shape of the region Ω and the chosen meshes some of the elements have large aspect ratios, (i.e. the ratio between the 'width' and 'height' of the element), causing some angles to be very small. Although the minimal angle is small, the maximal angle is $90°$, which is a sufficient criterion for convergence; see Remark 4.10.

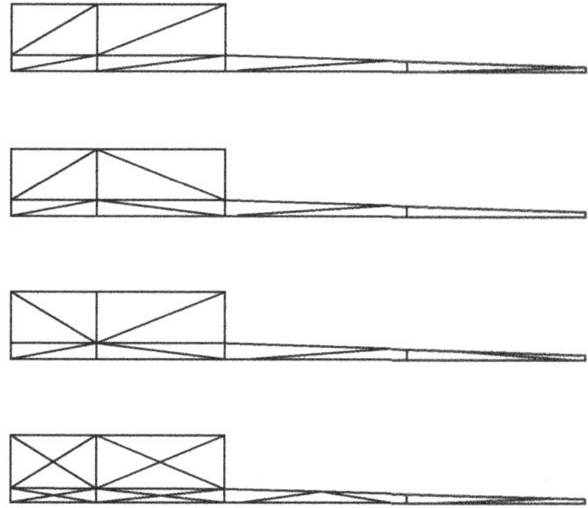

Fig. 3.27 Various mesh patterns for 2D Eng Problem 2.

The accuracy of the finite element solution will be related to N, the number of degrees of freedom in the computations, where $N = 0(h^{-2})$, with h being the size of the smallest element of the mesh. The number of degrees of freedom for the different meshes with levels of refinement $1, \dots, 7$ for $p = 1$ and 2 are given in Table 3.9.

One major difficulty in computing errors for the finite element solution of the 2D Eng Problem is that we do not know the exact solution of the problem. Whilst we shall discuss *a posteriori* error estimation (i.e. computed error estimates) in later chapters, we shall here use a simple Richardson extrapolation technique to produce accurate approximations to Q_i, $i = 1, 2, 3$ that can be used for comparison purposes. Richardson extrapolation has many entries in Google®; the original reference is Richardson (1910). We use a generalized form of the method with computed β; see e.g. Szabo and Babuška (1991), p.62. In each case, the 'extrapolated Q_i' is calculated by assuming that we can write

$$Q_i - Q_i^{\Delta} = CN^{-\beta}, \tag{3.81}$$

where Q_i, C and β have to be determined. The three coefficients are calculated using always the last three computed values of Q_i^{Δ} with the appropriate value for N, and hence the extrapolated Q_i value is obtained. In (3.81) the coefficient β characterizes the rate of convergence of Q_i^{Δ} to Q_i. We note first that in practical computations with non-uniform meshes the convergence is always taken with respect to N, and secondly

Table 3.9 The number of degrees of freedom N for the four mesh patterns and refinements. In each case the top number is for $p = 1$ and the lower number is for $p = 2$.

	Regular	Chevron	Union Jack	Criss-Cross
1	13	13	13	14
	37	37	37	61
2	37	37	37	61
	121	121	121	217
3	121	121	121	217
	433	433	433	817
4	433	433	433	817
	1 633	1 633	1 633	3 169
5	1 633	1 633	1 633	3 169
	6 337	6 337	6 337	12 481
6	6 337	6 337	6 337	12 481
	24 961	24 961	24 961	49 537
7	99 073	99 073	99 073	197 377

that the 'extrapolated Q_i' values and the errors $\mathrm{rel}(Q^{\Delta})$ are not exact. However, they are usually reliable if the meshes are fine enough to ensure that the calculations are in the asymptotic range.

Tables 3.10 and 3.11 show the values the temperature $Q_i^{\Delta}(A)$ and the values of extrapolated $Q_i(A)$, respectively, for $p = 1$ and 2. Similarly, Figs. 3.28 and 3.29 show $\mathrm{rel}\left(Q_i^{\Delta}(A)\right)$ v $\ln N$, respectively, for $p = 1$ and 2. From these we see that the convergence rates for $p = 1$ and 2 differ slightly. This is caused by imprecise knowledge of the exact solutions and also particularly, that we are not completely in the asymptotic range. It is slightly surprising that the rate of convergence is the same for $p = 1$ and 2. This is caused by the effect of the presence of a re-entrant comer at point (6) in Fig. 2.15, similarly as in Benchmark Problem 2. Analogous effects will occur in all the other QoI under consideration. Note that practical accuracy of 1% is obtained for refinement 3 when $p = 1$ and for refinement 1 when $p = 2$. Comparison of the computed values for the various N of the different refinements shows that $p = 2$ is much the more effective.

Tables 3.12 and 3.13 show the value of $Q_i^{\Delta}(B)$ and the values of extrapolated $Q_i(B)$, respectively, for $p = 2$ and 2, whilst Figs. 3.30 and 3.31 show $\mathrm{rel}\left(Q_i^{\Delta}(B)\right)$ v–$\ln N$, respectively, for $p = 1$ and 2. We see here that:

Table 3.10 Values of $Q_1^{\Delta}(A)$, for various mesh patterns and refinements (upper table), values of extrapolated $Q_1(A)$ and computed rates of convergence (lower table) for $p = 1$.

	Pattern			
Refinement	Regular	Chevron	Union Jack	Criss-Cross
1	257.8460	256.2494	256.7525	253.8330
2	251.2471	250.7757	250.9620	248.9207
3	247.4538	247.3284	247.4182	246.4585
4	245.8793	245.8511	245.8730	245.5476
5	245.3439	245.3381	245.3467	245.2298
6	245.1552	245.1540	245.1570	245.1136
7	245.0856	245.0854	245.0865	245.0700
8	245.0593	245.0593	245.0597	245.0534

Pattern	Extrapolated $Q_1(A)$	β
Regular	245.0435	0.7081
Chevron	245.0435	0.7053
Union Jack	245.0435	0.7057
Criss-Cross	245.0435	0.7097

Table 3.11 Values of $Q_1^{\Delta}(A)$, for various mesh patterns and refinements (upper table), values of extrapolated $Q_1(A)$ and computed rates of convergence (lower table) for $p = 2$.

Refinement	Pattern: Regular	Chevron	Union Jack	Criss-Cross
1	247.8097	247.7672	247.7733	247.1506
2	245.8764	245.8602	245.9042	245.6689
3	245.3287	245.3267	245.3376	245.2510
4	245.1556	245.1554	245.1594	245.1216
5	245.0872	245.0872	245.0887	245.0734
6	245.0603	245.0603	245.0609	245.0548
7	245.0498	245.0498	245.0500	245.0477

Pattern	Extrapolated $Q_1(A)$	β
Regular	245.0432	0.6912
Chevron	245.0432	0.6931
Union Jack	245.0432	0.6839
Criss-Cross	245.0432	0.7052

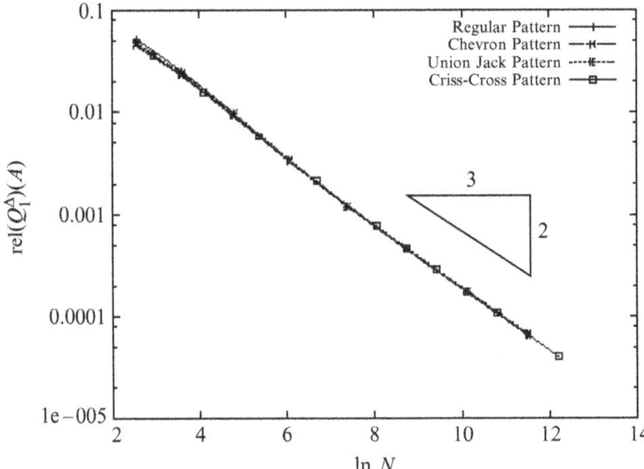

Fig. 3.28 rel $\left(Q_1^{\Delta}(A)\right)$ v $\ln N$ for $p = 1$ using extrapolated $Q_1(A)$ as exact value.

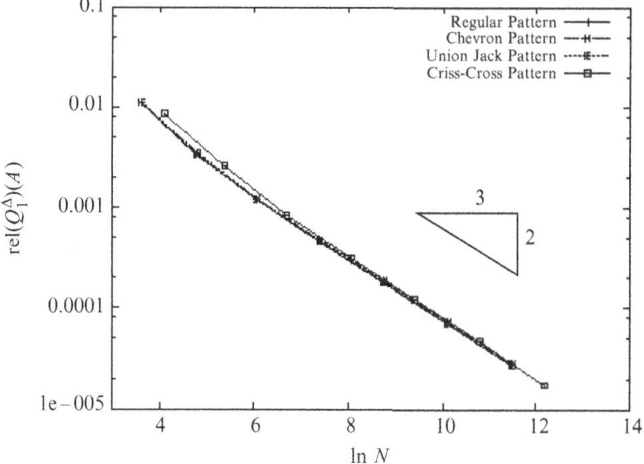

Fig. 3.29 rel $\left(Q_1^\Delta(A)\right)$ v $\ln N$ for $p = 2$ using extrapolated $Q_1(A)$ as exact value.

Table 3.12 Values of $Q_1^\Delta(B)$, for various mesh patterns and refinements (upper table), values of extrapolated $Q_1(B)$ and computed rates of convergence (lower table) for $p = 1$.

	Pattern			
Refinement	Regular	Chevron	Union Jack	Criss-Cross
1	72.5376	72.1001	72.2258	71.5177
2	75.5273	75.3928	75.4186	74.0460
3	75.8136	75.7777	75.7917	75.5824
4	75.6994	75.6905	75.6955	75.5986
5	75.5986	75.5964	75.6955	75.5986
6	75.5461	75.5456	75.5466	75.5296
7	75.5226	75.5226	75.5230	75.5162
8	75.5129	75.5129	75.5130	75.5103

Pattern	Extrapolated $Q_1(B)$	β
Regular	75.5062	0.6494
Chevron	75.5062	0.6494
Union Jack	75.5062	0.6265
Criss-Cross	75.5062	0.5970

Table 3.13 Values of $Q_1^\Delta(B)$, for various mesh patterns and refinements (upper table), values of extrapolated $Q_1(B)$ and computed rates of convergence (lower table) for $p = 2$.

	Pattern			
Refinement	Regular	Chevron	Union Jack	Criss-Cross
1	75.4252	76.5033	76.5053	75.1244
2	75.8635	75.8747	75.8794	75.7539
3	75.6436	75.6454	75.6485	75.5997
4	75.5579	75.5580	75.5596	75.5413
5	75.5260	75.5260	75.5266	75.5197
6	75.5139	75.5139	75.5141	75.5115
7	75.5092	75.5092	75.5093	75.5083

Pattern	Extrapolated $Q_1(B)$	β
Regular	75.5063	0.6882
Chevron	75.5063	0.6942
Union Jack	75.5063	0.6956
Criss-Cross	75.5063	0.6909

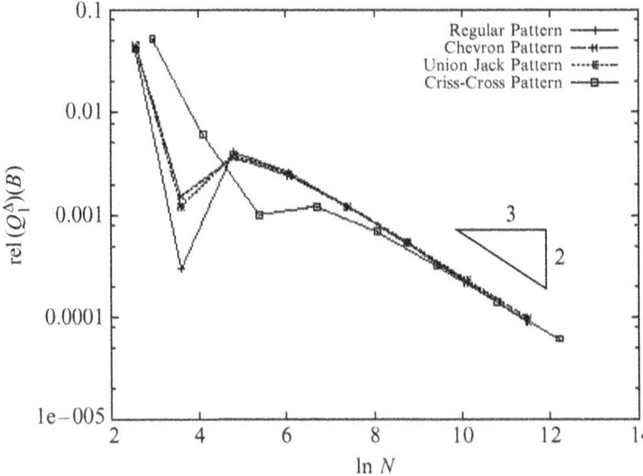

Fig. 3.30 rel $\left(Q_1^\Delta(B)\right)$ v $\ln N$ for $p = 1$ using extrapolated $Q_1(B)$ as exact value.

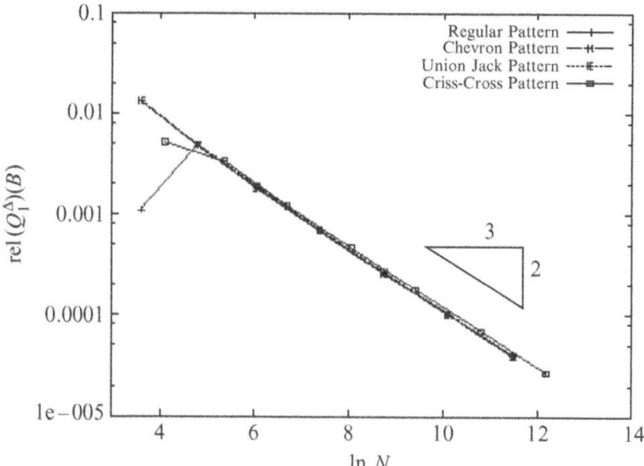

Fig. 3.31 rel $\left(Q_1^\Delta(B)\right)$ v $\ln N$ for $p = 2$ using extrapolated $Q_1(B)$ as exact value.

(1) the difference between the results for $p = 1$ and 2 is larger for $Q_i^\Delta(B)$ than for $Q_i^\Delta(A)$.
(2) Pre-asymptotic behaviour is clearly visible in each case but is larger for $p = 1$ than for $p = 2$. The basic mesh (refinement 1) produces a poor approximation, but in spite of this practical accuracy of 1% is in both cases achieved for refinement 2.

We next consider the QoI Q_2, the heat loss from the pipe to the environment giving, in Tables 3.14–3.17 the values of $Q_2^{\Delta,1}$and $Q_2^{\Delta,2}$ for $p = 1$ and 2, using the two strategies as described in Section 3.3.5 for the 1D Eng Problems, see (3.59) and (3.60). In the current two-dimensional case we have that, see Fig. 2.15,

$$Q_2 = 16 \int_{(4)}^{(7)} \frac{\partial \mathrm{u}}{\partial n_c}(s)ds,$$

so that in the first strategy we use $\dfrac{\partial u_\Delta^{[p]}}{\partial n_c}$ directly, giving

$$Q_2^{\Delta,1} = 16 \int_{(4)}^{(7)} \frac{\partial u_\Delta^{[p]}}{\partial n_c}(s)ds.$$

In the second strategy we utilize the boundary condition (2.90c) and define

$$Q_2 = 16 \int_{(4)}^{(7)} -\beta \left(\mathrm{u}(s) - T_1\right) ds,$$

giving

$$Q_2^{\Delta,1} = 16 \int_{(4)}^{(7)} -\beta \left(u_\Delta^{[p]}(s) - T_1\right) ds.$$

Table 3.14 Values of $Q_2^{\Delta,1}$, for various mesh patterns and refinements (upper table), values of extrapolated Q_2 and computed rates of convergence (lower table) for $p = 1$.

	Pattern			
Refinement	Regular	Chevron	Union Jack	Criss-Cross
1	6152.5265	4979.4140	5911.4517	5515.2115
2	5762.4936	4813.3172	5559.3451	5406.9402
3	5543.3064	4922.8677	5479.7837	5373.0423
4	5428.2453	5048.1365	5415.7585	5351.6649
5	5369.2931	5137.7452	5373.1713	5335.6937
6	5338.1720	5196.6671	5345.7675	5323.8590
7	5321.1461	5234.3224	5328.2523	5315.3610
8	5311.5387	5258.0878	5317.0825	5309.4348

Pattern	Extrapolated Q_2	β
Regular	5299.3026	0.4204
Chevron	5297.9959	0.3389
Union Jack	5297.7897	0.3313
Criss-Cross	5295.9290	0.2632

Table 3.15 Values of $Q_2^{\Delta,1}$, for various mesh patterns and refinements (upper table), values of extrapolated Q_2 and computed rates of convergence (lower table) for $p = 2$.

	Pattern			
Refinement	Regular	Chevron	Union Jack	Criss-Cross
1	5399.3843	5007.9831	5425.6462	5361.2598
2	5382.4554	5211.5261	5375.1542	5341.3472
3	5355.7041	5257.3550	5342.8687	5324.4923
4	5336.4158	5273.6769	5324.6941	5314.0073
5	5322.8193	5282.6506	5314.1227	5307.7414
6	5313.7029	5288.2448	5307.8117	5303.9354
7	5307.7944	5291.7725	5303.9699	5301.5905

Pattern	Extrapolated Q_2	β
Regular	5297.1206	0.3196
Chevron	5297.6830	0.3395
Union Jack	5298.0988	0.3652
Criss-Cross	5297.8618	0.3549

Table 3.16 Values of $Q_2^{\Delta,2}(A)$, for various mesh patterns and refinements (upper table), values of extrapolated Q_2 and computed rates of convergence (lower table) for $p = 1$.

	Pattern			
Refinement	Regular	Chevron	Union Jack	Criss-Cross
1	5382.2397	5371.4927	5376.4129	5361.9980
2	5318.2578	5315.2465	5316.0164	5307.7193
3	5292.2439	5291.4920	05291.8536	5287.1384
4	5281.4900	5281.3168	5281.4696	5279.2441
5	5277.1490	5277.1123	5277.1743	5276.2111
6	5275.4236	5275.4190	5275.4417	5275.0428
7	5274.7438	5274.7431	5274.7529	5274.5917
8	5274.4474	5274.4777	5274.4815	5274.4174

Pattern	Extrapolated Q_2	β
Regular	5274.3081	0.6856
Chevron	5274.3085	0.6845
Union Jack	5274.3075	0.6819
Criss-Cross	5274.3084	0.6913

Table 3.17 Values of $Q_2^{\Delta,2}$, for various mesh patterns and refinements (upper table), values of extrapolated Q_2 and computed rates of convergence (lower table) for $p = 2$.

	Pattern			
Refinement	Regular	Chevron	Union Jack	Criss-Cross
1	5294.5241	5294.5346	5294.4978	5288.6079
2	5282.4977	5282.4460	5282.7479	5279.9950
3	5277.5087	5277.4980	5277.6108	5276.5107
4	5275.5493	5275.5473	5275.5909	5275.1582
5	5274.7887	5274.7883	5274.8053	5274.6362
6	5274.4938	5274.4937	5274.5003	5274.4345
7	5274.3795	5274.3795	5274.3821	5274.3565

Pattern	Extrapolated Q_2	β
Regular	5274.3084	0.6936
Chevron	5274.3082	0.6935
Union Jack	5274.3083	0.6942
Criss-Cross	5274.3077	0.6898

Figures 3.32–3.35 show rel $\left(Q_2^{\Delta,j}\right)$ v–ln N, $j = 1, 2$, for $p = 1$ and 2. The results for $Q_2^{\Delta,1}$ are much less accurate than those for $Q_2^{\Delta,2}$, and there is much larger dispersion for $Q_2^{\Delta,1}$ and for $Q_2^{\Delta,2}$. Practical accuracy of 1% is not obtained at all for $Q_2^{\Delta,1}$ with $p = 1$, but is achieved for refinement 3 with $p = 2$.

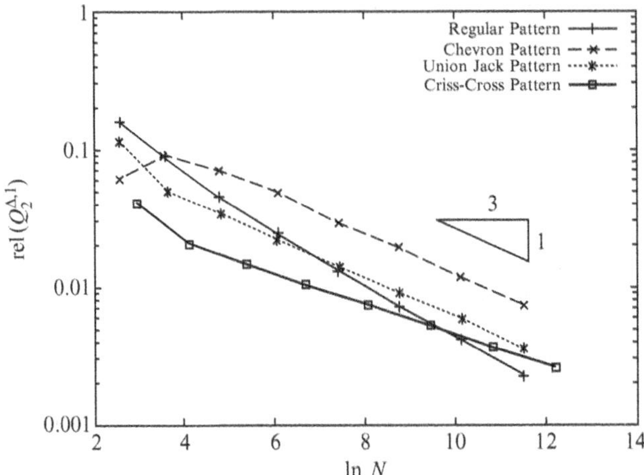

Fig. 3.32 rel $\left(Q_2^{\Delta,1}\right)$ v ln N for $p = 1$ using extrapolated Q_2 as exact value.

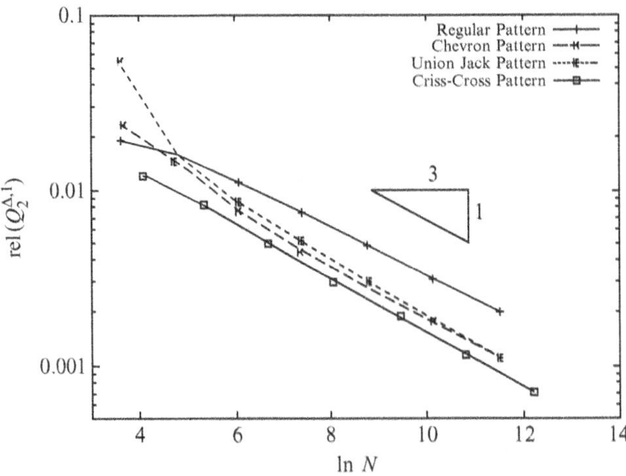

Fig. 3.33 rel $\left(Q_2^{\Delta,1}\right)$ v ln N for $p = 2$ using extrapolated Q_2 as exact value.

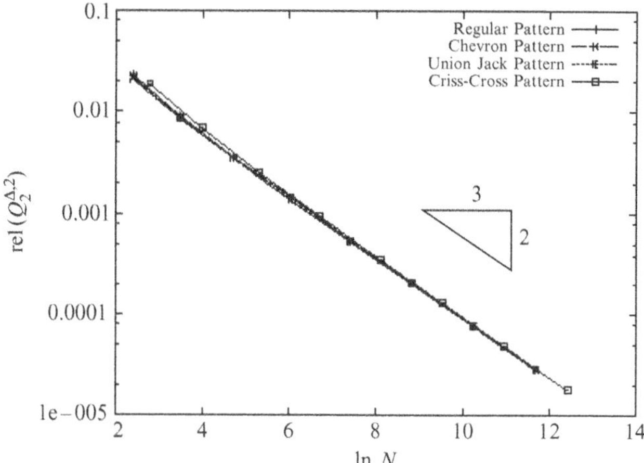

Fig. 3.34 rel $\left(Q_2^{\Delta,2}\right)$ v $\ln N$ for $p = 1$ using extrapolated Q_2 as exact value.

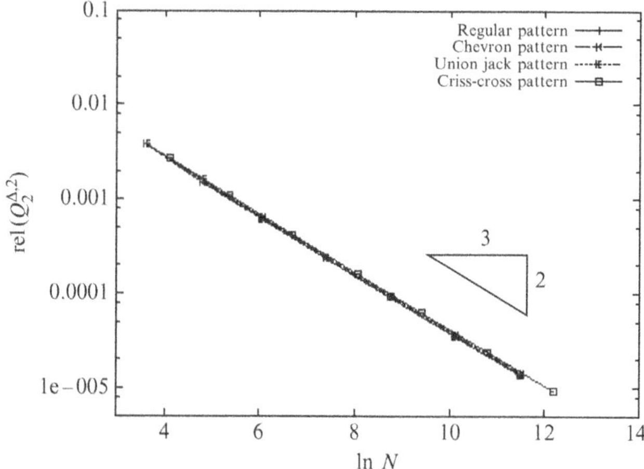

Fig. 3.35 rel $\left(Q_2^{\Delta,2}\right)$ v $\ln N$ for $p = 2$ using extrapolated Q_2 as exact value.

By contrast, practical accuracy of 1% is achieved for $Q_2^{\Delta,2}$ for refinement 2 with $p = 1$ and for refinement 1 with $p = 2$. These results show that direct use of $\dfrac{\partial u}{\partial n_c}$ is not a correct strategy. This will be discussed further in later chapters.

Finally, we consider the QoI Q_3. Tables 3.18 and 3.19 and Figs. 3.36 and 3.37 show the values of Q_3^{Δ} and the values of extrapolated Q_3 with the various mesh patterns and refinements, respectively, for $p = 1$ and 2. Comparison of the extrapolated values

Table 3.18 Values of Q_3^{Δ}, for various mesh patterns and refinements (upper table), values of extrapolated Q_3 and computed rates of convergence (lower table) for $p = 1$.

	Pattern			
Refinement	Regular	Chevron	Union Jack	Criss-Cross
1	1 285 200.0410	1 282 620.7581	1 283 801.6072	1 280 342.0319
2	1 269 844.3772	1 269 121.6740	1 269 306.4345	1 267 315.1453
3	1 263 601.0535	1 263 420.5803	1 263 507.3609	1 262 375.7110
4	1 261 020.1098	1 260 978.5396	1 261 015.2056	1 260 481.0855
5	1 259 978.2682	1 259 969.4679	1 259 984.3347	1 259 753.1707
6	1 259 564.1790	1 259 562.5931	1 259 568.5257	1 259 472.7727
7	1 259 401.0210	1 259 400.8619	1 259 403.2037	1 259 364.5237
8	1 259 337.0929	1 259 337.1452	1 259 338.0630	1 259 322.6746

Pattern	Extrapolated Q_3	β
Regular	1 259 296.2001	0.6828
Chevron	1 259 296.0190	0.6788
Union Jack	1 259 296.0078	0.6787
Criss-Cross	1 259 296.4856	0.6906

Table 3.19 Values of Q_3^{Δ}, for various mesh patterns and refinements (upper table), values of extrapolated Q_3 and computed rates of convergence (lower table) for $p = 2$.

	Pattern			
Refinement	Regular	Chevron	Union Jack	Criss-Cross
1	1 264 148.2959	1 264 150.8110	1 264 141.9880	1 262 728.3931
2	1 261 261.9557	1 261 249.5512	1 261 322.0073	1 260 661.3144
3	1 260 064.6052	1 260 062.0346	1 260 089.1070	1 259 825.0839
4	1 259 594.3475	1 259 593.8653	1 259 604.3261	1 259 500.4740
5	1 259 411.7841	1 259 411.7012	1 259 415.7679	1 259 375.1979
6	1 259 341.0126	1 259 341.0012	1 259 342.5809	1 259 326.7971
7	1 259 313.5912	1 259 313.5909	1 259 314.2037	1 259 308.0768

Pattern	Extrapolated Q_3	β
	1 259 296.4872	0.6940
	1 259 296.4770	0.6936
	1 259 296.4833	0.6935
	1 259 296.3530	0.6903

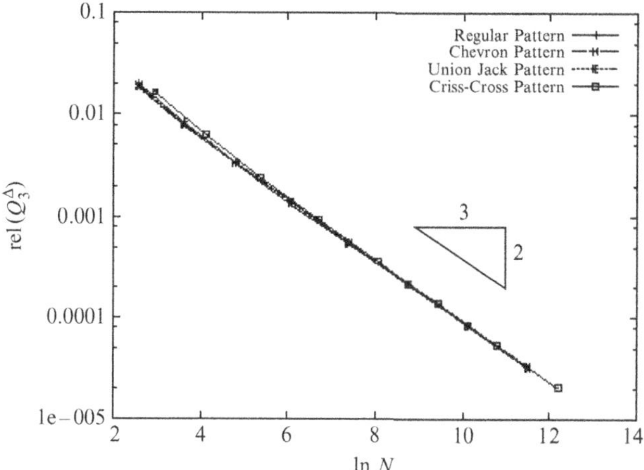

Fig. 3.36 rel $\left(Q_3^\Delta\right)$ v ln N for $p = 1$ using extrapolated Q_3 as exact value.

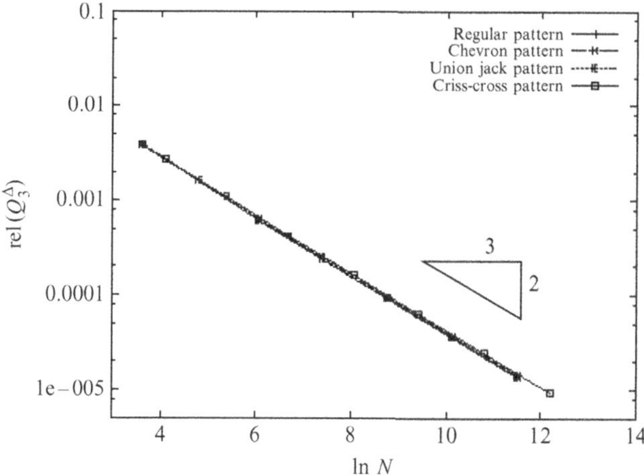

Fig. 3.37 rel $\left(Q_3^\Delta\right)$ v ln N for $p = 2$ using extrapolated Q_3 as exact value.

shows a difference of less than 0.7×10^{-5} between $p = 1$ and 2, which indicates the accuracy of the extrapolation. We see once more that the rate of convergence is the same for $p = 1$ and 2. Note that $Q_3^\Delta \approx \left\| u_\Delta^{[p]} \right\|_{\mathcal{U}}^2$.

Summary of results. We now give some general comments for all the results obtained for the 2D Eng Problem.

(1) The behaviour for this more complex practical problem of the results is less 'clean' than for the benchmark problems. This indicates that understanding the

results in these cases is more difficult, and affects the confidence that one has in the correctness of the program producing the results.

(2) Use of quadratic elements ($p = 2$), is much more effective than using linear elements ($p = 1$), as practical accuracy of 1% is in this case obtained with much cruder meshes.

(3) The direct use of the derivatives in Newton-type boundary conditions should be avoided.

(4) Computation using a well-selected sequence of meshes leads to very reliable accuracy checks for all computed QoI, but can be costly!

(5) All the computational results shown in this chapter have been presented in order to illustrate and moderate the theory given in subsequent chapters. The reader is invited to analyze the results in the light of the subsequent theory and to seek interesting features.

3.5 Best approximation property of the finite element solutions

In the previous sections of this chapter we have been concerned with developing methods that allow us to calculate Galerkin and finite element solutions to the weak problem (3.2). A common feature of the finite element problems of Sections 3.3 and 3.4 is that $S^{[p]}_{\Delta}, S^{[p]}_{\Delta,D}, S^{[p],\Gamma}_{\Delta,D} \subset \mathcal{U}$ and $S^{[p]}_{\Delta,0} \subset \mathcal{U}_0, p = 1, 2$. This feature is called the *conforming condition* and is important for enabling the error in the calculated solutions to be estimated.

Consider weak problem (3.2) for the case where $\mathfrak{u} \in \mathcal{U}_0$, i.e. with homogeneous Dirichlet boundary conditions on Γ. Correspondingly, for the finite element problems we have $u^{[p]}_{\Delta} \in S^{[p]}_{\Delta,g} \subset \mathcal{U}_g, p = 1, 2$, thus allowing us to write the weak problem and the finite element problem, respectively, as:

$$\begin{gathered}\text{find } \mathfrak{u} \in \mathcal{U}_g \text{ such that}\\ B(\mathfrak{u}, v) = F(v)\ \forall v \in \mathcal{U}_0,\end{gathered} \tag{3.82}$$

and

$$\begin{gathered}\text{find } u^{[p]}_{\Delta} \in S^{[p]}_{\Delta,g} \text{ such that}\\ B(u^{[p]}_{\Delta}, v) = F(v)\ \forall v \in S^{[p]}_{\Delta,0}.\end{gathered} \tag{3.83}$$

On subtracting (3.83) from (3.82) we obtain

$$B(\mathfrak{u} - u^{[p]}_{\Delta}, v) = 0 \quad \forall v \in S^{[p]}_{\Delta,0}, \tag{3.84}$$

which is known as the *Galerkin orthogonality condition.*

Equation (3.84) indicates that the *error*

$$e^{[p]}_{\Delta} \equiv \mathfrak{u} - u^{[p]}_{\Delta} \in \mathcal{U}_0 \tag{3.85}$$

in the finite element solution is orthogonal to any function v from $S^{[p]}_{\Delta,0}$ in terms of the bilinear form $B(\cdot,\cdot)$ defined on $\mathcal{U} \times \mathcal{U}$.

Remark 3.10 *Galerkin orthogonality is a very important property and will be used over and over again in the error analysis of the finite element methods in later chapters.*

We now use (3.75) in proving a fundamental property of the finite element solution; namely that the finite element solution is the **best approximation** to $\mathfrak{u}$ from $S^{[p]}_{\Delta,g}, p = 1, 2$ in terms of the energy norm $(B(u,u))^{\frac{1}{2}} \equiv \|u\|_{\mathfrak{u}}$ of (2.22).

Theorem 3.1 *Let $\mathfrak{u} \in \mathcal{U}$ be the solution of (3.82) and let $u^{[p]}_{\Delta} \in S^{[p]}_{\Delta,g}$ be the solution of (3.83). Then,*

$$\left\|e^{[p]}_{\Delta}\right\|_{\mathcal{U}} = \min_{\mathcal{X}\in S^{[p]}_{\Delta,g}} \|\mathfrak{u} - \mathcal{X}\|_{\mathcal{U}}, p = 1, 2. \tag{3.86}$$

Proof Let $\mathcal{X} \in S^{[p]}_{\Delta,0}$ be the best approximation to $\mathfrak{u}$ in the energy norm and let $\Psi \in S^{[p]}_{\Delta,0}$ be an arbitrary function. Then,

$$\|\mathfrak{u} - \mathcal{X}\|_{\mathcal{U}} = \min_{\Psi\in S^{[p]}_{\Delta,g}} \|\mathfrak{u} - \Psi\|_{\mathcal{U}}. \tag{3.87}$$

We can write $\mathfrak{u} - \Psi = \mathfrak{u} - u^{[p]}_{\Delta} + u^{[p]}_{\Delta} - \Psi = e^{[p]}_{\Delta} + (u^{[p]}_{\Delta} - \Psi) \in \mathcal{U}_0$, and hence

$$\begin{aligned}
\|\mathfrak{u} - \Psi\|^2_{\mathcal{U}} &= \left\|e^{[p]}_{\Delta} + \left(u^{[p]}_{\Delta} - \Psi\right)\right\|^2_{\mathcal{U}} \\
&= B\left(e^{[p]}_{\Delta} + \left(u^{[p]}_{\Delta} - \Psi\right), e^{[p]}_{\Delta} + \left(\mathcal{U}^{[p]}_{\Delta} - \Psi\right)\right) \\
&= B\left(e^{[p]}_{\Delta}, e^{[p]}_{\Delta}\right) + B\left(u^{[p]}_{\Delta} - \Psi, u^{[p]}_{\Delta} - \Psi\right) + 2B\left(e^{[p]}_{\Delta}, u^{[p]}_{\Delta} - \Psi\right) \\
&= \left\|e^{[p]}_{\Delta}\right\|^2_{\mathcal{U}} + \left\|u^{[p]}_{\Delta} - \Psi\right\|^2_{\mathcal{U}},
\end{aligned} \tag{3.87}$$

where the $B\left(e^{[p]}_{\Delta}, u^{[p]}_{\Delta} - \Psi\right)$ term vanishes due to the Galerkin orthogonality. Thus, from (3.87) we have that

$$\begin{aligned}
\|\mathfrak{u} - \mathcal{X}\|^2_{\mathcal{U}} &= \min_{\Psi\in S^{[p]}_{\Delta,g}} \|\mathfrak{u} - \Psi\|^2_{\mathcal{U}} \\
&= \left\|e^{[p]}_{\Delta}\right\|^2_{\mathcal{U}} + \min_{\Psi\in S^{[p]}_{\Delta,g}} \left\|u^{[p]}_{\Delta} - \Psi\right\|^2_{\mathcal{U}}.
\end{aligned}$$

Recalling the properties of the norm $\|\cdot\|_{\mathcal{U}}$ from Chapter 2, it follows that $\min_{\Psi\in S^{[p]}_{\Delta,0}} \left\|u^{[p]}_{\Delta} - \Psi\right\|_{\mathcal{U}} = 0$, which gives (3.86). □

Theorem 3.2 *Let $\mathfrak{u} \in \mathcal{U}_0$ be the solution of (3.82), where $\mathfrak{u}(\mathbf{x}) = g(\mathbf{x}) = 0$, $\mathbf{x} \in \Gamma_D$ and let $u^{[p]}_{\Delta} \in S^{[p]}_{\Delta,0}, p = 1, 2$ be the solution of the finite element problem, then*

$$\|\mathfrak{u}\|_{\mathcal{U}} \geqq \left\|u^{[p]}_{\Delta}\right\|_{\mathcal{U}} \tag{3.88a}$$

$$\left\|e_\Delta^{[p]}\right\|_{\mathcal{U}} = \left(\|\mathfrak{u}\|_{\mathcal{U}}^2 - \left\|u_\Delta^{[p]}\right\|^2\right)^{\frac{1}{2}}. \tag{3.88b}$$

Proof We have again that $e_\Delta^{[p]} \equiv \mathfrak{u} - u_\Delta^{[p]}$ or that $\mathfrak{u} = u_\Delta^{[p]} + e_\Delta^{[p]}$. Hence,

$$\begin{aligned}\|\mathfrak{u}\|_{\mathcal{U}}^2 = B(\mathfrak{u},\mathfrak{u}) &= B\left(u_\Delta^{[p]} + e_\Delta^{[p]}, u_\Delta^{[p]} + e_\Delta^{[p]}\right)\\ &= B\left(u_\Delta^{[p]}, u_\Delta^{[p]}\right) + 2B\left(e_\Delta^{[p]}, u_\Delta^{[p]}\right) + B\left(e_\Delta^{[p]}, e_\Delta^{[p]}\right)\\ &= \left\|u_\Delta^{[p]}\right\|_{\mathcal{U}}^2 + \left\|e_\Delta^{[p]}\right\|_{\mathcal{U}}^2.\end{aligned}$$

□

Remark 3.11 *Theorem 3.2 does not hold for $g(x) \neq 0$.*

Exercise 3.11 Work through the proof of Theorem 3.2 and identify where the condition $g = 0$ was used.

Exercise 3.12 Let the energy of Π be defined as in (2.42).
Show that

$$\Pi(\mathfrak{u}) \leqq \Pi\left(u_\Delta^{[p]}\right),$$

and that

$$\left\|e_\Delta^{[p]}\right\|_{\mathcal{U}}^2 = 2\left(\Pi\left(u_\Delta^{[p]}\right) - \Pi(\mathfrak{u})\right), \tag{3.89a}$$

irrespective of whether or not $g = 0$. Show that, if $g = 0$, then

$$2\left(\Pi\left(u_\Delta^{[p]}\right) - \Pi(\mathfrak{u})\right) = \|\mathfrak{u}\|_{\mathcal{U}}^2 - \left\|u_\Delta^{[p]}\right\|^2. \tag{3.89b}$$

Theorem 3.2 and Exercise 3.12 are very important for benchmark computations that are used to assess the performance of the finite element method. In the benchmark case the computations are performed on a problem for which the exact solution u is known, so that the error $e_\Delta^{[p]} = \mathfrak{u} - u_\Delta^{[p]}$ and the energy norm of the error,

$$\left\|e_\Delta^{[p]}\right\|_{\mathcal{U}} = \left(\int_\Omega \sum_{i,j=1} a_{ij}\frac{\partial e_\Delta^{[p]}}{\partial x_i}\frac{\partial e_\Delta^{[p]}}{\partial x_j}dx\right)^{\frac{1}{2}}, \tag{3.90}$$

can be computed.

Remark 3.12 *In benchmark computations the error should always be computed using (3.88b) or (3.89a) and not by direct computation of $\left\|e_\Delta^{[p]}\right\|_{\mathcal{U}}$. The reason for this is that numerical integration can be very inaccurate, especially when $\mathfrak{u}$ is singular in the neighbourhood of a corner.*

Remark 3.13 *Note that we have assumed in the above that all the integrations are exact. If this were not so the theorems would not hold. Nevertheless, in practice when quadratures are used to approximate the integrals, they are so accurate that in effect the results hold.*

4
Interpolation and its error

Summary

- **We wish to exploit the best approximation property of the finite element solution over the approximating space.**
- **Piecewise polynomial interpolants to $\mathfrak{u}$ from the space are candidate functions for use in the best approxiation inequality, (3.86).**
- **Interpolants $(\mathcal{L}^{[p]}(\tau)\mathfrak{u})$, $p = 1, 2$ to $\mathfrak{u}$ over a single element τ of Δ in one and two dimensions are considered and, for these, interpolation error estimates in various norms are derived.**

In Theorem 3.1 we saw that the error in the finite element solution of problem (3.2), for both the one- and the two-dimensional cases, is less than the error in the approximation to $\mathfrak{u}$ by any other function from $S_\Delta^{[p]}$, $p = 1, 2$. We now analyze the error in the particular case where the approximation to $\mathfrak{u}$ is its *interpolant* from $S_\Delta^{[p]}$. In this chapter, this analysis is done in the context of interpolation to $\mathfrak{u}$ on a single element, respectively in one and two dimensions.

These single-element interpolation results will be utilized in Chapter 5 for estimating the error in the global finite element solution over Ω.

4.1 Estimate of the error of linear and quadratic interpolants on a single element in one dimension

For ease of notation we shall consider the element $\tau \equiv (0, h)$ on the physical line and its closure $\bar{\tau} \equiv [0, h]$. This element involves the mesh size h. As in Section 3.3.2 we shall also use the master element $\tilde{\tau}$ on the ξ-line.

Let $u(x), x \in \bar{\tau}$, be a continuous function. The linear and quadratic interpolants to u on τ are denoted, respectively, by $(\mathcal{L}^{[1]}(\tau)u)$ and $(\mathcal{L}^{[2]}(\tau)u)$. This means that $\mathcal{L}^{[1]}(\tau)u$ is a linear function in τ such that

$$(\mathcal{L}^{[1]}(\tau)u)(0) = u(0), \tag{4.1a}$$

$$(\mathcal{L}^{[1]}(\tau)u)(h) = u(h), \tag{4.1b}$$

and $\mathcal{L}^{[2]}(\tau)(u)$ is correspondingly a quadratic function on τ such that

$$(\mathcal{L}^{[2]}(\tau)u)(0) = u(0), \tag{4.2a}$$

$$(\mathcal{L}^{[2]}(\tau)u)(h) = u(h), \tag{4.2b}$$

$$(\mathcal{L}^{[2]}(\tau)u)(\frac{h}{2}) = u(\frac{h}{2}). \tag{4.2c}$$

Equations (4.1) and (4.2) ensure that $\mathcal{L}^{[p]}(\tau)u, p = 1,2$ are uniquely defined functions on τ. We shall use the notation that $\mathcal{L}^{[p]}(\tau)u \in S^{[p]}[\tau]$, where

$$S^{[p]}[\tau] \equiv \left\{u(x), x \in \tau \mid u \in \mathbb{P}^{[p]}(\tau)\right\},$$

with $\mathbb{P}^{[p]}(\tau)$ the set of polynomials of degree p defined on τ.

Let us now define the respective *interpolation errors* $e_I^{[p]}(\tau)(x)$ on τ, where

$$e_I^{[p]}(\tau)(x) \equiv \left(u(x) - (\mathcal{L}^{[p]}(\tau)u)(x)\right), \; x \in \tau, \; p = 1,2. \tag{4.3}$$

Note that the finite element errors in the element τ will be denoted by $e_\Delta^{[p]}(\tau)(x)$.

In Section 3.3.2 we used the master element $\tilde{\tau} \equiv (-1,1)$ on the ξ-line, and introduced the linear mapping (3.30) of the master element onto the physical element τ_k. In the present particular case of the element $\tau = (0,h), h < h_0$, on the x-axis the mapping (3.31) becomes

$$x(\xi) = (\xi + 1)h/2, \tag{4.4}$$

and the inverse mapping of τ onto $\tilde{\tau}$ is

$$\xi(x) = 2x/h - 1, \;\; x \in \tau, \; \xi \in \tilde{\tau}. \tag{4.5}$$

Let $\tilde{u}(\xi)$ be defined on $\tilde{\tau}$ by

$$\tilde{u}(\xi) = u(x(\xi)),$$

and let the linear and quadratic interpolants of $\tilde{u}(\xi)$ on $\tilde{\tau}$ be

$$(\tilde{\mathcal{L}}^{[p]}(\tilde{\tau})\tilde{u})(\xi), \;\; p = 1,2. \tag{4.6}$$

For simplicity of notation in what follows we shall not keep indicating the dependence of $\tilde{\mathcal{L}}^{[p]}$ on $\tilde{\tau}$, but will write $\tilde{\mathcal{L}}^{[p]}\tilde{u}$ instead of $\tilde{\mathcal{L}}^{[p]}(\tilde{\tau})\tilde{u}$.

It is easy to see that

$$((\mathcal{L}^{[p]}u)(x)) = (\tilde{\mathcal{L}}^{[p]}\tilde{u})(\xi(x)), \;\; p = 1,2,$$

and that

$$e_I^{[p]}(x) = \tilde{e}_I^{[p]}(\xi(x)), \;\; p = 1,2,$$

where, following (4.3),

$$\tilde{e}_I^{[p]}(\xi) = \tilde{u}(\xi) - (\tilde{\mathcal{L}}^{[p]}\tilde{u})(\xi). \tag{4.7}$$

This notation allows us to analyze first the error in the interpolant on the master element $\tilde{\tau}$, and then to relate it to the error on the physical element τ.. We therefore first address the case of interpolation on the master element, and for this we need two lemmas, the results of which will be needed in the proof of our first main estimate, Theorem 4.1.

Lemma 4.1 *Let $\tilde{u}(\xi) \in C^2(\tilde{\tau})$ be such that $\tilde{u}(-1) = \tilde{u}(1) = 0$. Then,*

$$\|\tilde{u}\|_{C^2(\tilde{\tau})} \leqq C\,|\tilde{u}|_{C^2(\tilde{\tau})}, \tag{4.8}$$

where the constant C is independent of $\tilde{u}$.

Proof Let us first recall the definitions of $\|\tilde{u}\|_{C^k(\tilde{\tau})}$ and $|\tilde{u}|_{C^k(\tilde{\tau})}$, see (2.34) and (2.35), which for the current context are,

$$\|\tilde{u}\|_{C^k(\tilde{\tau})} \equiv \max_{j=0,\dots,k}\left(\max_{\xi\in[-1,1]}\left|\frac{d^j\tilde{u}}{d\xi^j}(\xi)\right|\right),$$

and

$$|\tilde{u}|_{C^k(\tilde{\tau})} \equiv \max_{\xi\in[-1,1]}\left|\frac{d^k\tilde{u}}{d\xi^k}(\xi)\right|,$$

with $\frac{d^j\tilde{u}}{d\xi^j}$, $j=0,\dots,k$ continuous on $[-1,1]$.

As $\tilde{u}(-1)=\tilde{u}(1)=0$, due to the mean-value theorem, see Rektorys, Vol.1, p.387, (1994), there exists an $\eta\in\tilde{\tau}$ such that

$$\frac{d\tilde{u}}{d\xi}(\eta)=0.$$

Hence, we have that

$$\frac{d\tilde{u}}{d\xi}(\xi)=\int_\eta^\xi \frac{d^2\tilde{u}}{d\gamma^2}(\gamma)d\gamma,$$

and

$$\left|\frac{d\tilde{u}}{d\xi}(\xi)\right| \leqq 2\max_{\xi\in\tilde{\tau}}\left|\frac{d^2\tilde{u}}{d\gamma^2}(\gamma)\right| = 2\,|\tilde{u}|_{C^2(\tilde{\tau})},$$

because $|\xi-\eta|<2$. Thus,

$$|\tilde{u}|_{C^1(\tilde{\tau})} \leqq 2\,|\tilde{u}|_{C^2(\tilde{\tau})}.$$

Further, as $\tilde{u}(-1)=0=\tilde{u}(1)$

$$\tilde{u}(\xi)=\int_{-1}^\xi \frac{d\tilde{u}}{d\gamma}(\gamma)d\gamma=\int_\xi^1 \frac{d\tilde{u}}{d\gamma}(\gamma)d\gamma,$$

and hence

$$|\tilde{u}|_{C^0(\tilde{\tau})} \leqq |\tilde{u}(\xi)|_{C^1(\tilde{\tau})} \leqq 2\,|\tilde{u}(\xi)|_{C^2(\tilde{\tau})},$$

from which we get inequality (4.8). □

Remark 4.1 *From the proof of Lemma 4.1 we see that, if $\tilde{u}\in C^2(\tilde{\tau})$ and $\tilde{u}(-1)=\tilde{u}(1)=0$, then*

$$\|\tilde{u}\|_{C^1(\tilde{\tau})} \leqq 2\,|\tilde{u}|_{C^1(\tilde{\tau})}, \tag{4.9a}$$

$$\|\tilde{u}\|_{C^0(\tilde{\tau})} \leqq |\tilde{u}|_{C^0(\tilde{\tau})}. \tag{4.9b}$$

Note that in (4.9a), and, respectively, (4.9b), we actually need that $\tilde{u}\in C^1(\tau)$, respectively, $\tilde{u}\in C^0(\tau)$, and not that $\tilde{u}\in C^2(\tau)$.

Clearly, we can write (4.8) and (4.9) compactly as

$$\|\tilde{u}\|_{C^j(\tilde{\tau})} \leqq C\,|\tilde{u}|_{C^j(\tilde{\tau})}\,, \quad j=0,1,2. \tag{4.10}$$

Note that because $\|\tilde{u}\|_{C^i(\tilde{\tau})} \leqq \|\tilde{u}\|_{C^j(\tilde{\tau})}\,, i \leqq j$, *we have for example that* $\|\tilde{u}\|_{C^1(\tilde{\tau})} \leqq \|\tilde{u}\|_{C^2(\tilde{\tau})}\,.$

In an analogous manner we may prove

Lemma 4.2 *Let* $\tilde{u}(\xi) \in C^3(\tilde{\tau})$ *and* $\tilde{u}(-1)=\tilde{u}(0)=\tilde{u}(1)=0$, *then*

$$\|\tilde{u}\|_{C^3(\tilde{\tau})} \leqq C\,|\tilde{u}|_{C^3(\tilde{\tau})}\,. \tag{4.11}$$

Remark 4.2 *In the same vein we can see that, if* $\tilde{u}(-1)=\tilde{u}(0)=\tilde{u}(1)=0$, *then*

$$\|\tilde{u}\|_{C^2(\tilde{\tau})} \leqq C\,|\tilde{u}|_{C^2(\tilde{\tau})}\,, \tag{4.12a}$$

$$\|\tilde{u}\|_{C^1(\tilde{\tau})} \leqq C\,|\tilde{u}|_{C^1(\tilde{\tau})}, \tag{4.12b}$$

$$\|\tilde{u}\|_{C^0(\tilde{\tau})} \leqq C\,|\tilde{u}|_{C^0(\tilde{\tau})}\,. \tag{4.12c}$$

Thus, for $\tilde{u}(\xi)$ *as in Lemma 4.2 we have in compact form*

$$\|\tilde{u}\|_{C^j(\tilde{\tau})} \leqq C\,|\tilde{u}|_{C^j(\tilde{\tau})}\,, \quad j=0,1,2,3. \tag{4.13}$$

Note that we have made different assumptions on $\tilde{u}(\xi)$ *in Remark 4.2 from those made in Remark 4.1.*

Exercise 4.1 Determine the value of C in (4.8).

Exercise 4.2 Determine the value of C in (4.11).

Exercise 4.3 Prove the results of (4.9) and (4.12).

Lemmas 4.1 and 4.2 will now be utilized in the proof of an error estimate of the interpolation error $\tilde{e}^{[p]}(\xi)$ on the master element, which is the first milestone of the present chapter.

Theorem 4.1 *Let* $\tilde{u}(\xi) \in C^2(\tilde{\tau})$, *then*

$$\left\|\tilde{e}_I^{[1]}\right\|_{C^2(\tilde{\tau})} \equiv \left\|\tilde{u} - \tilde{\mathcal{L}}^{[1]}\tilde{u}\right\|_{C^2(\tilde{\tau})} \leqq C\,|\tilde{u}|_{C^2(\tilde{\tau})}\,, \tag{4.14}$$

where C *in independent of* $\tilde{u}$.

Proof By the definition of the interpolant $\tilde{\mathcal{L}}^{[1]}\tilde{u}$ to $\tilde{u}$ we know that $\tilde{e}_I^{[1]}(\tau)(1) = \tilde{e}_I^{[1]}(\tau)(-1) = 0$. Further, because $(\tilde{\mathcal{L}}^{[1]}\tilde{u})(\xi)$ is a linear function, we have that $\left|\tilde{\mathcal{L}}^{[1]}\tilde{u}\right|_{C^2(\tilde{\tau})} = 0.$

Hence, $\left|\tilde{e}_I^{[1]}\right|_{C^2(\tilde{\tau})} = |\tilde{u}|_{C^2(\tau)}\,,$ and application of Lemma 4.1 gives (4.14). □

Similarly, using (4.9a) in an analogous way for $\tilde{e}_I^{[1]}$ we get:

Theorem 4.2 *Let $\tilde{u}(\xi) \in C^1(\tilde{\tau})$, (respectively, $C^0(\tilde{\tau})$). Then,*

$$\left\| \tilde{e}_I^{[1]} \right\|_{C^j(\tilde{\tau})} \leqq \mathcal{C} \left| \tilde{u} \right|_{C^j(\tilde{\tau})}, \quad j = 0, 1, \tag{4.15}$$

where again $\mathcal{C}$ is independent of $\tilde{u}$.

Proof In the proof of Theorem 4.1 we used the fact that $\left| \mathcal{L}^{[1]} \tilde{u} \right|_{C^2(\tilde{\tau})} = 0$. This is not useful in the present case. Instead, using (4.9a) and the triangle inequality we have that

$$\left| \tilde{e}_I^{[1]} \right|_{C^1(\tilde{\tau})} \leqq \left| \tilde{u} \right|_{C^1(\tilde{\tau})} + \left| \tilde{\mathcal{L}}^{[1]} \tilde{u} \right|_{C^1(\tilde{\tau})}. \tag{4.16}$$

Thus, we have to estimate $\left| \mathcal{L}^{[1]} \tilde{u} \right|_{C^1(\tilde{\tau})}$. Now,

$$\frac{d\left(\tilde{\mathcal{L}}^{[1]} \tilde{u}\right)}{d\xi} = \frac{1}{2} \left(\tilde{u}(1) - \tilde{u}(-1) \right) = \frac{1}{2} \int_{-1}^{1} \frac{d\tilde{u}}{d\xi}(\xi) d\xi,$$

and hence

$$\left| \frac{d\tilde{\mathcal{L}}^{[1]} \tilde{u}}{d\xi} \right| \leqq \left| \tilde{u} \right|_{C^1(\tilde{\tau})},$$

from which inequality (4.16) follows and the theorem is proved for $j = 1$.

In a similar way, using (4.9b) we need to estimate $\left| \tilde{e}_I^{[1]} \right|_{C^0(\tilde{\tau})}$. As $\tilde{\mathcal{L}}^{[1]} \tilde{u}$ is a linear function in $\tilde{\tau}$ we have that

$$\left| \tilde{\mathcal{L}}^{[1]} \tilde{u} \right|_{C^0(\tilde{\tau})} = \max \left\{ \left| \tilde{u}(-1) \right|, \left| \tilde{u}(1) \right| \right\}, \leqq \left| \tilde{u} \right|_{C^0(\tilde{\tau})},$$

so that using the triangle inequality and (4.9b) we obtain (4.15) for the case $j = 0$. □

Bounds similar to those of Theorems 4.1 and 4.2 can be established for the error in the quadratic interpolant $\tilde{\mathcal{L}}^{[2]} \tilde{u}$ on $\tilde{\tau}$. We state these as theorems.

Theorem 4.3 *Let $\tilde{u}(\xi) \in C^3(\tilde{\tau})$. Then,*

$$\left\| \tilde{e}_I^{[2]} \right\|_{C^3(\tilde{\tau})} \equiv \left\| \tilde{u} - \tilde{\mathcal{L}}^{[2]} \tilde{u} \right\|_{C^3(\tilde{\tau})} \leqq \mathcal{C} \left| \tilde{u} \right|_{C^3(\tilde{\tau})}, \tag{4.17}$$

with $\mathcal{C}$ independent of $\tilde{u}$.

Proof The proof is similar to that of Theorem 4.1, except that we now use Lemma 4.2 and the fact that $\left| \mathcal{L}^{[2]} \tilde{u} \right|_{C^3(\tilde{\tau})} = 0$. □

Exercise 4.4 Prove inequality 4.17.

Theorem 4.4 *Let $\tilde{u}(\xi) \in C^2(\tilde{\tau})$, respectively, $C^1(\tilde{\tau})$ and $C^0(\tilde{\tau})$, then*

$$\left\| \tilde{e}_I^{[2]} \right\|_{C^2(\tilde{\tau})} \leqq C \left| \tilde{u} \right|_{C^2(\tilde{\tau})}, \tag{4.18a}$$

$$\left\| \tilde{e}_I^{[2]} \right\|_{C^1(\tilde{\tau})} \leqq C \left| \tilde{u} \right|_{C^1(\tilde{\tau})}, \tag{4.18b}$$

$$\left\| \tilde{e}_I^{[2]} \right\|_{C^0(\tilde{\tau})} \leqq C \left| \tilde{u} \right|_{C^0(\tilde{\tau})}. \tag{4.18c}$$

Proof The proof of (4.18) follows that of Theorem 4.2. □

Exercise 4.5 Prove inequalities (4.18a–c).

The results of Theorems 4.3 and 4.4 can now be combined into the single compact estimate

$$\left\| \tilde{e}_I^{[p]} \right\|_{C^j(\tilde{\tau})} \leqq C \left| \tilde{u} \right|_{C^{\min(p+1,j)}} (\tilde{\tau}) \tag{4.19}$$

$$p = 1, 2,\ j = 0, 1, \ldots, p+1.$$

As was stated earlier, we now proceed to utilize the error estimates for interpolation on the master element $(\tilde{\tau})$ to derive similar estimates on the physical element τ.

Theorem 4.5 *Let $u \in C^k(\tau)$, $k = 0, 1, 2$. Then, we have that*

$$\left\| e_I^{[1]} \right\|_{C^j(\tau)} \leqq C h^{k-j} \left| u \right|_{C^k(\tau)}, \tag{4.20a}$$

and

$$\left\| e_I^{[1]} \right\|_{H^j(\tau)} \leqq C h^{k-j+\frac{1}{2}} \left| u \right|_{C^k(\tau)}, \tag{4.20b}$$

$$0 \leqq j \leqq 1,\ k \geqq j,\ k \leqq 2,$$

where C is independent of u and h, but depends on k and j.

Proof Using (4.4) we can easily see that

$$\left| \tilde{u} \right|_{C^j[\tilde{\tau}]} = \left(\frac{h}{2} \right)^j \left| u \right|_{C^j[\tau]},\quad j = 0, 1, 2,$$

and so from Theorems 4.1 and 4.2 we have immediately that

$$\left\| \tilde{e}_I^{[1]} \right\|_{C^j(\tilde{\tau})} \leqq C h^j \left| u \right|_{C^j(\tau)},\quad j = 0, 1, 2. \tag{4.21}$$

Further, using (4.5) we see that

$$\left\| e_I^{[1]} \right\|_{C^j(\tau)} = \left(\frac{2}{h} \right)^j \left| \tilde{e}_I^{[1]} \right|_{C^j(\tilde{\tau})},\quad j = 0, 1, 2. \tag{4.22}$$

Combining (4.20) with (4.21) and (4.22), and using the definitions of $|\cdot|_{C^j}$ and $\|\cdot\|_{C^j}$, we obtain (4.20a). Inequality (4.20b) is obtained by noting that $\|\cdot\|_{H^j(\tau)} \leqq h^{\frac{1}{2}} \|\cdot\|_{C^j(\tau)}$.

□

Theorem 4.5 produced an estimate for the error of the linear interpolant to u on the element τ. We have an analogous result for the quadratic interpolant; this is now stated as a theorem.

Theorem 4.6 *Let* $u \in C^k(\tau), k = 0, 1, 2, 3,$ *then*

$$\left\| e_I^{[2]} \right\|_{C^j(\tau)} \leqq C h^{k-j} |u|_{C^k(\tau)}, \tag{4.23a}$$

and

$$\left\| e_I^{[2]} \right\|_{H^j(\tau)} \leqq C h^{k-j+\frac{1}{2}} |u|_{C^k(\tau)}, \tag{4.23b}$$

$$0 \leqq j \leqq 1,\ k \geqq j,\ k \leqq 3,$$

where C is independent of u and h but can depend on k and j.

Exercise 4.6 Prove inequalities (4.23).

Remark 4.3 *Clearly, Theorems 4.5 and 4.6 can be written in a combined form that covers both linear and quadratic interpolation ($p = 1, 2$) simultaneously. This combined form is*

$$\left\| e_I^{[p]} \right\|_{C^j(\tau)} \leqq C h^{\min(k,p+1)-j} |u|_{C^{\min(k,p+1)}(\tau)}, \tag{4.24a}$$

$$\left\| e_I^{[p]} \right\|_{H^j(\tau)} \leqq C h^{\min(k,p+1)-j+\frac{1}{2}} |u|_{C^{\min(k,p+1)}(\tau)}, \tag{4.24b}$$

$$0 \leqq j \leqq 1,\ k \geqq j,\ p = 1, 2.$$

The combined estimates (4.24) for the linear and quadratic interpolation errors in the norm $\|\cdot\|_{C^j}$ on the element τ are particularly significant. The element estimates $\left\| e_I^{[p]} \right\|_{C^j(\tau)}$ with $j = 0, 1$ allow us to compute the norm $\left\| e_I^{[p]} \right\|_{\mathcal{U}(\tau)}$ for any $\tau \in \Delta$, and from this it is possible to compute $\left\| e_I^{[p]} \right\|_{\mathcal{U}(\Omega)}$.

This global interpolation result can thus be used in association with the best approximation results (3.86) to produce global estimates of the finite element errors for problem (3.2) for $p = 1, 2$. Chapter 5 contains the derivation of these error estimates.

Note that in order to obtain $\left\| e_I^{[p]} \right\|_{\mathcal{U}(\tau)}$ in this way we made the (stronger) assumption that $u \in C^k(\tau)$ rather than $u \in H^k(\tau)$. This was done so that we could utilize elementary interpolation theory.

Remark 4.4 *We note that the results of Theorems 4.5 and 4.6 can be generalized so that we have for $p = 1, 2$*

$$\left\|e_I^{[p]}\right\|_{H^j(\tau)} \leqq C h^{\min(k,p+1)-j} |u|_{H^{\min(k,p+1)}(\tau)},$$

$$0 \leqq j \leqq 1,\ k \geqq j,\ k \geqq 1.$$

Note also that we use here only $k \geqq 1$, because we need the continuity of the approximated function to be such that the interpolant makes sense.

Exercise 4.7 Derive the inequalities of (4.24).

Exercise 4.8 Why would inequality (4.24a) be wrong if, instead of $|u|_{C^{\min(k,p+1)}(\tau)}$ we were to use $|u|_{C^k(\tau)}$? Illustrate this with an example.

Exercise 4.9 Deduce the values of C in Theorems 4.5 and 4.6.

Example 4.1 Let us now analyze the accuracy of estimate (4.20) for the case $j = 0$. In deriving (4.20) we have shown (see the proof) that

$$\begin{aligned}\left\|e_I^{[1]}\right\|_{C^0(\tau)} &= \left\|\tilde{e}_I^{[1]}\right\|_{C^0(\tilde{\tau})} \leqq 2\left|\tilde{e}_I^{[1]}\right|_{C^2(\tilde{\tau})}\\ &= 2|\tilde{u}|_{C^2(\tilde{\tau})} \leqq \frac{1}{2}h^2|u|_{C^2(\tau)}.\end{aligned}$$

Similarly,

$$\begin{aligned}\left|e_I^{[1]}\right|_{C^1(\tau)} &= \frac{2}{h}\left|\tilde{e}_I^{[1]}\right|_{C^1(\tilde{\tau})} \leqq \frac{4}{h}\left|\tilde{e}_I^{[1]}\right|_{C^2(\tilde{\tau})}\\ &\leqq \frac{4}{h}|\tilde{u}|_{C^2(\tilde{\tau})} \leqq h|u|_{C^2(\tau)}.\end{aligned}$$

Now, using (4.12) we have that

$$\begin{aligned}\left|e_I^{[1]}\right|_{C^1(\tau)} &= \frac{2}{h}\left|\tilde{e}_I^{[1]}\right|_{C^1(\tilde{\tau})}\\ &\leqq \frac{2}{h}|\tilde{u}|_{C^1(\tilde{\tau})} + \frac{2}{h}\left|\tilde{\mathcal{L}}^{[1]}\tilde{u}\right|_{C^1(\tilde{\tau})}\\ &\leqq \frac{4}{h}|\tilde{u}|_{C^1(\tilde{\tau})} \leqq 2|u|_{C^1(\tau)},\end{aligned}$$

and

$$\begin{aligned}\left|e_I^{[1]}\right|_{C^0(\tau)} &= \left|\tilde{e}_I^{[1]}\right|_{C^0(\tilde{\tau})} \leqq \left|\tilde{e}_I^{[1]}\right|_{C^1(\tilde{\tau})}\\ &\leqq \frac{h}{2}|e|_{C^1(\tau)}\\ &\leqq h|u|_{C^1(\tau)}.\end{aligned}$$

Note that in deriving these estimates we have used the mean value theorem and have assumed that the derivative at the particular point is the maximum over the interval. This will occur only in special cases. Hence, the estimates are sharp with respect to the power of h, but not with respect to the constant.

We now apply these estimates where $u(x)$ has the form $u(x) = e^{cx}$. Specifically, we consider the case where $u = e^{5x}$ with $\tau \equiv (0, h)$. For $u = e^{5x}$ we obtain

$$e_I^{[1]} = e^{5x} - x\frac{e^{5h} - 1}{h} - 1,$$

and

$$\begin{aligned}
|\mathfrak{u}|_{C^2(\tau)} &= 25\, e^{5h} \\
|\mathfrak{u}|_{C^1(\tau)} &= 5\, e^{5h} \\
\left|e^{[1]}\right|_{C^0} &= \frac{e^{5h} - 1}{5h}\left(1 - \ln\left(\frac{e^{5h-1}}{5h}\right)\right) - 1 \\
\left|e^{[1]}\right|_{C^1} &= 5\, e^{5h} - \frac{e^h - 1}{h}.
\end{aligned}$$

Specific values of $\left|e^{[1]}\right|_{C^i}$ are given in Table 4.1 for $u(x) = e^{5x}$.

From these values we can see that

(1) As $h \to 0$ the estimates converge to specific values. However, these values are not those that we derived, which are more pessimistic. The reason for this is that our estimates have to be valid for any function.

(2) The theoretical estimates derived above are the 'worst scenario' estimates that are valid for the entire class of admissible functions characterized by the information given; e.g. that only first derivatives of u are bounded. It is unlikely that any particular case will be the worst case, so that the estimates will usually be pessimistic.

(3) In the estimates of this section we have used throughout a general constant $\mathcal{C}$ that depends on various factors, but not on the data of the problem nor on $\mathfrak{u}$. These constants could be determined by analysing the proofs in which

Table 4.1 Interpolation errors and norm of interpolated function.

h	$\left\|e_I^{[1]}\right\|_{C^0}$	$\left\|e_I^{[1]}\right\|_{C^1}$	$\|u\|_{C^2}$	$\|u\|_{C^1}$
0.01	0.00032	0.1292	26.21	5.256
0.2	0.21186	5.000	67.93	13.59
h	$\left\|e_I^{[1]}\right\|_{C^0} / \frac{1}{2}h^2 \|u\|_{C^2}$	$\left\|e_I^{[1]}\right\|_{C^0} / h\|u\|_{C^1}$	$\left\|e_I^{[1]}\right\|_{C^1} / h\|u\|_{C^2}$	$\left\|e_I^{[1]}\right\|_{C^1} / \|u\|_{C^1}$
0.01	0.1223	0.6095	0.4929	0.0246
0.2	0.0779	0.0779	0.3682	0.3679

they appear, but the values obtained would not necessarily be the minimal ones. The task of obtaining the 'best' constants is complicated, and will not be pursued here.

Theorems 4.5 and 4.6 relate the error in the interpolation on τ to powers of the size of the element. The question that arises immediately is, "Are these estimates pessimistic in the sense that the exponent in the powers of the element size h could be higher than those in the theorems?" The answer to this question is 'no'. The exponents are the largest possible.

In order to illustrate this, we consider the case of Theorem 4.5 with $j = 0$ and $k = 2$ and assume to the contrary that (4.20) can be replaced, for $\epsilon > 0$, by

$$\left\|e_I^{[1]}\right\|_{C^0(\tau)} \leqq Ch^{2+\epsilon} |u|_{C^2(\tau)} . \tag{4.25}$$

Suppose that we choose $u = x^2$, $x \in \tau$ as a special case. Then, $|u|_{C^2(\tau)} = 2$ and

$$\left\|e_I^{[1]}\right\|_{C^0(\tau)} = \max_{x \in \tau} \left|x^2 - h^2 \frac{x}{h}\right| = \max_{x \in \tau} \left|hx - x^2\right| .$$

The maximum occurs at $x_0 = \frac{h}{2}$, at which point the value is $hx_0 - x_0^2 = \frac{h^2}{2}$, which contradicts (4.25).

It can similarly be shown that the exponent of h is in general maximal in all the estimates occurring in Theorems 4.5 and 4.6. The word general here means that there is a function for which the exponent is maximum. This does not mean that this exponent is maximum for all functions; for example, if u is a polynomial of degree 2 then $e_I^{[2]} = 0$.

Exercise 4.10 Show that if $\left|\dfrac{d^{p+1}u}{dx^{p+1}}\right| \geqq \delta$, then

$$\left(\min_{\chi \in \mathbb{P}^p(\tau)} \|u - \chi\|_{U(\tau)}\right) / h^{p-\varepsilon} \to \infty \text{ as } h \to 0 \text{ for any } \varepsilon > 0.$$

Example 4.2 We now analyze the interpolation of the function $u = x^\alpha$, $0 < \alpha < 1$ on $\tau = (0, h)$. We note first that $u \in C^0(\tau)$ but $u \notin C^1(\tau)$; i.e.

$$\max_{x \in \tau} \frac{du}{dx}(x) = \alpha \max_{x \in \tau} \left|x^{\alpha-1}\right| = \infty.$$

Now, we have

$$e_I^{[1]} = u - \mathcal{L}^{[1]}u = x^\alpha - h^{\alpha-1}x,$$

and so

$$\left|e_I^{[1]}\right|_{C^0(\tau)} = \bar{x}^\alpha - h^{\alpha-1}\bar{x},$$

where

$$\bar{x} = h\alpha^{1/(1-\alpha)}.$$

Hence,

$$\left|e_I^{[1]}\right|_{C^0(\tau)} = h^\alpha \alpha^{\alpha/(1-\alpha)} - h^\alpha \alpha^{1/(1-\alpha)}$$

$$= h^\alpha \left(\alpha^{\alpha/(1-\alpha)} - \alpha^{1/(1-\alpha)}\right).$$

Let us estimate $\left\|e_I^{[1]}\right\|_{L^2(\tau)}$ and $\left\|\frac{de_I^{[1]}}{dx}\right\|_{L^2(\tau)}$. Because $e^{[1]} \notin C^1(\tau)$, we have to make direct estimates via the master element. We have that $\tilde{u} = (\xi+1)^\alpha\, h^\alpha\, 2^{-\alpha}$. Further, consider

$$\left|(\xi+1)^\alpha - \tilde{\mathcal{L}}^{[1]}\,(\xi+1)^\alpha\right| \leqq |(\xi+1)^\alpha|_{C^0(\tilde{\tau})} + \left|\tilde{\mathcal{L}}^{[1]}\,(\xi+1)^\alpha\right|_{C^0(\tilde{\tau})}$$

$$\leqq 2\cdot 2^\alpha.$$

Hence,

$$\left|\tilde{u} - \mathcal{L}^{[1]}\tilde{u}\right| \leqq 2\,h^\alpha \leqq C\,h^\alpha.$$

If we also estimate $\left|e_I^{[1]}\right|_{L^2(\tau)}$, we find that

$$\left|e_I^{[1]}\right|_{L^2(\tau)} \leqq C\,h^{\alpha+\frac{1}{2}}.$$

We now estimate $\frac{de_I^{[1]}}{dx} = \alpha x^{\alpha-1} - h^{\alpha-1}$ with $\alpha > \frac{1}{2}$, so that

$$\int_0^h \left(\frac{de_I^{[1]}}{dx}\right)^2 dx = h^{2\alpha-1}\left(\frac{\alpha^2}{2\alpha-1} - 1\right) \leqq C\left(h^{\alpha-\frac{1}{2}}\right)^2.$$

The results of Example 4.2 will play an important role later, because this is a one-dimensional model of the two-dimensional situation that occurs when a domain has a corner and the solution is singular in the neighbourhood of this corner. We see that the error is $0(h^{\alpha-\frac{1}{2}})$ independent of p, and also that for convergence we need $\alpha > \frac{1}{2}$. Note that $x^\alpha \in H^1(\tau)$ only when $\alpha > \frac{1}{2}$.

Exercise 4.11 Consider the function $u = x^\beta, x \in \tau = (0,h)$ and estimate $\left(\int_0^h \left(e_I^{[2]}\right)^2 dx\right)^{\frac{1}{2}} \leqq C\,h^{\tilde{\beta}}$. For what values of $\tilde{\beta}$ is this estimate valid?

Exercise 4.12 Find the values of β_i such that

$$\left(\int_0^h \left(\frac{de^{[1]}}{dx}\right)^2 dx\right)^{\frac{1}{2}} \leqq C_1 h^{\beta_1},$$

$$\left(\int_0^h \left(\frac{de^{[1]}}{dx}\right)^2 dx\right)^{\frac{1}{2}} \leqq C_2 h^{\beta_2},$$

where the constants C_i, $i = 1, 2$, are independent of h.

Exercise 4.13 Assume that we are interested in the estimate

$$\left(\int_0^h a(x)\left(\frac{de^{[1]}}{dx}\right)^2 dx\right)^{\frac{1}{2}} \leqq C\, h^{\beta},$$

with $0 < a_{\min} \leqq a(x) \leqq a_{\max} < \infty$. Assume also that you know the exponent β_1 and the constant C_1 of Exercise 4.11. What will the values of C and β be here?

Exercise 4.14 Find $\alpha > 0$ such that

$$h\left(\int_0^h \left(\frac{de^{[2]}}{dx}\right)^2 dx\right)^{\frac{1}{2}} + \left(\int_0^h \left(e^{[2]}\right)^2 dx\right)^{\frac{1}{2}} \leqq C\, h^{\alpha},$$

with C independent of h.

4.2 Estimate of the error of linear and quadratic interpolants on a single element in two dimensions

We proceed in the two-dimensional setting in a manner similar to the one-dimensional context of Section 4.1. The single element is now a triangle τ in the physical (x_1, x_2) -plane and, correspondingly, we also have the master triangle $\tilde{\tau}$ in the (ξ, η) -plane.

The triangle τ is taken as in Fig. 4.1 and has vertices z_1, z_2, z_3, sides ${}^3e \equiv z_1z_2$, ${}^1e \equiv z_2z_3$, ${}^2e \equiv z_3z_1$, and internal angles $\alpha_1, \alpha_2, \alpha_3$ as shown. The midpoints of the sides are, respectively, z_4, z_5 and z_6. As before, $\mathbf{x} \equiv (x_1, x_2)$ and the diameter of τ is h.

As in Section 4.1, we let $u(\mathbf{x})$, $\mathbf{x} \in \bar{\tau}$ be a continuous function. We again denote by $(\mathcal{L}^{[1]}(\tau)u)$ and $(\mathcal{L}^{[2]}(\tau)u)$, respectively, the linear and quadratic interpolants to $u(\mathbf{x})$, $\mathbf{x} \in \bar{\tau}$. This means that $\mathcal{L}^{[1]}(\tau)u$ is a linear function on τ and has the form

$$\mathcal{L}^{[1]}(\tau)u = a_{0,0}^{(1)} + a_{1,0}^{(1)}x_1 + a_{0,1}^{(1)}x_2, \; (x_1, x_2) \in \tau, \tag{4.26}$$

where $a_{i,j} \in \mathbb{R}$, and is (similarly to (4.1)) such that

$$\left(\mathcal{L}^{[1]}(\tau)u\right)(z_i) = u(z_i), \; i = 1, 2, 3. \tag{4.27}$$

Similarly, $\mathcal{L}^{[2]}(\tau)u$ is a quadratic function on τ and has the form

$$\mathcal{L}^{[2]}(\tau)u = a_{0,0}^{(2)} + a_{1,0}^{(2)}x_1 + a_{0,1}^{(1)}x_2 + a_{2,0}^{(2)}x_1^2 + a_{1,1}^{(2)}x_1x_2 + a_{0,2}^{(2)}x_2^2, \; (x_1, x_2) \in \tau,$$

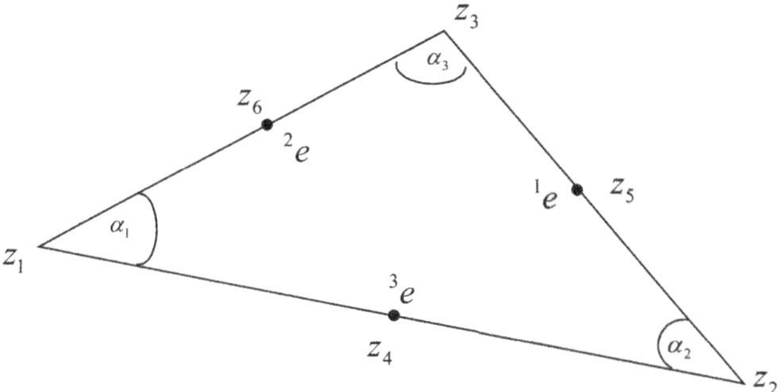

Fig. 4.1 The element τ with vertices, sides and side midpoints.

where $a_{i,j} \in \mathbb{R}$, and is such that

$$\left(\mathcal{L}^{[2]}(\tau)u\right)(z_i) = u(z_i),\ i = 1, 2, 3, \tag{4.28a}$$

$$\left(\mathcal{L}^{[2]}(\tau)u\right)(z_i) = u(z_i),\ i = 4, 5, 6. \tag{4.28b}$$

We shall see below that the interpolants as above exist and are uniquely defined on τ. Similarly to the one-dimension case in (4.3) we now define the respective interpolation errors to be

$$e_I^{[p]}(\tau) \equiv \left(u(\mathbf{x}) - (\mathcal{L}^{[p]}(\tau)u(\mathbf{x})\right),\ \mathbf{x} \in \tau.$$

In Section 3.4 we introduced the three- and six-node master triangle elements $\tilde{\tau}$, see Figs. 3.19 and 3.20; in the (ξ_1, ξ_2) -plane. For ease of use we give these elements again as Fig. 4.2.

Note that in Figs. 4.1 and 4.2 the numbering of the points ${}^i\tilde{z}$ is in each case counterclockwise. As before, τ is an arbitrary element of the mesh.

Again, as in Chapter 3, we introduce the linear mapping: $\mathbf{x}(\xi) = ((x_1(\xi_1, \xi_2), x_2(\xi_1, \xi_2))$ that maps $\tilde{\tau}$ onto τ, and has the form of (3.67). Specifically, we have for the element τ that

$$\begin{aligned} x_j(\xi_1, \xi_2) = x_j^{(1)} & \left(\frac{1}{2}(1 - \xi_1) - \frac{1}{2\sqrt{3}}\xi_2\right) \\ & + x_j^{(2)}\left(\frac{1}{2}(1 + \xi_1) - \frac{1}{2\sqrt{3}}\xi_2\right) + x_j^{(3)}\frac{\xi_2}{\sqrt{3}}, j = 1, 2, \end{aligned} \tag{4.29}$$

where $\left(x_1^{(k)}, x_2^{(k)}\right)$ are the coordinates of the vertices z_k, $k = 1, 2, 3$ numbered locally for τ. This mapping maps the points ${}^i\tilde{z}$ into z_i, $i = 1, \ldots, 6$.

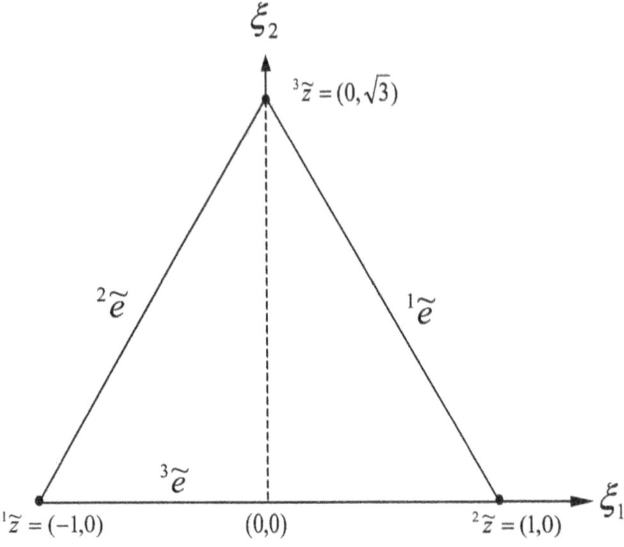

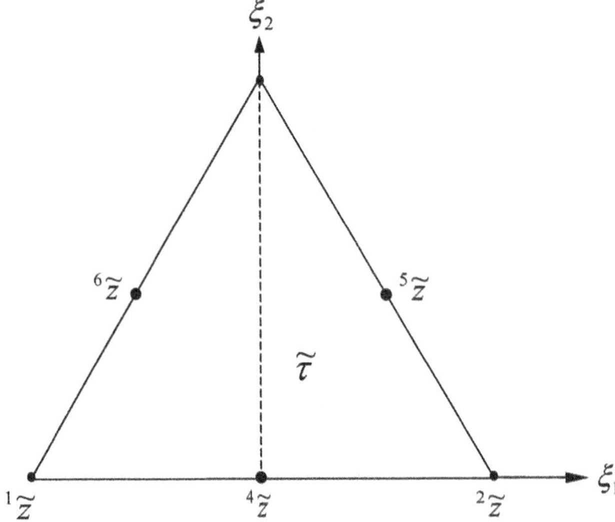

Fig. 4.2 Master triangle showing the locations of the nodes ${}^i\tilde{z}, i = 1, \ldots, 6$ and the sides ${}^i\tilde{e}, i = 1, 2, 3$.

The Jacobian of the mapping is

$$J = \begin{bmatrix} \frac{1}{2}\left(x_1^{(2)} - x_1^{(1)}\right) & \frac{1}{\sqrt{3}}\left(x_1^{(3)} - \frac{\left(x_1^{(1)} + x_1^{(2)}\right)}{2}\right) \\ \frac{1}{2}\left(x_1^{(2)} - x_1^{(1)}\right) & \frac{1}{\sqrt{3}}\left(x_2^{(3)} - \frac{\left(x_2^{(1)} + x_2^{(2)}\right)}{2}\right) \end{bmatrix}. \tag{4.30}$$

We now make the assumption that the diameter of τ is h, with the result that all the entries in J are smaller than $\mathcal{C}h$. We recall from Section 3.4.1 that, because the mapping $\mathbf{x}(\boldsymbol{\xi})$ is linear and the nodal points z_k (respectively, ${}^k\tilde{z}$) are not collinear, there exists the inverse mapping

$$\xi = (\xi_1(x_1, x_2), \xi_2(x_1, x_2)),$$

which is again linear, and for which the Jacobian is $[J]^{-1}$.

From computation we can see that the (constant) entries in $[J]^{-1}$ are bounded by Kh^{-1}, where K depends on $\alpha_{\text{min}} \equiv \min(\alpha_1, \alpha_2, \alpha_3)$ with α_i the vertex angles of τ, as shown in Fig. 4.1. The constant $K \to \infty$ as $\alpha_{\text{min}} \to 0$.

Exercise 4.15 Prove that the determinant of the Jacobian in (4.30) is positive because the z_k are not collinear. Show that this guarantees that the mapping is one-to-one.

Remark 4.5 *The constant K plays an important role in the interpolation estimates for the two-dimensional case. This did not occur for the one-dimensional estimates, as we had (see (3.33)) $J = dx/d\xi = h$, whilst $[J]^{-1} = d\xi/dx = 1/h$; i.e. exact reciprocals.*

As in the one-dimensional case, we now define the function $\tilde{u}(\boldsymbol{\xi})$ on the master triangle $\tilde{\tau}$ as $\tilde{u}(\boldsymbol{\xi}) \equiv u(\mathbf{x}(\boldsymbol{\xi}))$, and now denote the linear and quadratic interpolants to $\tilde{u}(\boldsymbol{\xi})$ on $\tilde{\tau}$, respectively, by $(\tilde{\mathcal{L}}^{[p]}(\tilde{\tau})\tilde{u})(\boldsymbol{\xi})$, $p = 1, 2$. Similarly to (4.3) and (4.5) we also define the respective interpolation errors on τ and $\tilde{\tau}$ as

$$e_I^{[p]}(\tau) = \left(u(\mathbf{x}) - (\mathcal{L}^{[p]}(\tau)u)(\mathbf{x})\right), \tag{4.31a}$$

$$\tilde{e}_I^{[p]}(\tilde{\tau}) = \left(\tilde{u}(\boldsymbol{\xi}) - (\mathcal{L}^{[p]}(\tilde{\tau})\tilde{u})(\boldsymbol{\xi})\right), \tag{4.31b}$$

and get

$$e_I^{[p]}(\tau) = \tilde{e}_I^{[p]}(\tilde{\tau}), \quad p = 1, 2. \tag{4.31c}$$

Equations (4.31) show (again analogously to one dimension) that we can analyze the interpolation error on τ by considering the interpolation error on the master element $\tilde{\tau}$. We assume that the interpolant exists and is uniquely determined as will be shown in Remark 4.8, and proceed quite analogously to the one-dimensional analysis proving:

Lemma 4.3 *Let $\tilde{u}(\boldsymbol{\xi}) \in C^2(\tilde{\tau})$ be such that $\tilde{u}({}^i\tilde{z}) = 0$, $i = 1, 2, 3$. Then,*

$$\|\tilde{u}\|_{C^2(\tilde{\tau})} \leqq \mathcal{C}\, |\tilde{u}|_{C^2(\tilde{\tau})}, \tag{4.32}$$

where $\mathcal{C}$ is independent of $\tilde{u}$.

Proof First, recollect the definitions of $\|\tilde{u}\|_{C^k(\tilde{\tau})}$ and $|\tilde{u}|_{C^k(\tilde{\tau})}$. These are

$$\|\tilde{u}\|_{C^2(\tilde{\tau})} \equiv \max_{0 \leqq l_1+l_2 \leqq 2} \left(\max_{\xi \in \tilde{\tau}} \left| \frac{\partial^{l_1+l_2} \tilde{u}}{\partial \xi_1^{l_1} \partial \xi_2^{l_2}} \right| \right),$$

$$|\tilde{u}|_{C^2(\tilde{\tau})} \equiv \max_{l_1+l_2=2} \left(\max_{\xi \in \tilde{\tau}} \left| \frac{\partial^{l_1+l_2} \tilde{u}}{\partial \xi_1^{l_1} \partial \xi_2^{l_2}} \right| \right),$$

$$l_i \geqq 0, \; i = 1, 2.$$

Using Lemma 4.1 we have on each side ${}^i\tilde{e}$, $i = 1, 2, 3$ of $\tilde{\tau}$

$$\|\tilde{u}\|_{C^2({}^i\tilde{e})} \leqq C \, |\tilde{u}|_{C^2({}^i\tilde{e})}. \tag{4.33}$$

□

Note that in (4.33) the derivatives under consideration are with respect to the edge ${}^i\tilde{e}$, and clearly

$$|\tilde{u}|_{C^2({}^i\tilde{e})} \leqq C \, |\tilde{u}|_{C^2(\tau)}, \tag{4.34}$$

see Fig. 4.2. Consider now the vertex ${}^2\tilde{z}$ that is common to the sides ${}^1\tilde{e}$ and ${}^3\tilde{e}$, which have length parameters s_1 and s_3.

We have that

$$\begin{bmatrix} \dfrac{\partial \tilde{u}}{\partial \tilde{s}_1}({}^2\tilde{z}) \\ \dfrac{\partial \tilde{u}}{\partial \tilde{s}_2}({}^2\tilde{z}) \end{bmatrix} = \begin{bmatrix} \cos \xi_1 \tilde{s}_1 & \cos \xi_2 \tilde{s}_1 \\ \cos \xi_1 \tilde{s}_3 & \cos \xi_2 \tilde{s}_3 \end{bmatrix} \begin{bmatrix} \dfrac{\partial \tilde{u}}{\partial \xi_1}({}^2\tilde{z}) \\ \dfrac{\partial \tilde{u}}{\partial \xi_2}({}^2\tilde{z}) \end{bmatrix},$$

from which by inverting the Jacobian we find that

$$\left| \frac{\partial \tilde{u}}{\partial \xi_1}({}^2\tilde{z}) \right| \leqq C \, |u|_{C^2(\tilde{\tau})},$$

$$\left| \frac{\partial \tilde{u}}{\partial \xi_2}({}^2\tilde{z}) \right| \leqq C \, |u|_{C^2(\tilde{\tau})}.$$

Analogously, it is clear that

$$\left| \frac{\partial \tilde{u}}{\partial \xi_i}({}^j\tilde{z}) \right| \leqq C \, |u|_{C^2(\tilde{\tau})}, \quad i = 1, 2, \quad j = 1, 2, 3. \tag{4.35}$$

Suppose that $\xi \equiv (\xi_1, \xi_2) \in \tilde{\tau}$ is a general point, see Fig. 4.3. Let us denote the straight lines connecting ξ with ${}^j\tilde{z}$ by ${}^j\tilde{L} \equiv \xi^j\tilde{z}$.

Let us resolve that for the set of points $\xi \in \omega_j, j \neq k, l$ we can define the angles $\theta_{k,l}$ between the lines ${}^k\tilde{L}$ and ${}^l\tilde{L}$ so that there exist constants γ_1 and γ_2 independent of ξ such that $0 < \gamma_1 < \theta_{k,l} < \gamma_2 < \pi$. For example, let $\omega_3 \equiv \left\{ \xi \mid \xi_2 > \frac{2}{\sqrt{3}} \right\}$ and select $k = 1, \; l = 2$. By rotation of coordinates we can obtain the k and l for $\xi \in \omega_2$ and

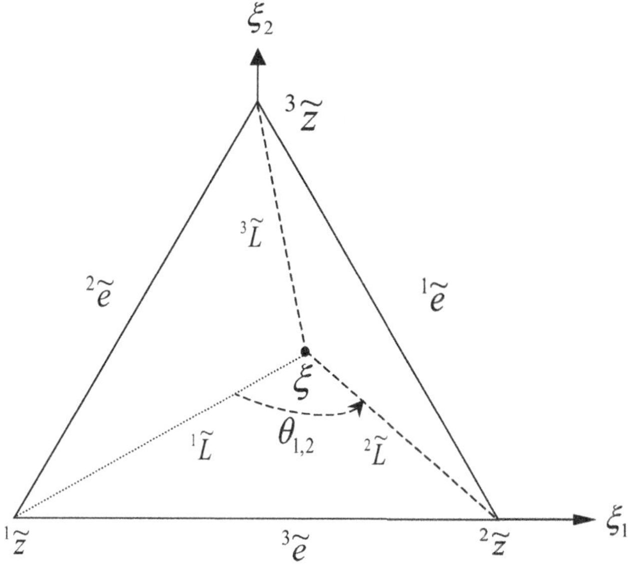

Fig. 4.3 Master element with general point ξ.

$\xi \in \omega_1$, where ω_1 and ω_2 are analogous to the set ω_3. We note that $\cup\omega_i = \tilde{\tau}$. Again, assuming that $\xi \in \omega_3$ we have

$$\frac{\partial \tilde{u}}{\partial^1 \tilde{L}}(\xi) = \int_{^1\tilde{z}}^{\xi} \frac{\partial^2 \tilde{u}}{\partial^1 \tilde{L}^2} d^1\tilde{L} + \frac{\partial \tilde{u}}{\partial^1 \tilde{L}}(^1z).$$

From (4.35) we have that

$$\left|\frac{\partial \tilde{u}}{\partial^1 \tilde{L}}(\xi)\right| \leqq \mathcal{C}\, |u|_{C^2(\tilde{\tau})}, \tag{4.36a}$$

and similarly that

$$\left|\frac{\partial \tilde{u}}{\partial^2 \tilde{L}}(\xi)\right| \leqq \mathcal{C}\, |u|_{C^2(\tilde{\tau})}. \tag{4.36b}$$

Because the angle between $^1\tilde{L}$ and $^2\tilde{L}$ is bounded from above and below we obtain

$$\left|\frac{\partial \tilde{u}}{\partial \xi_j}(\xi)\right| \leqq \mathcal{C}\left[\left|\frac{\partial \tilde{u}}{\partial^1 \tilde{L}}(\xi)\right| + \left|\frac{\partial \tilde{u}}{\partial^2 \tilde{L}}(\xi)\right|\right] \leqq \mathcal{C}\, |\tilde{u}|_{C^2(\tilde{\tau})}, \; j = 1, 2.$$

Hence,

$$|\tilde{u}|_{C^1(w_j)} \leqq \mathcal{C}\, |\tilde{u}|_{C^2(\tilde{\tau})},$$

and also that

$$|\tilde{u}|_{C^1(\tilde{\tau})} \leqq C\,|\tilde{u}|_{C^2(\tilde{\tau})}\,. \tag{4.37}$$

We must also estimate $|u(\xi)|$. For $\xi \in \omega_3$

$$|\tilde{u}(\xi)| \leqq \int_{{}^1\tilde{z}} \left|\frac{\partial \tilde{u}}{\partial^1 \tilde{L}}\right| d^1\tilde{L} < C\,|\tilde{u}|_{C^2(\tilde{\tau})}\,,$$

using (4.37) and the lemma is proven.

There are other proofs of Lemma 4.3; this one was used on account of its elementarity.

Remark 4.6 *From the proof of Lemma 4.3 we see immediately that for $\tilde{u} \in C^j(\tilde{\tau})$ and $\tilde{u}(z_i) = 0$,*

$$\|\tilde{u}\|_{C^1(\tilde{\tau})} \leqq C\,|\tilde{u}|_{C^1(\tilde{\tau})}, \tag{4.38a}$$

$$\|\tilde{u}\|_{C^0(\tilde{\tau})} \leqq C\,|\tilde{u}|_{C^0(\tilde{\tau})}, \tag{4.38b}$$

which is the analogue of Remark 4.1. Further, we can write analogously, as in (4.10), in compact form

$$\|\tilde{u}\|_{C^j(\tilde{\tau})} \leqq C\,|\tilde{u}|_{C^j(\tilde{\tau})}\,,\ j = 0, 1, 2. \tag{4.39}$$

In an analogous manner we have:

Lemma 4.4 *Let $\hat{u} \in C^3(\tilde{\tau})$ and suppose that $u({}^jz) = 0,\ j = 1, 2, \ldots, 6$. Then,*

$$\|\tilde{u}\|_{C^3(\tilde{\tau})} \leqq C\,|\hat{u}|_{C^3(\tilde{\tau})}\,, \tag{4.40}$$

where C is independent of $\tilde{u}$.

The proof of this lemma is omitted because it is analogous to that of Lemma 4.1.

Remark 4.7 *For $\tilde{u} \in C^2(\tilde{\tau})$ satisfying $\tilde{u}({}^iz) = 0,\ i = 1, 2, \ldots, 6$, it is immediate that also*

$$\|\tilde{u}\|_{C^2(\tilde{\tau})} \leqq C\,|\tilde{u}|_{C^2(\tilde{\tau})}, \tag{4.41}$$

$$\|\tilde{u}\|_{C^1(\tilde{\tau})} \leqq C\,|\tilde{u}|_{C^1(\tilde{\tau})}, \tag{4.42}$$

$$\|\tilde{u}\|_{C^0(\tilde{\tau})} \leqq C\,|\tilde{u}|_{C^0(\tilde{\tau})}, \tag{4.43}$$

and we can once more write, as in (4.12) for the one-dimensional case,

$$\|\tilde{u}\|_{C^j(\tilde{\tau})} \leqq C\,|\tilde{u}|_{C^j(\tilde{\tau})},\ \ j = 0, 1, 2, 3. \tag{4.44}$$

Exercise 4.16 Prove (4.41a and b)

Our next task is, as before, to estimate the error $\tilde{e}^{[p]}(\tilde{\tau})$ of the interpolant $\mathcal{L}^{[p]}(\tilde{\tau})u$, $p = 1, 2$, but now of course in the two-dimensional master element $\hat{\tau}$.

Theorem 4.7 *Let $\hat{u} \in C^2(\tilde{\tau})$, then*

$$\left\| \tilde{e}_I^{[1]} \right\|_{C^2(\hat{\tau})} \equiv \left\| \tilde{u} - \mathcal{L}^{[1]}\tilde{u} \right\|_{C^2(\tilde{\tau})} \leqq C \left| \tilde{u} \right|_{C^2(\tilde{\tau})}, \tag{4.45}$$

where C is independent of $\tilde{u}$.

Proof Again, by the definition of the interpolant $\mathcal{L}^{[1]}\tilde{u}$ to $\tilde{u}$, we know that $\tilde{e}_I^{[1]}({}^j z) = 0$, $j = 1, 2, 3$. Further, because $(\mathcal{L}^{[1]}\tilde{u})(\boldsymbol{\xi})$ is a linear function, we have that $\left|\mathcal{L}^{[1]}\tilde{u}\right|_{C^2(\tilde{\tau})} = 0$. Hence, $\left\| \tilde{e}_I^{[1]} \right\|_{C^2(\hat{\tau})} = \|\tilde{u}\|_{C^2(\tilde{\tau})}$, and application of Lemma 4.3 gives (4.45). □

we also have:

Theorem 4.8 *Let $\tilde{u} \in C^1(\tilde{\tau})$ (respectively, $C^0(\tilde{\tau})$). Then,*

$$\left\| \tilde{e}_I^{[1]} \right\|_{C^1(\hat{\tau})} \leqq C \left| \tilde{u} \right|_{C^1(\tilde{\tau})}, \tag{4.46}$$

and

$$\left\| \tilde{e}_I^{[1]} \right\|_{C^0(\hat{\tau})} \leqq C \left| \tilde{u} \right|_{C^1(\tilde{\tau})}, \tag{4.47}$$

while C is independent of $\tilde{u}$.

Proof In the proof of Theorem 4.7 we used the fact that $\left|\tilde{\mathcal{L}}^{[1]}\tilde{u}\right|_{C^2(\tilde{\tau})} = 0$. Now, analogously as in Theorem 4.2, we have to estimate $\left|\tilde{\mathcal{L}}^{[1]}\tilde{u}\right|_{C^1(\tilde{\tau})}$. As $\tilde{\mathcal{L}}^{[1]}\tilde{u}$ is linear, $\dfrac{\partial \tilde{\mathcal{L}}^{[1]}\tilde{u}}{\partial \xi_i}$, $i = 1, 2$, is a constant function, and hence we can estimate these two constants from the derivatives along the sides ${}_j\tilde{e}$, $j = 1, 2, 3$ of $\tilde{\tau}$. This immediately gives

$$\left|\tilde{\mathcal{L}}^{[1]}\tilde{u}\right| \leqq C \left| \tilde{u} \right|_{C^1(\tilde{\tau})},$$

and (4.46) is proven. Because $\tilde{\mathcal{L}}^{[1]}\tilde{u}$ is maximal on $\tilde{S}_j$, (4.47) follows. □

Theorems similar to 4.7 and 4.8 hold for the error in the quadratic interpolant $\tilde{\mathcal{L}}^{[2]}(\tilde{\tau})\tilde{u}$.

Theorem 4.9 *Let $\tilde{u} \in C^3(\tau)$. Then,*

$$\left\| \tilde{e}_I^{[2]} \right\|_{C^3(\hat{\tau})} \equiv \left\| \tilde{u} - \tilde{\mathcal{L}}^{[2]}\tilde{u} \right\|_{C^3(\tau)} \leqq C \left| \tilde{u} \right|_{C^3(\tilde{\tau})}, \tag{4.48}$$

where C is independent of $\hat{u}$.

Proof The proof of this theorem is completely similar to that of Theorem 4.7. We use Lemma 4.4 and the fact that $\left|\tilde{\mathcal{L}}^{[2]}\tilde{u}\right|_{C^3(\tilde{\tau})} = 0$. □

We now state without proof Theorem 4.10, which is the direct two-dimensional analogy to the one-dimensional case, Theorem 4.4.

Theorem 4.10 *Let $\tilde{u} \in C^2(\tau)$ (respectively, $C^1(\tilde{\tau})$ and $C^0(\tilde{\tau})$). Then,*

$$\left\| \tilde{e}_I^{[2]} \right\|_{C^2(\hat{\tau})} \leqq C \left| \tilde{u} \right|_{C^2(\tilde{\tau})}, \tag{4.49}$$

$$\left\| \tilde{e}_I^{[2]} \right\|_{C^1(\hat{\tau})} \leqq C \left| \tilde{u} \right|_{C^1(\tilde{\tau})}, \tag{4.50}$$

$$\left\| \tilde{e}_I^{[2]} \right\|_{C^0(\hat{\tau})} \leqq C \left| \tilde{u} \right|_{C^0(\tilde{\tau})}. \tag{4.51}$$

We note here that in Theorems 4.7–4.10 we have assumed that the interpolants exist and are unique.

Remark 4.8 *We now show that the interpolants $\tilde{\mathcal{L}}^{[1]}\tilde{u}$ and $\tilde{\mathcal{L}}^{[2]}\tilde{u}$ exist and are unique. Because the mapping of $\tilde{\tau}$ onto τ is linear and invertible, it is sufficient to address only the master elements. The coefficient in (4.26) and (4.28) can be computed from a system of linear algebraic equations. Hence, it is sufficient to show that, if the polynomials are zero at the nodal points, these are identically zero. Because $\left|\mathbb{P}^{[1]}(\mathbf{x})\right|_{C^2(\tilde{\tau})} = 0$ and $\left|\mathbb{P}^{[2]}(\mathbf{x})\right|_{C^3(\tilde{\tau})} = 0$, the desired result follows from Lemmas 4.3 and 4.4.*

As a result of Theorems 4.7–4.10 we are now in a position to estimate the interpolation errors in the (physical) element τ.

Lemma 4.5 *If $u \in C^j(\tau_k)$ and $\tilde{u} \in C^j(\tilde{\tau})$, $j = 0, 1, 2$, then*

$$\left| \tilde{u} \right|_{C^j(\tilde{\tau})} \leqq Ch^j \left| u \right|_{C^j(\tau)}, \quad j = 0, 1, 2\,, \tag{4.52}$$

where C is independent of u and h is the diameter of τ.

Proof This lemma follows immediately from the fact that the mapping ${}^k x(\boldsymbol{\xi})$ is linear and that all the entries, see (4.29), are bounded by Ch, where h is the diameter of τ. □

Exercise 4.17 Prove inequality (4.47)

Lemma 4.6 *If $u \in C^j(\tau)$ and $\tilde{u} \in C^j(\tilde{\tau})$, $j = 0, 1, 2$, then*

$$\left| u \right|_{C^j(\tau)} \leqq Ch^{-j} K^j \left| \tilde{u} \right|_{C^j(\tilde{\tau})}, \quad j = 0, 1, 2, \tag{4.53}$$

where K is a constant that depends on the minimum angle of τ.

Proof This lemma follows immediately from the fact that the inverse mapping $\boldsymbol{\xi}(x)$ of $\boldsymbol{\xi}(\mathbf{x})$ is linear and $\left|J^{-1}\right| \leqq K/h$. □

Exercise 4.18 Prove (4.48) in detail.

Theorem 4.11 *Let $u \in C^k(\tau)$, $k = 0, 1, 2$. Then,*

$$\left\| e_I^{[1]} \right\|_{C^j(\tau)} \leqq Ch^{k-j} K^j \left| u \right|_{C^k(\tau)}, \tag{4.54a}$$

and

$$\left\| e_I^{[1]} \right\|_{H^j(\tau)} \leqq C h^{k-j+1} K^j \left| u \right|_{C^k(\tau)}, \tag{4.54b}$$

$$0 \leqq j \leqq 1,\ j \leqq k,\ k \leqq 2,$$

where C is independent of τ, h and u, and K depends on the minimal angle of τ.

Proof The proof of this theorem is completely analogous to that of Theorem 4.5, when (4.20) and (4.21) are replaced by (4.47) and (4.48) and Theorems 4.1 and 4.2 by Theorems 4.7 and 4.8, respectively. □

Exercise 4.19 Prove (4.49) in detail.

Theorem 4.12 *Let $u \in C^k(\tau)$, $k = 0, 1, 2, 3$, with $0 \leqq j \leqq k$, $k \leqq 3$, then*

$$\left\| e_I^{[2]} \right\|_{C^j(\tau)} \leqq C h^{k-j} K^j \left| u \right|_{C^k(\tau)}, \tag{4.55a}$$

and

$$\left\| e_I^{[2]} \right\|_{H^j(\tau)} \leqq C h^{k-j+1} K^j \left| u \right|_{C^k(\tau)}, \tag{4.55b}$$

where C is independent of τ, h and u, but depends on the minimal angle of τ.

Proof The proof is very similar to that of Theorem 4.10. □

Exercise 4.20 Prove (4.54) and (4.55) in detail.

Remark 4.9 *As in the one-dimensional case, the statements of Theorems 4.11 and 4.12 can be written in the combined form (see also (4.24))*

$$\left\| e_I^{[p]} \right\|_{C^j(\tau)} \leqq C\, h^{\min(k,p+1)-j} K^j \left| u \right|_{C^{\min(k,p+1)}(\tau)}, \tag{4.56a}$$

and

$$\left\| e_I^{[p]} \right\|_{H^j(\tau)} \leqq C\, h^{\min(k,p+1)-j+1} K^j \left| u \right|_{C^{\min(k,p+1)}(\tilde{\tau})}, \tag{4.56b}$$

$$0 \leqq j \leqq \min(k, p+1),\ p = 1, 2,$$

where C is independent of τ, h and u, but depends on the minimum angle of τ.

In what follows we shall assume that the minimum angle condition is uniformly satisfied for all elements. We will therefore include the K in the constant C, rather than display it explicitly.

Exercise 4.21 Prove (4.56b) in detail.

We note the similarity between the one-dimensional and two-dimensional cases; the main difference being the use of the constant K in the two-dimensional setting.

As in the one-dimensional case with (4.24) the combined estimates (4.54) and (4.55) are significant for the two-dimensional context. Again, the estimates $\left\| e_I^{[p]} \right\|_{C^j(\tau)}$ with

$j = 0, 1$ allow us to compute $\left\| e_I^{[p]} \right\|_{\mathcal{U}(\tau)}$ and hence $\left\| e_I^{[p]} \right\|_{\mathcal{U}(\Omega)}$. The use of (3.86) then allows global error estimates on the finite element errors $\left\| e^{[p]} \right\|_{\mathcal{U}(\Omega)}$ to be produced for problem (3.2) in the two-dimensional case for $p = 1, 2$.

Note also that (4.54) and (4.55) show that the rate of convergence, as defined in the numerical experiments, which is related directly to the exponent of h, is governed by the degree of the polynomials in the elements (p) and the smoothness of the function u. It can be shown that the exponents in (4.56a) and (4.56b) cannot be improved.

Remark 4.10 *The* ***minimal*** *angle condition of Theorem 4.11 is only a* ***sufficient*** *condition for the estimates (4.48)–(4.56). The necessary condition is the* ***maximal*** *angle condition, i.e. that* $\pi - \omega_{\max} > \beta > 0$*, rather that* $\omega_{\min} > \beta > 0$*. This is because we used the mapping of* τ *onto* $\tilde{\tau}$ *and a resulting simple argument. If we were to use a master element of diameter 1 but with the same angles as in the physical plane, we would get the maximal-angle condition; for the proof of this see Babuška and Aziz, (1976). The miminum angle does not influence the approximation errors, but it does influence the condition number of the stiffness matrix.*

Exercise 4.22 Consider the triangle τ where

$$\tau \equiv \left\{ \begin{array}{l} (x_1, x_2) \mid -\frac{h}{2} < x_1 < \frac{h}{2},\ 0 < x_2 < \alpha \left(x_1 + \frac{h}{2}\right), x_1 \leqq 0, \\ \qquad 0 < x_2 < \alpha \left(\frac{h}{2} - x_1\right), x_1 > 0 \end{array} \right\},$$

and show that if $u = x_1^2$, then

$$\frac{\left| e_I^{[1]} \right|_{C^1(\tau)}}{h \left| u \right|_{C^2(\tau)}} \to \infty \text{ as } \alpha \to 0.$$

Exercise 4.23 Consider the triangle

$$\tau \equiv \left\{ (x_1, x_2) \mid \frac{-h}{2} < x_1 < 0,\ 0 < x_2 < \alpha \left(x_1 + \frac{h}{2} \right) \right\};$$

i.e. half of that of Exercise 4.22. If $u = \left((\frac{h}{2})^2 - x_1^2 \right)$, show that on τ we have

$$\left| e_I^{[1]} \right|_{C^1(\tau)} \leqq C\, h \left| u \right|_{C^2(\tau)},$$

with C independent of α.

In Chapter 5 we shall need to estimate the error $e^{[p]}, p = 1, 2$ of the interpolant in the H^1- semi-norm and the L_2 - norm; see Chapter 2.

That is, we need to estimate

$$\left| e^{[p]} \right|_{H^1(\tau)} = \left(\int\!\!\int_\tau \left(\frac{\partial e_I^{[p]}}{\partial x_1} \right)^2 + \left(\frac{\partial e_I^{[p]}}{\partial x_2} \right)^2 d\mathbf{x} \right)^{\frac{1}{2}},$$

and

$$\left|e^{[p]}\right|_{L^2(\tau)} = \left(\int\int_\tau \left(e^{[p]}\right)^2 d\mathbf{x}\right)^{\frac{1}{2}}, \quad p = 1, 2.$$

We have the following theorem:

Theorem 4.13 *Let τ_k satisfy the minimal angle condition and $u \in C^2(\tau_k)$ (respectively, $C^3(\tau_k)$). Then,*

$$\left|e_I^{[1]}\right|_{H^1(\tau)} \leqq C\,h^2\,|u|_{C^2(\tau)} \tag{4.57a}$$

$$\left\|e_I^{[1]}\right\|_{L^2(\tau)} \leqq C\,h^3\,|u|_{C^2(\tau)}, \tag{4.57b}$$

and

$$\left\|e_I^{[2]}\right\|_{H^1(\tau)} \leqq C\,h^3\,|u|_{C^3(\tau)} \tag{4.58a}$$

$$\left\|e_I^{[2]}\right\|_{L^2(\tau)} \leqq C\,h^4\,|u|_{C^3(\tau)}, \tag{4.58b}$$

where C is independent of u and h, but depends on the minimum angle condition. Note that on the right-hand sides of (4.57b) and (4.58b) we get one higher power of h than the order of the norm being used. This happens because we integrate over τ that has area h^2.

Proof The results of this theorem follow immediately from those of Theorems 4.11 and 4.12. □

Exercise 4.24 Prove the results of (4.52).

In Section 2.2.2, in the context of problems of the form (2.64) in polygonal domains as in Fig. 2.11, we considered the regularity conditions that the solutions $\mathrm{u}(\mathbf{x})$ satisfy in neighbourhoods of boundary vertices. These were exhibited in (2.86) and (2.87). We now state and prove a theorem concerning interpolation error near such a vertex, which utilizes analogues of Example 4.2.

Theorem 4.14 *Suppose that $\mathrm{u}(\mathbf{x})$ satisfies (2.89c), i.e. that*

$$\left|\frac{\partial^l \mathrm{u}}{\partial x_1^{l_1}\partial x_2^{l_2}}(\mathbf{x})\right| \leqq QC(\varepsilon)\,|\mathbf{x}|^{\beta_j-\varepsilon-l},\ \beta_j > 0, l_1 + l_2 = l \geqq 1, \tag{4.59}$$

in the neighbourhood ω_1 of the vertex A_1 of Γ, coinciding with the origin, and let τ be an element of Δ with vertex ${}^1z$ at A_1, as in Fig. 4.4, then

$$\int_\tau \left|\nabla\left(\mathrm{u} - \mathcal{L}^{[p]}\mathrm{u}\right)\right|^2 dx \leqq C\,h^{2(\beta_j-\varepsilon)}, \tag{4.60}$$

where C depends on $\beta_j - \varepsilon$ but is independent of h.

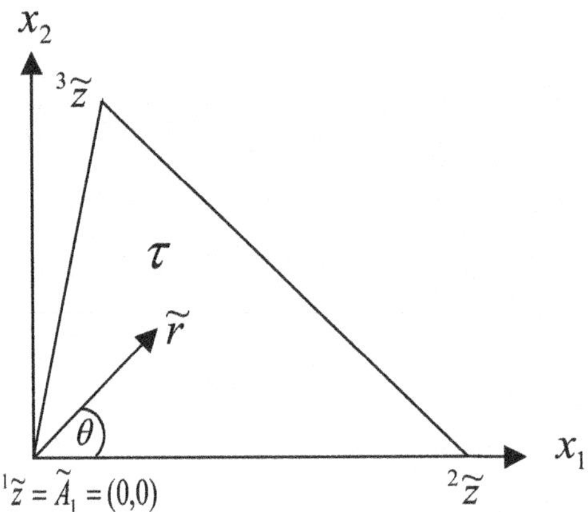

Fig. 4.4 Element τ with one vertex at the origin of coordinates, coinciding with the vertex A_1.

Proof Clearly,

$$\begin{aligned}\int_\tau \left|\nabla\left(u - \mathcal{L}^{[p]}u\right)\right|^2 dx &= \int_\tau |\left(\nabla u - u(0)\right) - \mathcal{L}^{[p]}\left(u - u\left(0\right)\right)|^2 d\mathbf{x} \\ &\leqq 2\left(\int_\tau |\nabla\left(u - u\left(0\right)\right|^2 d\mathbf{x} + \int_\tau |\nabla\left(\mathcal{L}^{[p]}\left(u - u(0)\right)\right|^2 d\mathbf{x}\right) \\ &\equiv 2\left(\mathcal{J}_1 + \mathcal{J}_2\right). \end{aligned} \quad (4.61)$$

Using (4.59) we get $|\mathcal{J}_1| \leqq C\, h^{2\beta_j - \varepsilon}$. From (4.59) we see that the values of $\left(u - u(0)\right)\left({}^i z\right) \leqq Ch^{\left(\beta_j - \varepsilon\right)}$, from which it follows that $|\mathcal{J}_2| \leqq C\, h^{2\beta_j - \varepsilon}$, and the theorem is proven. □

5

a priori estimates of the error of the finite element solution in the energy norm

Summary

- **The single element interpolation error estimates of Chapter 4 are used to produce global *a priori* error estimates for finite element solutions.**
- **The effect on the global error of using different types of mesh is discussed.**
- **For problems in two dimensions the effects of boundary singularaties on the errors are illustrated, and mesh refinement methods are presented.**

5.1 Introduction to *a priori* error analysis

Having in Chapter 4 considered the case of the interpolation error on a single element of the finite element mesh, we now recall the analysis of Section 3.5. There, we showed that over the whole partition of the region Ω of a problem the error in the finite element solution measured in the energy norm (see (2.21c) and (2.66)) can be bounded in terms of the error of any function χ from the space $S^{[p]}_{\Delta,0}$ or $S^{[p]}_{\Delta}$.

In this chapter we shall use global (piecewise linear and piecewise quadratic) interpolants to the weak solutions of the problems ((2.14) and (2.63)) over the partitions Δ of Ω as the function $\chi \in S^{[p]}_{\Delta,0}$ or $S^{[p]}_{\Delta}$. These *global* interpolants and global error estimates over Δ are constructed from the *local* interpolants and their corresponding *local* error estimates on each element. The error estimates of this chapter are of *a priori* type, because they are based on *a priori* (known) information about the exact (weak) solutions and in particular on measures of these exact solutions in specific norms.

The results of this chapter contrast with those of Chapter 7, where we shall obtain error estimates by utilising information of *a posteriori* type, in the sense that this information concerns *computed* results.

Remark 5.1 *In this chapter we shall derive error estimates primarily for problems with homogeneous Dirichlet boundary conditions; i.e.* $g(\mathbf{x}) = 0$, $\mathbf{x} \in \Gamma_D$, *see (2.14) and (2.63). The reason for this is that, for two-dimensional problems with polygonal boundaries* Γ, *and with triangular mesh partitions of* Ω *where the boundary segments*

Γ_j always coincide with element edges, and where the boundary vertices A_j are always element vertices, when $g(\mathbf{x}) = 0$, $\mathbf{x} \in \Gamma_D$, the finite element solution is identically zero on Γ_D, so that no boundary error is introduced as a result of the Dirichlet condition.

We feel that treatment of boundary error is not central to finite element error estimation, so we limit our consideration to contexts where it does not occur.

Note that the Dirichlet boundary conditions are prescribed only on Γ_{D} and not on the entire Γ.

5.1.1 Error of the finite element solution in one dimension

We first consider again the one-dimensional boundary value problem of (2.14) in which, for simplicity, we take $g_1 = 0$, giving a homogeneous Dirichlet boundary condition at $\{0\} = \Gamma_{\mathrm{D}}$. Assumptions (2.20) on $a(x), c(x)$, and $f(x)$ are again assumed to hold. With $\mathcal{U}_0$ as in (2.21b) the weak form of (2.14) is that of (2.31), namely:

find $\mathrm{u} \in \mathcal{U}_0$ such that

$$B(\mathrm{u}, v) \equiv \int_0^l \left(a(x) \frac{d}{dx}\mathrm{u}(x) \frac{d}{dx} v(x) + c(x)\mathrm{u}(x)v(x) \right) dx + \alpha_2 \mathrm{u}(l)v(l)$$

$$= \int_0^l f(x)v(x)dx + G_2 v(l) \equiv F(v) \ \ \forall v \in \mathcal{U}_0. \tag{5.1}$$

In (2.40) and (2.41) we stated that for $a(x) \in C^{k-1}(I)$, and $c(x), f(x) \in C^{k-2}[I]$ we have $\mathrm{u}(x) \in C^k(I)$, $k = 2, 3$. In Section 3.3.1 we defined the mesh Δ on I for which the elements are τ_q, $q = 1, \ldots, M$, $\tau = (x_q, x_{q+1})$ and have size $h_q = x_{q+1} - x_q$; $h^{\Delta} \equiv \max h_q^{\Delta}$, $q = 1, \ldots, M$. We further in (3.17) and (3.49) introduced for $p = 1, 2$, the finite element spaces

$$S_{\Delta}^{[p]} \equiv \left\{ v(x), v \in C^0(\bar{I}) \mid \ v|_{\tau_q} \in \mathbb{P}^{[p]}(\bar{\tau}_q) \right\}, \tag{5.2}$$

$$S_{\Delta,0}^{[p]} \equiv \left\{ v(x), v \in S_{\Delta}^{[p]} \mid \ v(x) = 0, \ x \in \Gamma_{\mathrm{D}} \right\}. \tag{5.3}$$

Although they are not needed immediately we also have

$$S_{\Delta,D}^{[p]} \equiv \left\{ v(x) \in S_{\Delta}^{[p]} \mid v(x) = 0, x \in \tau_q, \bar{\tau}_q \cap \Gamma_D = \emptyset \right\}, \tag{5.4}$$

$$S_{\Delta,g_1}^{[p]} \equiv \{ v \in \mathcal{U}_{\Gamma_{\mathrm{D}}} \mid v(0) = g_1 \}. \tag{5.5}$$

A family of meshes Δ is said to be *quasiuniform* if there exists a constant $\nu > 0$ such that

$$\max_{q=1,\ldots,M} h_q^{\Delta} / \min_{q=1,\ldots,M} h_q^{\Delta} \leq \nu \tag{5.6}$$

for any Δ in the family. The constant ν is called the *quasiuniformity constant.* If the mesh is quasiuniform then, again recalling from Chapter 3 that $h^{\Delta} \equiv \max h_q^{\Delta}$, $q = 1, \ldots, M$, it follows immediately from (5.4) that

$$\frac{1}{\nu} h^{\Delta} \leqq h_q^{\Delta} \leqq h^{\Delta}, \; q = 1, \ldots, M. \tag{5.7}$$

The mesh Δ is *uniform* if $h_q^{\Delta} = h^{\Delta}$, $q = 1, \ldots, M$, i.e. $\nu = 1$.

Remark 5.2 *A quasiuniform mesh is simple to construct and works well for problems where the solutions do not contain singularities.*

The finite element solution $u_{\Delta}^{[p]} \in S_{\Delta,0}^{[p]}$ to problem (5.1), with homogeneous Dirichlet conditions on Γ_{D}, for $p = 1, 2$, satisfies

$$B\left(u_{\Delta}^{[p]}, v\right) = F(v) \;\; \forall v \in S_{\Delta,0}^{[p]}. \tag{5.8}$$

As $S_{\Delta,0}^{[p]} \subset \mathcal{U}_0$ we have that

$$B(\mathrm{u}, v) = F(v) \;\; \forall v \in S_{\Delta,0}^{[p]}. \tag{5.9}$$

On subtracting (5.8) from (5.9) and defining the error

$$e_{\Delta}^{[p]} \equiv \mathrm{u} - u_{\Delta}^{[p]}, \; p = 1, 2, \tag{5.10}$$

we obtained the best approximation property, (3.86),

$$\left\| e_{\Delta}^{[p]} \right\|_{\mathcal{U}} \leqq \|\mathrm{u} - \chi\|_{\mathcal{U}} \;\text{ for any } \chi \in S_{\Delta,0}^{[p]}. \tag{5.11}$$

As noted earlier, the approach to estimating $\left\| e_{\Delta}^{[p]} \right\|_{\mathcal{U}}$ is, for $p = 1, 2$, to construct χ as the piecewise pth-order interpolant to u over Δ using the local interpolant $\mathcal{L}^{[p]}(\tau_q)$ as in Chapter 4 on every element $\tau_q, q = 1, \ldots, M$. The following theorem ensures that the global interpolant constructed in this way lies in $S_{\Delta,0}^{[p]}, p = 1, 2$.

Theorem 5.1 *Let $\chi(x)$ be a piecewise pth-order polynomial function, $p = 1, 2$, of x over Δ such that, when $\mathrm{u} \in \mathcal{U}_0$,*

$$\chi \mid_{\tau_q^{\Delta}} \equiv \mathcal{L}^{[p]}(\mathrm{u})(\tau_q^{\Delta}) \equiv \mathcal{L}^{[p]}(\tau_q)\mathrm{u}, \; \tau_q \in \Delta, \tag{5.12}$$

then $\chi(x) \in S_{\Delta,0}^{[p]}$.

Proof Given the definition of $\chi(x)$ on $\tau_q \in \Delta$, we need to show that χ is continuous over the partition of Δ, and that $\chi(x) = 0, x \in \Gamma_D$; i.e. that in this case $\chi(0) = 0$. The continuity condition is clearly satisfied, because $\left(\chi \mid_{\tau_q^{\Delta}} (x_q)\right) = \left(\chi \mid_{\tau_{q-1}^{\Delta}} (x_q)\right) = u(x_q), q = 1, \ldots, M$ and the function $u(x)$ is continuous. Further, $\chi(0) = \mathrm{u}(0) = 0$. □

In the following, we shall refer to $\mathcal{L}_{\Delta}^{[p]}(\mathrm{u})$ as the global piecewise pth-order polynomial interpolant to u over the partition Δ of Ω. As before, when no misunderstanding can occur, we shall adopt the notation in which we omit the Δ.

The next theorem gives the estimate of the global error $\left\|e^{[p]}\right\|_{\mathcal{U}}$, as in (5.11), using the global interpolant $\mathcal{L}^{[p]}(\mathrm{u})$, which is constructed by taking the local error on each element τ_q and summing over the partition. In the proof, we invoke Theorem 4.5 for $p=1$ and Theorem 4.6 for $p=2$.

Note that from (5.11) and Theorem 5.1,

$$\left\|e_\Delta^{[p]}\right\|_{\mathcal{U}} \equiv \left\|\mathrm{u}-u_\Delta^{[p]}\right\|_{\mathcal{U}} \leqq \|\mathrm{u}-\chi\|_{\mathcal{U}} = \left\|\mathrm{u}-\mathcal{L}^{[p]}\mathrm{u}\right\|_{\mathcal{U}}. \tag{5.13}$$

Inequality (5.13) shows that the error of the interpolant in the energy norm on the domain I majorizes the error in the finite element solution $u_\Delta^{[p]}$, in the energy norm. We emphasize that the phenomenon is true only when these errors are measured in the *energy norm.*

We now prove a theorem that provides a bound in the energy norm for the error $e_\Delta^{[p]}$. This is proved jointly for $p=1$ and 2, and then the specific cases are discussed.

Theorem 5.2 *Let* $\mathrm{u}\in C^k(\tau_q), k=1,2,3,\ \tau_q\in\Delta$, *then for* $p=1,2$

$$\left\|e_\Delta^{[p]}\right\|_{\mathcal{U}}^2 \leqq C\left(\textstyle\sum_{q=1}^M \left(h_q^{2(\min(k-1,p))+1} + h_q^{2(\min(k,p+1))+1}\right)|\mathrm{u}|^2_{C^{\min(k,p+1)}(\tau_q)}\right) \tag{5.14a}$$

and

$$\left\|e_\Delta^{[p]}\right\|_{\mathcal{U}}^2 \leqq C\left(\left(\sum_{q=1}^M h_q^{2(\min(k-1,p))+1}\right)|\mathrm{u}|^2_{C^{\min(k,p+1)}(I)}\right), \tag{5.14b}$$

where C *depends on* $k, a_{\max}, c_{\max}, \alpha_2$ *and* l, *but is independent of* u *and the mesh* Δ.

Proof We have that

$$\begin{aligned}\left\|\mathrm{u}-\mathcal{L}^{[k]}\mathrm{u}\right\|_{\mathcal{U}}^2 &= \sum_{q=1}^M \int_{\tau_q} a(x)\left(\frac{d\mathrm{u}}{dx}-\frac{d}{dx}\left(\mathcal{L}^{[p]}(\tau_q)\mathrm{u}\right)\right)^2 dx \\ &\quad+\int_{\tau_q} c(x)\left(\mathrm{u}-\mathcal{L}^{[p]}(\tau_q)\mathrm{u}\right)^2)dx + \alpha_2\left(\mathrm{u}(l)-(\mathcal{L}^{[p]}_{(\tau_M)}(\mathrm{u}))(l)\right)^2.\end{aligned}$$

Using now Theorems 4.5 and 4.6 we get

$$\begin{aligned}\left\|e_\Delta^{[p]}\right\|_{\mathcal{U}}^2 &\leq \left\|\mathrm{u}-u_\Delta^{[p]}\right\|_{\mathcal{U}}^2 \\ &\leqq C\left(\sum_{q=1}^M\left(h^{2\min(k-1,p))+1}|\mathrm{u}|^2_{C^{\min(k,p+1)}(\tau_q)} + h^{2(\min(k,p+1))+1}|\mathrm{u}|^2_{C^{\min(k,p+1)}(\tau_q)}\right)\right),\end{aligned}$$

which gives (5.14a).

If we use the fact that

$$|\mathrm{u}|_{C^{\min(k,p+1)}(\tau_q)} \leqq |\mathrm{u}|_{C^{\min(k,p+1)}(I)},$$

then we get

$$\left\| e_{\Delta}^{[p]} \right\|_{\mathcal{U}}^{2} \leqq \mathcal{C} \left(\sum_{q=1}^{M} \left(h_q^{2(\min(k-1,p))+1} + h_q^{2(\min(k,p+1))+1} \right) |\mathrm{u}|^2_{C^{\min(k,p+1)}(I)} \right).$$

Because the second term in the above summation is less that the first, we get (5.14b). □

Remark 5.3 *At first sight, the results of Theorem 5.2 may look rather daunting. However, we emphasize that all that is being done here is to amalgamate the single element interpolation error results of Chapter 4 into a single global interpolation error estimate* $e_{\Delta}^{[p]}$ *for the whole mesh, (5.14a and b).*

Remark 5.4 *Whereas Theorem 5.3 also covers the general case, for a specific example the estimates become much simplified. Suppose* $\mathrm{u} \in C^2[I]$ *so that* $k = 2, p = 1$ *and* Δ *is a uniform mesh with elements of size* h*, then* $\min(k-1,p) = 1$ *and* $\min(k, p+1) = 2$*, so that*

$$\left\| e_{\Delta}^{[1]} \right\|_{\mathcal{U}}^{2} \leqq \mathcal{C} \sum_{q=1}^{M} h^3 \left\| \mathrm{u} \right\|^2_{C^2(\tau_q)} \leqq \mathcal{C} h^2 \left\| \mathrm{u} \right\|^2_{C^2(I)}. \tag{5.15}$$

We note that in (5.15) we have used norms of u as opposed to semi-norms. Because $\|\cdot\|_{C^k} \geqq |\cdot|_{C^k}$, instead of the exponents in the inequalities above with $\min(k, p+1)$ terms, we are now able to use only k. This does of course affect the constants.

In Theorem 5.2 we have utilized only global information concerning $a(x), c(x), \alpha_2$; i.e. that they are all bounded, together with assumptions on the regularity of the solution.

The next theorem, exploits the properties of quasiuniform meshes. Here, all elements have essentially the same size. In this case it is more natural to use h as the size measure. In the non-quasiuniform context estimates with large ratio $h_{\max}/h_{\min}$ can be very pessimistic when expressed in terms of h.

Theorem 5.3 *Let the mesh* Δ *be quasiuniform with quasiuniformity constant* ν*. Then, if* $\mathrm{u} \in C^k(I)$*,* $k = 1, 2, 3$ *and* $p = 1, 2$*,*

$$\left\| e_{\Delta}^{[p]} \right\|_{\mathcal{U}} \leqq \mathcal{C}\, h^{\min(k-1,p)} \left| \mathrm{u} \right|_{C^{\min(k,p+1)}(I)}, \tag{5.16}$$

where $\mathcal{C}$ *depends on* $\mathrm{u}, a_{\max}, c_{\max}, \alpha, l$ *and* ν*.*

Proof Inequality (5.16) follows immediately from Theorem 5.2, because $\sum_{q=1}^{M} h_q = l$, and hence

$$\sum_{k=1}^{M} h_q^{2\min(k-1,p)} h_q \leqq h^{2\min(k-1,p)} l.$$

□

Although (5.16) was, for reasons of elementarity, proved for $\mathrm{u} \in C^{\min(k,p+1)}(I)$, in fact the result also holds if the C semi-norm is replaced by the $H^{\min(k,p+1)}$ semi-norm.

Theorems 5.2 and 5.3 give typical *a priori* asymptotic error estimates. They express the accuracy of the finite element solution as the size of the elements approaches zero. Then, the exponents of h_k, i.e. of the size of the elements, govern this accuracy. If the mesh is quasiuniform, then the exponent $\min(k-1,p)$ in (5.16) is called the *rate of convergence* of $u_\Delta^{[p]}$ to u; see also Section 3.6.

Remark 5.5 *Graphically, the behaviour of the error* $\left\| e_\Delta^{[p]} \right\|$ *as a function of* h *is best expressed on a* ln / ln *scale. This is illustrated in Fig. 5.1 where, referring to Theorem 5.3,* h *is the mesh size and the straight line has the form*

$$\ln(Ch^\beta \|\mathrm{u}\|_{C^{\min(k,p+1)}}) = \beta \ln h + D,$$

where C *and* D *are constants, and the slope of the line is the rate of convergence* β*. The position of the line is characterized by* D*.*

Note from Fig. 5.1 that in general the values of the actual errors for different values of h do not lie on a straight line, because (5.16) is an inequality and not an equality. Further, note that the actual error does not decrease monotonically; see Remark 5.8.

We say that the rate of convergence β is optimal if, for any $\bar{\beta} > \beta$,

$$\left\| e_{\Delta(h)}^{[p]} \right\|_{\mathcal{U}} / h^{\bar{\beta}} \to \infty \text{ as } h \to 0.$$

If the mesh is quasiuniform, then the size N_1 of the linear equation system (3.13) is of order $\frac{l}{h} \approx N_1$; recall that N_1 is the number of degrees of freedom of the finite element solution. Hence, we can express (5.16) in terms of N_1 as

$$\left\| e^{[p]} \right\|_{\mathcal{U}} \leqq C N_1^{-\min(k-1,p)} \|\mathrm{u}\|_{C^{\min(k,p+1)}(I)} . \tag{5.17}$$

Note that generically $N_1 \approx N$. The number of degrees of freedom in a finite element solution is a measure of the amount of *computational effort* needed to obtain the

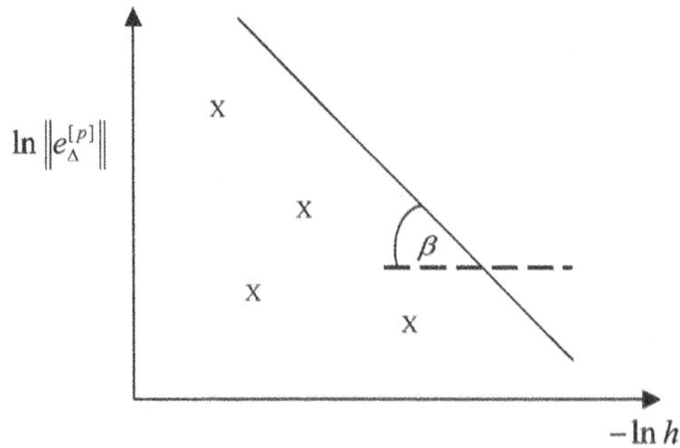

Fig. 5.1 Behaviour of $\left\| e_\Delta^{[p]} \right\|$ against h on a ln / ln scale. The line depicts the estimate, whilst the $\times$ are values of the actual errors for various h.

solution. From this point of view it is more informative to express the estimate in the form (5.17) instead of (5.16).

If the mesh is not quasiuniform then the rate of convergence should be expressed only through N, because as has been said using h, the maximum size of the elements, leads to very pessimistic estimates.

Remark 5.6 *We note that the rate of convergence obtained using the interpolant (see Section 4, Exercise 4.10) is optimal in the sense that the rate cannot be improved provided that*

$$\left| \frac{d^{p+1}\mathfrak{u}}{dx^{p+1}}(x) \right| \geqq \nu > 0,$$

for all $x \in I^* \subset I$ with $|I^*| > 0$. This is usually the case.

The examples that now follow illustrate this optimality.

Remark 5.7 *The estimate (5.16) is an upper bound on the error and holds for any h. Nevertheless, it does **not** mean that $\left\|e^{[p]}\right\|_{\mathcal{U}}$ has to behave as $h^{\min(k-1,p)}$ for large h.*

Remark 5.8 *If the input data of the problem are sufficiently smooth, then for a uniform mesh and $p = 1, 2$ for h small.*

$$\left\|e^{[p]}\right\|_{\mathcal{U}} = C\ h^{\min(k-1,p)}\left(1 + 0(h)\right).$$

Remark 5.9 *If we have a sequence of meshes such that the associated space $S_{\Delta}^{[p]}$ of one is contained in the associated space of the next, then $\left\|e^{[p]}\right\|_{\mathcal{U}}$ decreases from mesh to mesh monotonically as $h \to 0$. If the meshes do not satisfy this inclusion property, then the convergence is not necessarily monotonic, as seen in the numerical results for the 1D Eng Problem 1 in Section 3.3.5.*

From Theorems 5.2 and 5.3 we see that less computation is required to achieve the prescribed accuracy if piecewise quadratic functions are used rather than piecewise linear ones, provided the data is sufficiently smooth. For this reason, quadratic elements are in general much more popular for practical computations.

The improved accuracy that results from the use of quasiuniform meshes occurs in general only when $\mathfrak{u}$ is smooth, see (5.17). *For example*, if $k = 2$ (respectively, 3) then the rates of convergence for $p = 1$ (respectively, 2), linear (respectively, quadratic) elements, are 1 (respectively, 2). In the non-smooth case it will be seen later that the situation is different.

The situation is more complicated in two dimensions, where there can be a specific lack of smoothness in the solution locally, see Chapter 2. In this case, if the two-dimensional mesh is properly designed, see later in this chapter, in terms of N the rate of convergence remains the same as for the smooth case, with the result that it is again better to use piecewise quadratic functions. In order to illustrate many of the above estimates and remarks we now consider in the form of examples and exercises a number of two-point boundary value problems of the type as in (2.14), which give rise to the weak form (5.1).

The examples are designed to illustrate rates of convergence, whilst the exercises should enable the reader to confirm his/her understanding of the results above.

Example 5.1 Consider the problem of type (2.14) for which $I \equiv (0, \frac{\pi}{2})$, in which $\mathrm{u}(x)$ satisfies,

$$-\frac{d^2\mathrm{u}}{dx^2}(x) + c\mathrm{u}(x) = \sin kx, \ x \in I, \tag{5.18a}$$

$$\mathrm{u}(0) = 0, \tag{5.18b}$$

$$\frac{d\mathrm{u}}{dx}(x) = 0, \ x = \frac{\pi}{2}, \tag{5.18c}$$

where c is a non-negative constant and k is an odd positive integer. In this case, it can easily be determined that the exact solution u to (5.18) is

$$\mathrm{u}(x) = \sin kx \,/\, (k^2 + c). \tag{5.19}$$

Let us now determine the finite element solution for linear elements on a uniform mesh $\Delta = \{x_q\}$, where $x_q = (q-1)h, q = 1, 2, \ldots, M+1$, with $Mh = \pi/2$. Using Example 3.3, we determine the $[K]$ for $p = 1$ to be

$$\frac{1}{h}\begin{bmatrix} 2 & -1 & & & \\ -1 & 2 & -1 & & \\ & & & & \\ & & -1 & 2 & -1 \\ & & & -1 & 1 \end{bmatrix} + ch\begin{bmatrix} \frac{2}{3} & \frac{1}{6} & & & \\ \frac{1}{6} & \frac{2}{3} & \frac{1}{6} & & \\ & & & & \\ & & \frac{1}{6} & \frac{2}{3} & \frac{1}{6} \\ & & & \frac{1}{6} & \frac{1}{3} \end{bmatrix}.$$

The augmented right-hand vector $\hat{\mathbf{f}}$ is

$$\left[\hat{f}_2, \ldots, \hat{f}_{M+1}\right]^T,$$

where

$$\hat{f}_q = \frac{2}{k^2}\sin kx_q(1 - \cos kh), \ q = 2, \ldots, M+1$$

and

$$\hat{f}_{M+1} = \frac{1}{k^2}\sin kx_{M+1}(1 - \cos kh).$$

We can easily check that

$$u_\Delta(x_q) = Q\sin kx_q,$$

where

$$Q = \frac{2}{k^2}(1 - \cos kh) \,/\, \left(2(1 - \cos kh) + \frac{2ch^2}{3}(1 + \frac{1}{2}\cos kh)\right)$$

$$= \left(\frac{1}{k^2} + \frac{ch^2k^2}{3}\left(\frac{1 + \frac{1}{2}\cos kh}{1 - \cos kh}\right)\right)^{-1},$$

and for $c = 0$ we have that $u_\Delta(x_q) = \mathfrak{u}(x_q)$, independent of k. Further, using Taylor's series expansions we get for $h \to 0$ that

$$Q = \frac{1}{k^2 + c}\left(1 + \mathcal{C}(k)(h^2)\right). \tag{5.20}$$

Note that Q is independent of x_q. Hence,

$$e_\Delta^{[1]}(x_q) \equiv \left(\mathfrak{u}(x_q) - u_\Delta^{[1]}(x_q)\right) = \sin kx_q \left(0(h^2)\right). \tag{5.21}$$

It can further be seen that the error has a complicated character when k is large.

Exercise 5.1 For the problems and numerical scheme of Example 5.1 compute $Q \cdot (k^2 + c)$ for various values of h, k and c.

We see that the error in an element τ_q can be written in the form

$$\begin{aligned} e_\Delta^{[1]}(x) &\equiv \left(\mathfrak{u}(x) - u_\Delta^{[1]}(x)\right) \\ &\equiv \left(\mathfrak{u}(x) - \mathcal{L}_\Delta^{[p]}(u)(x)\right) + \chi_q(x), \end{aligned}$$

where χ_q is the linear function on τ_q that has values $(\mathfrak{u} - u_\Delta^{[1]}(x_l),\ l = q, q+1$. Thus,

$$\chi_q(x) = \left(\sin(x_q + \frac{1}{2}) + k^2 \cos(x_{q+1})(x - x_q + \frac{1}{2})\right) 0(h^2).$$

We can now in τ_q easily compute the error $\left\|\mathfrak{u}(x) - u_\Delta^{[1]}(x)\right\|_{\mathcal{U}(I)}$ and we will see that

$$\left\|\chi_q\right\|_{\mathcal{U}} \,/\, \left\|e_\Delta^{[1]}\right\|_{\mathcal{U}} \leqq Ch,$$

i.e. that χ_q can be asymptotically neglected so that the error is $e_\Delta^{[1]}$, which is $0(h)$. Nevertheless, we emphasize that this happens for this example only where h is small with respect to k and to c.

Exercise 5.2 For the problem and numerical scheme of Example 5.1

(1) For fixed h compute the error for several large values of c. What happens as $c \to \infty$?
(2) For fixed h compute the error for various (odd) values as k gets very large.

Exercise 5.3 Repeat all the calculations of Example 5.1 for the case $p = 2$.

Exercise 5.4 Consider the problem of type (2.14) for which $I \equiv (0,1)$ and $\mathrm{u}(x)$ satisfies

$$-\frac{d^2\mathrm{u}}{dx^2}(x) = \begin{cases} 1, 0 < x < \frac{1}{\pi}, \\ 0, \frac{1}{\pi} < x < 1, \end{cases} \tag{5.22a}$$

$$\mathrm{u}(0) = \frac{d\mathrm{u}}{dx}(1) = 0. \tag{5.22b}$$

(1) Determine the exact solution of (5.22).
(2) Compute the finite element solution using piecewise linear and quadratic approximations over various partitions of I.
(3) Using the results of (1) and (2) evaluate

$$\left\| e_\Delta^{[p]} \right\|_{\mathcal{U}} \; / \; \|\mathrm{u}\|_{\mathcal{U}},$$

and make a graph of this on a ln / ln scale.

You will observe from your graph behaviour of *chaotic* type, for $p = 2$, very different from that of Examples 5.1 and 5.2. Why does this occur?

Hint: Use the fact that $e_\Delta^{[p]}, p = 1, 2$, is zero at the vertices of Δ.

Exercise 5.5 Consider the problem of type (2.14) for which $I \equiv (0,1)$ and $\mathrm{u}(x)$ satisfies

$$-\frac{d^2\mathrm{u}}{dx^2}(x) = f(x), \; x \in I, \tag{5.23a}$$

$$\mathrm{u}(0) = 0, \tag{5.23b}$$

$$\frac{d\mathrm{u}}{dx}(1) = 0. \tag{5.23c}$$

Taking a uniform mesh with element size h and quadratic elements, construct a function f such that

$$\mathrm{u} \,|_{\tau_k} \in C^3(\tau_k), k = 1, \ldots, M,$$

for any k and M even, but with the $\mathrm{u} \notin C^3(I)$. Analyze the difference between the estimates (5.14a and b)

We now return to the lack of smoothness of the solution $\mathrm{u}(x)$ of the boundary value problem, and discuss how this can affect the rate of convergence and the accuracy of the finite element solution.

Example 5.2 Consider the problem of type (2.14) in which $I \equiv (0,1)$ and $\mathrm{u}(x)$ is given by

$$\mathrm{u}(x) = x^\alpha, \; 0 < x < 1, \tag{5.24}$$

where $\alpha = \frac{1}{2} + \theta, \theta > 0$, is not an integer. The finite element solution will be computed using a uniform mesh Δ, where $x_j = (j-1)/h_j, j = 1, \ldots, (1/h + 1)$. The weak formulation of the problem is: find $\mathfrak{u} \in \mathcal{U}_0$ such that

$$B(\mathfrak{u}, v) = F(v) \;\; \forall v \in \mathcal{U}_0, \tag{5.25}$$

with

$$B(\mathfrak{u}, v) = \int_0^1 \frac{du}{dx}(x)\frac{dv}{dx}(x)dx,$$

and $F(v)$ has the form that arises when $\mathfrak{u}$ has the form (5.24).

We now seek to estimate the error in the interpolant $\mathcal{L}^{[p]}\mathfrak{u}$, $p = 1, 2$ that leads to the finite element error estimate (5.14).

We shall estimate the interpolation error separately in τ_1 and $\tau_q, q \neq 1$. First, using the solution $\mathfrak{u}$ of problem (5.24) we have by direct computation that

$$\int_0^h \frac{d}{dx}\left(\mathfrak{u} - \mathcal{L}^{[p]}(\tau_1)\mathfrak{u}\right)^2 dx \leqq Ch^{2(\min(\theta,p))}, \quad p = 1, 2, \tag{5.26a}$$

where C takes a different value depending whether $p = 1, 2$. Further, we see that for $q > 1$

$$|u|_{C^s(\tau_q)} \leqq Cx_q^{\alpha - s}, \; \forall s. \tag{5.26b}$$

Hence, applying Theorem 4.7 we have for $p = 1$ with $s = 2$

$$\left\|e_I^{[1]}\right\|_{\mathcal{U}(\tau_q)} \leqq Ch^{1+\frac{1}{2}}x_q^{\alpha-2},$$

and for $p = 2$ with $s = 3$ that

$$\left\|e_I^{[2]}\right\|_{\mathcal{U}(\tau_q)} \leqq Ch^{2+\frac{1}{2}}x_q^{\alpha-3}.$$

Hence, for $p = 1$

$$\sum_{q=2}^{1/h}\left\|e_I^{[1]}\right\|_{\mathcal{U}(\tau_q)}^2 \leqq Ch^3\sum_{q=2}^{1/h}x_q^{2(\alpha-2)} \leqq Ch^{3+2\alpha-4} = Ch^{2\theta} \tag{5.27}$$

for all α such that $2(\alpha - 2) + 1 < 0$, which $\Rightarrow \theta < 1$ and for $p = 2$

$$\sum_{q=2}^{1/h}\left\|e_I^{[2]}\right\|_{\mathcal{U}(\tau_q)} \leqq Ch^5\sum_{q=2}^{1/h}x_q^{2(\alpha-3)} \leqq Ch^{2\theta}, \tag{5.28}$$

for all α such that $2(\alpha - 3) + 1 < 0$, which $\Rightarrow \theta < 2$.

Combining (5.27) and (5.28) with (5.26) we obtain for $\alpha = 1 + \theta$, α not an integer
for $p = 1$: if $\theta < 1$, then the convergence rate is affected by the θ,
for $p = 2$: if $\theta < 2$, then the convergence rate is affected by the θ,
and for $\theta < 1$ both linear and quadratic elements give the same rate of convergence.

From the error estimate (5.24) we see that reduced level of regularity of the solution $\mathfrak{u}$ of Example 5.2 can reduce the rate of convergence of the finite element solution. For

$p = 1$ this is seen if $\theta < 1$ but not for $\theta > 1$, whilst for $p = 2$ the convergence rate is not reduced for $\theta > 2$.

The above behaviour is typical, and also occurs in the two-dimensional setting when a problem (see Section 2.2.2) has boundary vertex singularities.

Theorem 5.2, and in particular estimate (5.14), give a pointer to the form that a non-uniform mesh should take in a specific case.

We now consider the relative error of the interpolant $\mathcal{L}_\Delta^{[p]}$ to u over I in the energy norm, which is

$$\left\| \mathrm{u} - \mathcal{L}_\Delta^{[p]}(\mathrm{u}) \right\|_{\mathcal{U}} / \ \|\mathrm{u}\|_{\mathcal{U}}, \tag{5.29}$$

for sequences of meshes Δ of I that fall into three types and that in each case have the same number of elements N. The three types of mesh are (a) uniform, (b) equilibrated, (c) the optimal mesh that minimizes the relative interpolation error over all meshes that have N elements. In practice, equilibrated and optimal meshes are very similar, and it is not necessary to distinguish between them.

An *equilibrated* mesh is one for which the errors in all elements are the same. Such meshes give (in some sense) optimal accuracy, with convergence rates in N being the same whether a solution is smooth or non-smooth, and the errors are relatively insensitive to small perturbations of the mesh.

The *quality* of a mesh, for particular $\Delta, \mathcal{U}$ and degree p, can be determined by the ratio

$$\frac{\min\limits_q \left\| e_\Delta^{[p]} \right\|_{\mathcal{U}(\tau_q)}}{\max\limits_q \left\| e_\Delta^{[p]} \right\|_{\mathcal{U}(\tau_q)}} \equiv \rho(\Delta, \mathrm{u}, p). \tag{5.30}$$

For problems with smooth solutions and computations with quasiuniform meshes the ratio ρ is *reasonable*, in the sense that $\rho > \frac{1}{5}$ or $\frac{1}{10}$ but not less than $\frac{1}{100}$, hence such meshes with sufficiently large M produce efficient computations. We note that exceptional elements can occur when $e_\Delta^{[p]}$ is very small or zero, and should be excluded from the minimum.

If ρ is small then the mesh is of *low* quality. For a problem with a smooth solution, where we have $\gamma_1^{[p]}$ and $\gamma_2^{[p]}$, $p = 1, 2$, such that

$$0 < \gamma_1^{[1]} \leqq \left| \frac{d^2\mathrm{u}}{dx^2} \right| \leqq \gamma_2^{[1]} < \infty \ \ \forall x \in I,$$

and

$$0 < \gamma_1^{[2]} \leqq \left| \frac{d^3\mathrm{u}}{dx^3} \right| \leqq \gamma_2^{[2]} < \infty \ \ \forall x \in I,$$

then the *quality* of a uniform mesh will be good, because $\rho \approx \gamma_1^{[p]} / \gamma_2^{[p]}, p = 1, 2$.

A uniform mesh is also of very low quality when it is used for a problem for which the solution has a singularity. Such a mesh in this case gives poor results.

Exercise 5.6 Compute ρ for the problem of Example 5.1 with $k = 1$ and $c = 0$, using a uniform mesh.

Example 5.3 Consider problem (2.14) in which $I \equiv (0, \frac{\pi}{2})$ and $\mathfrak{u}(x)$ satisfies

$$-\frac{d^2\mathfrak{u}}{dx^2}(x) = \sin x,\ x \in I, \tag{5.31a}$$

$$\mathfrak{u}(0) = 0, \tag{5.31b}$$

$$\frac{d\mathfrak{u}}{dx}\left(\frac{\pi}{2}\right) = 0. \tag{5.31c}$$

The solution to this problem is $\mathfrak{u}(x) = \sin x$. Recall now that $\tau_q = (x_q, x_{q+1})$, then for sufficiently small h_q we have

$$|\mathfrak{u}|_{C^{(p+1)}(\tau_q)} \approx \left|\sin\ x_{q+\frac{1}{2}}\right|, \quad \mathit{for}\ p = 1,$$

$$\approx \left|\cos x_{q+\frac{1}{2}}\right|, \quad \mathit{for}\ p = 2,$$

where $x_{q+\frac{1}{2}} = \frac{1}{2}\left(x_q + x_{q+1}\right)$, which means that for $p = 1$ the equilibrated mesh should have smaller elements in the neighbourhood of $\frac{\pi}{2}$ than in the neighbourhood of 0, whilst for $p = 2$ we have the opposite situation.

From this example we see that the character of an optimal mesh depends not only on the smoothness of $\mathfrak{u}$, but also on the degree of the elements.

So far we have considered only uniform meshes, and now we discuss other classes of meshes. Let us have a *radical* mesh over $I = (0, 1)$, for which

$$x_q = \left(\frac{q-1}{M}\right)^{\gamma},\ q = 1, \ldots, M+1. \tag{5.32}$$

Note that for $\gamma = 1$ this gives a uniform mesh.

Example 5.4 Let us consider the problem of Example 5.2, for which the solution is $\mathfrak{u}(x) = x^{\alpha}, 0 < x < 1$, with $\alpha = \frac{1}{2} + \theta$, where $\theta > 0$ is not an integer. For this, we shall study the error $e_{\Delta}^{[p]}$ for the case of radical meshes; for a mathematical analysis we refer to Babuška and Strouboulis (2001), from which the illustrative results are taken.

In Example 5.2 we saw that with a uniform mesh the error is $0(h^{\alpha-\frac{1}{2}}) = 0(h^{\theta})$, $\frac{1}{h} = M$, the number of elements.

In this context we define

$$e_{\mathrm{Re}\,l}(\alpha, \gamma, M, p) \equiv \left\|e_{\Delta}^{[p]}\right\|_{\mathcal{U}} / \left\|\mathfrak{u}\right\|_{\mathcal{U}}, \tag{5.33}$$

and

$$C(\alpha, \gamma, M, p) = e_{\mathrm{Re}\,l}/M^{-\min(p, \gamma(\alpha-\frac{1}{2}))}, \tag{5.34}$$

where $C(\alpha, \gamma, M, p)$ is called the quality index. Then, for the case of a uniform mesh, $\gamma = 1$, and $\alpha = 0.7$ we obtain the results of Table 5.1.

Table 5.1 The relative error (5.33) and estimate (5.34) for $\alpha = 0.7$ with a uniform mesh.

		$p = 1$	$p = 2$
$M = 2$	e_{Rel}	0.3742	0.2853
	C_2	0.42982	0.32776
$M = 16$	e_{Rel}	0.2473	0.1882
	C_{16}	0.43064	0.32766
$M = 512$	e_{Rel}	0.1237	0.09412
	C_{512}	0.43078	0.32775

We see that as $M \to \infty$ the C converges to a constant; this shows that the estimate characterizes the behaviour of the method well. Note also that the relative error for $p = 1$ and 2 for $M = 512$ is, respectively, 12% and 9%.

We now illustrate the effect of using a radical mesh with $\alpha = 0.7$. It is possible to show, see Babuška and Strouboulis, (2001), that for $p = 2$ $\gamma_{\text{opt}} \approx 12.5$. The results of Table 5.2 give the ratio between the error achieved with differing refinements, characterized by γ, and the error achieved with the optimal mesh for quadratic elements. We see clearly the influence of the refinement; for $\gamma < 12.5$ it is clearly under-refined, whilst for $\alpha > 12.5$ it is over-refined. We also see that for the optimal mesh the accuracy improves by a factor of 4 when M is doubled, which is the rate when the solution is smooth. This would be expected for the case of quadratic elements

Table 5.2 Ratio of $e_{\text{Rel}}/e_{\text{Rel}}^{\text{opt}}$ for different radical meshes.

$M(\Delta)$	$e_{\text{Rel}}/e_{\text{Rel}}^{\text{opt}}$						$e_{\text{Rel}}^{\text{opt}}$
	$\gamma = 1$	$\gamma = 4$	$\gamma = 7$	$\gamma = 10$	$\gamma_{\text{opt}} = 12.5$	$\gamma = 15$	
2	0.98	0.69	0.74	0.90	1.0	1.06	2.915×10^{-1}
4	1.60	0.76	0.67	0.83	1.0	1.17	1.548×10^{-1}
8	3.84	1.21	0.77	0.83	1.0	1.22	0.5630×10^{-1}
16	11.19	2.33	1.02	0.89	1.0	1.20	0.1681×10^{-1}
32	35.85	4.92	1.44	0.96	1.0	1.17	0.4572×10^{-2}
64	119.81	10.84	2.12	1.05	1.0	1.15	0.1191×10^{-2}
128	408.82	24.41	3.15	1.14	1.0	1.13	0.3038×10^{-3}
256	1409.01	55.51	4.73	1.22	1.0	1.12	0.7672×10^{-4}
512	4881.74	126.87	7.14	1.31	1.0	1.12	0.1928×10^{-4}

on a uniform mesh for a problem with a smooth solution. Hence, we see that, even if the solution contains a singularity, use of a suitable mesh gives an error that is $0(M^{-2})$.

We now present an important example that demonstrates the influence that highly non-uniform meshes can have on the accuracy of the finite element solution. In this, we take a particular case for which the first element of the mesh is much bigger than all of the rest. The purpose is to show that the error in the finite element solution over the region need not in every element be close to that of the interpolant.

Example 5.5 Consider the problem of the type (2.14) for which $I \equiv (0,1)$ and $a(x) = (1+x), c(x) = 0, f(x) = 0, g_1 = 0, \alpha_2 = 2$ and $G_2 = 1$. We thus have that

$$-\frac{d}{dx}\left((1+x)\frac{du}{dx}(x)\right) = 0, \; x \in I, \tag{5.35a}$$

$$\mathfrak{u}(0) = 0, \tag{5.35b}$$

$$2\frac{du}{dx}(1) + \mathfrak{u}(1) = 1. \tag{5.35c}$$

The exact solution of (5.33) can be found by integrating (5.35a) to get first

$$(1+x)\frac{du}{dx}(x) = C,$$

and then

$$\mathfrak{u}(x) = C\ln(1+x) + D,$$

where C and D are constants to be found. From (5.35b) it follows immediately that $D = 0$, whilst from (5.35c) we then have that $C = \frac{1}{1+\ln 2}$, giving

$$\mathfrak{u}(x) = \frac{\ln(1+x)}{1+\ln 2}, \tag{5.36}$$

which is a very smooth function with $\mathfrak{u}(0) = 0, 2\frac{du}{dx}(1) + \mathfrak{u}(1) = 1$.

For the finite element method we take the mesh Δ with mesh points

$$0 = x_1 < x_2 < \ldots < x_{M+1} = 1.$$

We select $x_2 \equiv h_1, x_j = h_1 + (j-2)h, \; j = 3, \ldots, M+1$ for some h.

Consider the finite element solution $u_\Delta^{[1]}$ on Δ. Further, let $\hat{u}(x), x \in I$ be a solution to a modified problem, where $\hat{u}(x)$ satisifies the boundary condition (5.35c), the differential equation (5.35a) on $\hat{I} \equiv (h_1, 1)$, and is such that

$$\left|u_\Delta^{[1]}(h_1) - \hat{\mathfrak{u}}(h_1)\right| \leq Ch^2, \tag{5.37a}$$

$$\left\|u_\Delta^{[1]} - \hat{\mathfrak{u}}\right\|_{\mathcal{U}(\hat{I})} \leq Ch, \tag{5.37b}$$

$$\|\mathfrak{u} - \hat{\mathfrak{u}}\|_{\mathcal{U}(\hat{I})} \approx Ch_1^3, \tag{5.37c}$$

(see Babuška and Strouboulis, (2001), p.263). This modified problem eliminates the influence of the large element τ_1.

We consider the errors

$$\left\| e_{\Delta}^{[1]} \right\|_{\mathcal{U}(\hat{I})} \equiv \left\| \mathrm{u} - u_{\Delta}^{[1]} \right\|_{\mathcal{U}(\hat{I})},$$

$$\left\| \hat{e}_{\Delta}^{[1]} \right\|_{\mathcal{U}(\hat{I})} \equiv \left\| \widehat{\mathrm{u}} - u_{\Delta}^{[1]} \right\|_{\mathcal{U}(\hat{I})}.$$

If $\overline{\mathrm{u}}_{\Delta}^{[1]} \approx \mathcal{L}^{[1]}\mathrm{u}$ for the interval $\hat{I}$, then using (5.37) we get

$$Qh_1^3 \leqq \left\| \mathrm{u} - u_{\Delta}^{[1]} \right\|_{\mathcal{U}(\hat{I})} + \mathcal{C}h. \tag{5.38}$$

Suppose we choose $h_1 = h^{\beta}, \beta < 1$ and the specific value $\beta = 0.15$. Then, from (5.38) we get $\left\| \mathrm{u} - u_{\Delta}^{[1]} \right\|_{\mathcal{U}(\hat{I})} \approx 0(h^{0.45})$. The purpose of this example is to demonstrate that if the mesh is of low quality then the error in the interpolant is not close to the actual error.

$$\left\| e_{\Delta}^{[1]} \right\|_{\mathcal{U}(\hat{I})} \approx 0(h^{0.45}), \tag{5.39a}$$

and with $\hat{e}_{\Delta}^{[1]} = \mathrm{u} - \mathcal{L}^{[1]}(\mathrm{u})$ we have

$$\left\| \hat{e}_{\Delta}^{[1]} \right\|_{\mathcal{U}(\hat{I})} \approx 0(h). \tag{5.39b}$$

In Table 5.3 we present the computed values of

$$\left\| e_{\Delta}^{[1]} \right\|_{\mathcal{U}(\hat{I})}, h^{-0.45} \left\| e_{\Delta}^{[1]} \right\|_{\mathcal{U}(\hat{I})}, \left\| \hat{e}_{\Delta}^{[1]} \right\|_{\mathcal{U}(\hat{I})}, h^{-1} \left\| \hat{e}_{\Delta}^{[1]} \right\|_{\mathcal{U}(\hat{I})},$$

Table 5.3 The error $e_{\Delta}^{[1]}$ of the problem and $\hat{e}_{\Delta}^{[1]}$ of the modified problem.

$N(\Delta)$	h	$\Vert e^{[1]} \Vert_{\mathcal{U}(\tilde{I})}$	$h^{-.45} \Vert e^{[1]} \Vert_{\mathcal{U}(\tilde{I})}$	$\Vert \hat{e}^{[1]} \Vert_{\mathcal{U}(\tilde{I})}$	$h^{-1} \Vert \hat{e}^{[1]} \Vert_{\mathcal{U}(\tilde{I})}$
11	0.0386	0.2239×10^{-2}	0.9861×10^{-2}	0.1704×10^{-2}	0.4412×10^{-1}
21	0.0218	0.1646×10^{-2}	0.9203×10^{-2}	0.1050×10^{-2}	0.4809×10^{-1}
31	0.0155	0.1399×10^{-2}	0.9123×10^{-2}	0.7798×10^{-3}	0.5033×10^{-1}
41	0.0121	0.1255×10^{-2}	0.9147×10^{-2}	0.6282×10^{-3}	0.5189×10^{-1}
51	0.0100	0.1158×10^{-2}	0.9208×10^{-2}	0.5297×10^{-3}	0.5308×10^{-1}
61	0.0085	0.1087×10^{-2}	0.9282×10^{-2}	0.4600×10^{-3}	0.5404×10^{-1}
71	0.0074	0.1031×10^{-2}	0.9359×10^{-2}	0.4078×10^{-3}	0.5484×10^{-1}
81	0.0066	0.9863×10^{-3}	0.9437×10^{-2}	0.3671×10^{-3}	0.5553×10^{-1}

for different values of h. The values given in the table illustrate how the error on a specific subdomain can be heavily influenced by the choice of the size of the elements in I .

In Figs. 5.2 and 5.3 we compare the true error and the error of the modified problem. For $h = 0.01$, $h_1 = h^{0.15} = 0.5012$ the difference between these is very visible. It can be seen that the first large element adversely influences the error over the whole of I. In this case, we say that we have pollution caused by this first element, for more see

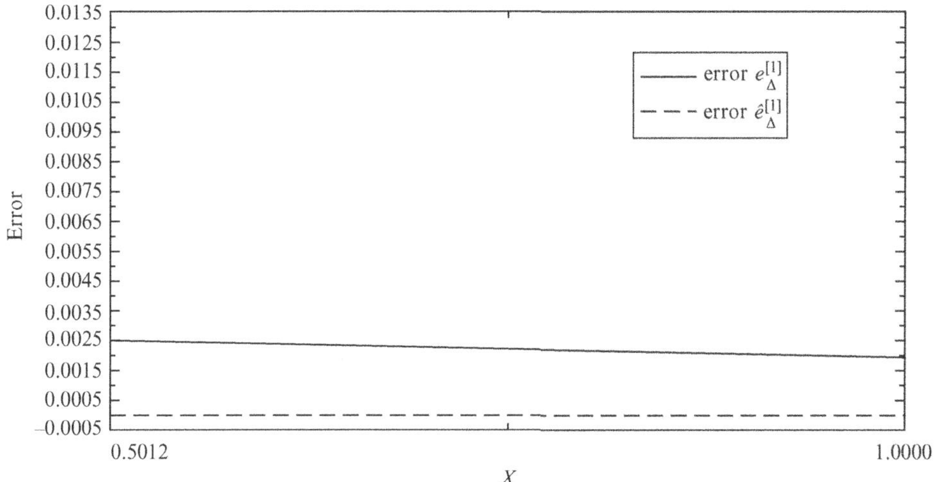

Fig. 5.2 The errors $e_\Delta^{[1]}$ and $\hat{e}_\Delta^{[1]}$, respectively, for the problem and the modified problem.

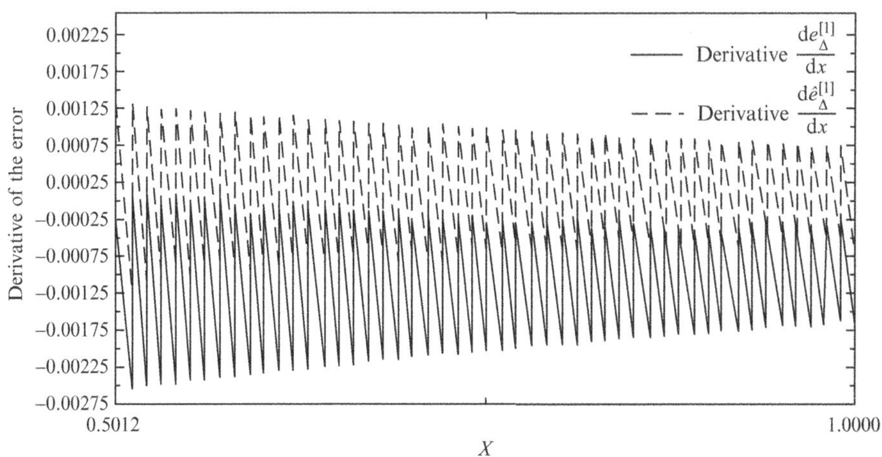

Fig. 5.3 The derivatives $\dfrac{de_\Delta^{[1]}}{dx}$ and $\dfrac{d\hat{e}_\Delta^{[1]}}{dx}$ for the problem and the modified problem.

Section 6.1.2. When, as in Example 5.1, there is no such adverse influence, we say that there is no pollution.

If the difference between the two solutions $u_\Delta^{[1]}$ and $\hat{u}_\Delta^{[1]}$ is negligible in comparison with the error of the finite element method for the modified problem, we say that the pollution on $\hat{I}$ is negligible. If this is not the case, then the pollution is significant. If the pollution error is negligible then the finite element error over $\hat{I}$ is very close to the interpolation error. For the present example it can be shown, see Babuška and Strouboulis, (2001), that the pollution is negligible if $\beta > \frac{1}{3}$. For $\beta < \frac{1}{3}$, as illustrated here, the pollution is significant throughout $\hat{I}$.

We shall return again to pollution effects in the two-dimensional context when the problems have boundary vertex singularities.

As was indicated initially in this chapter, we have so far in this section assumed that $g_1 = 0$ in problem (2.14) and we have been considering finite element solutions to the weak problem (2.31). Not withstanding Remark 5.1 we now take the case when $g_1 \neq 0$ so that we are considering problem (2.13). In order to treat this case for the partition Δ of I we define $\bar{u}^{[p]} \in S^{[p]}(\tau_1)$, $p = 1, 2$, such that $\bar{u}^{[p]}(x_1) = g_1$ and $\bar{u}^{[p]}(x_2) = 0$, where $\bar{u}^{[p]} \in S^{[p]}_{\Delta,g_1} \cap S^{[p]}_{\Delta,D}$. Then, to obtain the finite element solution $u_\Delta^{[p]} \equiv u_0^{[p]} + \bar{u}^{[p]}$ we seek $u_0^{[p]} \in S^p_{\Delta,0}$ such that

$$B\left(u_0^{[p]}, v\right) = F(v) - B\left(\bar{u}^{[p]}, v\right) \ \forall v \in S^{[p]}_{\Delta,0}. \tag{5.40}$$

The error of the interpolant from $S_\Delta^{[p]}$ to $\mathrm{u} - \bar{u}^{[p]}$ on τ_q, $q = 1, \ldots, M$ is the same as the error in the interpolant from S^p_Δ to u; i.e.

$$\mathrm{u} - \mathcal{L}^{[p]}(\tau_k)(\mathrm{u}) = \left(\mathrm{u} - \bar{u}^{[p]}\right) - \mathcal{L}^{[p]}(\tau_k)\left(\mathrm{u} - \bar{u}^{[p]}\right),$$

because

$$\bar{u}^{[p]} - \mathcal{L}^{[p]}(\tau_k)\bar{u}^{[p]} = 0, \ q = 1, \ldots, M + 1.$$

Thus, we can immediately use Theorems 5.2 and 5.3 to estimate the error of the finite element approximation to u in this case of the non-homogeneous Dirichlet boundary condition g_1 at $x = 0$. The construction of $\bar{u}^{[p]}, p = 1, 2$, is not unique, but we must have that $\bar{u}^{[p]} \in S_\Delta^{[p]}$.

Suppose now that in (2.14) we have $c(x) = 0$, $x \in I$ and $\alpha_2 = 0$, and that we take natural boundary conditions of type (2.14c) at both $x = 0$ and $x = l$. The weak form of our problem is thus now: find $\mathrm{u} \in \mathcal{U}$ such that

$$B(\mathrm{u}, v) = F(v), \ \ \forall v \in \mathcal{U}, \tag{5.41}$$

where, for this special case, $\mathcal{U}$ has become

$$\mathcal{U} \equiv \left\{ \mathrm{u} \mid \int_0^l a(x)\left(\frac{du}{dx}(x)\right)^2 dx \equiv \|\mathrm{u}\|^2_{\mathcal{U}} < \infty \right\}, \tag{5.42}$$

and $\|\cdot\|_{\mathcal{U}}$ as in (5.42) is a semi-norm. The solution of (5.41) exists only when the consistency condition is satisfied; i.e. when

$$F(v) = \int_0^l fv dx + G_1 v(0) + G_2 v(l) = 0. \tag{5.43}$$

We take $v(x) \in S_\Delta^{[p]}$ in (5.41) using $v(x) = 1$, and we find that (5.43) arises. Then, the exact solution exists up to a constant. In this case, the finite element solution $u_\Delta^{[p]} \in S_\Delta^{[p]}$ to

$$B(u_\Delta^{[p]}, v) = F(v) \;\; \forall v \in S_\Delta^{[p]} \tag{5.44}$$

exists and is unique up to a constant, and the error estimates of Theorems 5.2 and 5.3 hold as before. The free constant in $u_\Delta^{[p]}$ can be obtained by setting $u^{[p]}(0) = 0$. This means when we replace the Neumann boundary condition at $x = 0$ by a Dirichlet condition, so that the solution of the Dirichlet problem is $\hat{u}$, then $\mathfrak{u} \in \mathcal{U}_0$ and $\hat{u}_\Delta^{[p]} \in S_{\Delta,0}^{[p]}$. These solutions exist without (5.43) being satisfied, but nevertheless $\widehat{\mathfrak{u}} = \mathfrak{u}$ only if (5.43) holds.

Remark 5.10 *In practice, as noted in Chapter 3, we compute $\int_0^l fv dx$ in (5.43) using numerical integration. Thus, numerically (5.43) can only be satisfied up to the numerical integration error, which we assume to be small. For specific given data it is usually known in advance whether (5.43) is satisfied. In practice, this should always be verified, particularly as the computation is not costly!*

Exercise 5.7 Consider again problem (2.14) in which $I \equiv (0, 1)$ and $\mathfrak{u}(x)$ satisfies

$$-\frac{d^2\mathfrak{u}}{dx^2}(x) = x^2, \quad x \in I, \tag{5.45a}$$

$$\mathfrak{u}(0) = 0, \tag{5.45b}$$

$$\frac{d\mathfrak{u}}{dx}(1) = 1. \tag{5.45c}$$

(1) Derive the exact solution $\mathfrak{u}(x)$ to (5.45a).
(2) Write out the finite element formulation of the problem (5.45) as in (3.12) with $u_\Delta^{[1]} \in S_\Delta^{[1]}(I)$, for any mesh partition Δ of I.
(3) For a uniform mesh of size h and with linear elements derive the constrained global stiffness equation system and constrained load vector arising from (2).
(4) Observe that this system is in fact a system of finite difference equations but with different right-hand sides that can be solved analytically for the nodal values for both finite differences and finite elements.
(5) Derive the solution of this equation system.
(6) Compare the finite element nodal values (of $u_\Delta^{[1]}$) with the exact values of $\mathfrak{u}(x)$ at the nodal points and derive analytically the formula of the error as a function of h; what do you observe about the error at the nodal points?
(7) Compute analytically the error of the finite element solution in the energy norm as a function of h.

(8) Use Theorem 5.3 and compare the estimate $\left\|e_\Delta^{[p]}\right\|_{\mathcal{U}}$ with the true error in the energy norm.

(9) Write down the analytical formular for the strain energy $B(\mathfrak{u},\mathfrak{u})$ of the exact solution $\mathfrak{u}$ and the strain energy of the finite element solution $u_\Delta^{[1]}$. Compare these energies and relate their difference to the error in the finite element solution.

(10) Write down the formulae for the total energy Π (see (2.42)) for both the exact solution $\mathfrak{u}$ and the finite element solution $u_\Delta^{[1]}$. Compare these energies and relate them to the strain energies in (1). Relate the error in the total energy of the finite element solution to the error in the finite element solution.

5.1.2 Error analysis for the one-dimensional engineering problem of Section 3.3.5

The performance of the finite element method for computing the quantities of interest $Q_1^\Delta, Q_2^\Delta, Q_3^\Delta$ for the one-dimensional heat-transfer problems of Section 2.1.5 was demonstrated in Section 3.3.5. In the present section, we now analyze those numerical results in the light of the theory of the present chapter. Because the error in total energy is directly related to the error $e_\Delta^{[p]}$, we consider it most appropriate to analyze the accuracy of the quantity of interest Q_3^Δ for the two problems. Specifically, we can see this from the following.

For $\Pi(u)$ defined as in (2.42) we saw in Exercise 3.12 that

$$Q_3^\Delta - Q_3 = \Pi\left(u_\Delta^{[p]}\right) - \Pi(\mathfrak{u}) = \frac{1}{2}\left\|e_\Delta^{[p]}\right\|_{\mathcal{U}}^2. \tag{5.46}$$

Hence, the relative error $rel\left(Q_3^\Delta\right)$ in Q_3^Δ may be defined as

$$rel\left(Q_3^\Delta\right) \equiv \left|Q_3^\Delta - Q_3\right| / \left|Q_3\right| = \left\|e_\Delta^{[p]}\right\|^2 / 2\left|Q_3\right|, \tag{5.47}$$

and can be analyzed by considering $e_\Delta^{[p]}$. Values for $\mathrm{rel}\left(Q_3^\Delta\right)$ were given in Figs. 3.14 and 3.15 for 1D Eng Problems 1 and 2.

We consider first 1D Eng Problem 1 with a uniform mesh such that for some M the interface coincides with a vertex and refer to Tables 3.1 and 3.2. As was said in Section 3.3.5 the error does not decrease monotonically with h, and for certain values of M, namely when the vertex coincides with the interface, the accuracy is clearly superior to that seen generally. We first study this case and note that the solution $\mathfrak{u}$ is such that $\mathfrak{u} \in C^3[3, 3.5]$ and $\mathfrak{u} \in C^3[3.5, 6.5]$, but $\mathfrak{u} \notin C^3[3, 6.5]$. Now, by applying (5.14a), since $\mathfrak{u} \in C^3(\tau_q)$ for all $\tau_q \in \Delta$,

$$\left\|e_\Delta^{[1]}\right\|_{\mathcal{U}} \leqq Ch, \tag{5.48a}$$

$$\left\|e_\Delta^{[2]}\right\|_{\mathcal{U}} \leqq Ch^2, \tag{5.48b}$$

$$\left|Q_3^\Delta - Q_3\right| \leqq Ch^{2p}, p = 1, 2. \tag{5.48c}$$

In Fig. 3.14 using (5.47) we see that for ln(rel) the lower envelopes of the error are almost exactly on the straight lines with slopes $2p, p = 1, 2$.

We next consider values of M where the interface does not coincide with a vertex. This occurs when $M = 7k + s, k = 1, 2, \ldots, s = 1, \ldots 6$. Then, $h = 3.5 / (7k + s)$ and the interface is located in the element $\tau_q = (x_q, x_{q-1})$ where

$$x_q = \frac{3 + 3.5k}{7k + s},\ x_{q-1} = \frac{3 + 3.5(k+1)}{7k + s};$$

the relative position of the interface in the element is governed by the s.

The solution $\mathfrak{u}$ in the element τ_q can be written in the form $\mathfrak{u} = \mathfrak{u}_1 + \mathfrak{u}_2$, where $\mathfrak{u}_1 \in C^3(\tau_q)$ and $\mathfrak{u}_2 \in C(\tau_q)$ is piecewise linear. The jump $\mathcal{J}$ of the derivative of $\mathfrak{u}$ occurs at the interface $x = b$, where we have

$$\gamma_{\text{ss}} \frac{d\mathfrak{u}}{dx}(b - 0) = \gamma_{\text{cs}} \frac{d\mathfrak{u}}{dx}(b + 0) = \rho,$$

and hence that

$$\mathcal{J} = \frac{d\mathfrak{u}}{dx}(b + 0) - \frac{d\mathfrak{u}}{dx}(b - 0) = \rho \left(\frac{1}{\gamma_{\text{cs}}} - \frac{1}{\gamma_{\text{ss}}} \right).$$

Now, using (4.24b)

$$\left\| \mathfrak{u}_1 - \mathcal{L}^{[p]}(\tau_q)\, \mathfrak{u}_2 \right\|_{\mathcal{U}(\tau_q)} \leqq C h^{p+1}, \tag{5.49a}$$

and, by simple direct computations, we obtain

$$\left\| \mathfrak{u}_2 - \mathcal{L}^{[p]}(\tau_q)\, \mathfrak{u}_2 \right\|_{\mathcal{U}(\tau_q)} \approx C(s) h^{\frac{1}{2}}. \tag{5.49b}$$

Hence, we have

$$\left\| \mathfrak{u}_1 - \mathcal{L}^{[p]}(\tau_q)\, \mathfrak{u} \right\|_{\mathcal{U}(\tau_q)} \approx C(s) h^{\frac{1}{2}}, \tag{5.49c}$$

as (5.49a) is negligible by comparison to (5.49b), and further, from (5.46) and (5.48a and b) that

$$\left| Q_3^{\Delta} - Q_3 \right| \leqq C h^2 + C(s) h \approx C(s) h,\ p = 1, \tag{5.50a}$$

$$\left| Q_3^{\Delta} - Q_3 \right| \leqq C h^4 + C(s) h \approx C(s) h,\ p = 2. \tag{5.50b}$$

Considering again Fig. 3.14 we see exactly this behaviour. For every s (i.e. the same relative position of the interface in the element) the upper envelope of the errors lies on straight lines with slope 1 for both linear and quadratic elements.

Moral 1: It is not in general possible to say that as h decreases the errors lie on a straight line!

Moral 2: The interface should coincide with an element vertex.

Moral 3: Irrespective of the degree of the elements, if the problem has a non-smooth solution in some elements, then the rate of convergence is governed by the non-smoothness irrespective of p.

We next consider 1D Eng Problem 2, which differs from 1D Eng Problem 1 in that there are no values of M for which the interface coincides with a nodal point, and the position of the interface in an element is essentially random. Considering Fig. 3.15 we see the theory illustrated. There is now no straight-line lower-bound envelope, as in the previous case, as the behaviour is essentially chaotic, but the upper $|Q_3^\Delta - Q_3| \leqq Ch$ bound holds.

5.2 Two-dimensional problems

5.2.1 Error of the finite element solution in two dimensions

We now move to two-dimensional boundary value problems in domains with polygonal boundaries Γ and consider again problem (2.63) with homogeneous Dirichlet boundary conditions on Γ_{D} so that $g(\mathbf{x}) = 0, \mathbf{x} \in \Gamma_{\mathrm{D}}$. Assumptions (2.73) and (2.74) on $a_{ij}(\mathbf{x}), c(\mathbf{x}), \alpha(\mathbf{x}), f(\mathbf{x})$ and $G(\mathbf{x})$ are again assumed to hold. With $\mathcal{U}_0$ as in (2.69) the weak form of (2.63) with $g(\mathbf{x}) = 0$ on Γ_{D} is:

$$\text{find } \mathrm{u} \in \mathcal{U}_0 \text{ such that}$$

$$B(\mathrm{u}, v) = F(v) \quad \forall v \in \mathcal{U}_0, \tag{5.51}$$

where, as in (2.66) and (2.67),

$$B(\mathrm{u}, v) \equiv \int_\Omega \left(\sum_{i,j=1}^{2} a_{ij}(\mathbf{x}) \frac{\partial \mathrm{u}}{\partial x_i}(\mathbf{x}) \frac{\partial v}{dx_j}(\mathbf{x}) + c(\mathbf{x})\mathrm{u}(\mathbf{x})v(\mathbf{x}) \right) d\mathbf{x} + \int_{\Gamma_N} \alpha(\mathbf{x})\, \mathrm{u}(\mathbf{x})\, v(\mathbf{x})\, ds, \tag{5.52}$$

and

$$F(v) \equiv \int_\Omega f(\mathbf{x})\, v(\mathbf{x})\, d\mathbf{x} + \int_{\Gamma_N} G(\mathbf{x})\, v(\mathbf{x})\, ds. \tag{5.53}$$

Once again, Ω is a polygonal domain with boundary $\Gamma = \cup \Gamma_i$ as in Section 2.2.1.

As in Section 3.4.1 we define a mesh of triangular elements τ_q over Ω in which the uniform minimum angle conditions (see Remark 4.10) are satisfied. Again, as in one dimension, we consider quasiuniform meshes, in this case triangles, for which there exist constants $\nu > 0$ such that

$$\max_{\tau_q \in \Delta} h(\tau_q) \,/\, \min_{\tau_q \in \Delta} h(\tau_q) \leqq \nu. \tag{5.54}$$

We have for two dimensions the piecewise polynomial spaces $S_\Delta^{[p]}$ and $S_{\Delta,0}^{[p]}$, $p = 1, 2$, as in (3.59) and (3.73). The finite element problem for (5.51) is:

$$\text{find } u_\Delta^{[p]} \in S_{\Delta,0}^{[p]} \text{ such that}$$

$$B(u_\Delta^{[p]}, v) = F(v) \quad \forall v \in S_{\Delta,0}^{[p]}. \tag{5.55}$$

Denoting the error

$$e_{\Delta}^{[p]} \equiv \mathfrak{u} - u_{\Delta}^{[p]},\ p = 1, 2, \tag{5.56}$$

for this two-dimensional context we again have inequalities (5.11) and (5.13), where $\mathcal{L}_{\Delta}^{[p]}(\mathfrak{u})$ is now the piecewise pth-order polynomial interpolant to $\mathfrak{u}(\mathbf{x})$ over Δ, and $\mathcal{L}_{\Delta}^{[p]}(\mathfrak{u}) \in S_{\Delta,0}^{[p]}$.

Exercise 5.8 Show that $\mathcal{L}_{\Delta}^{[p]}(\mathfrak{u}) \in S_{\Delta,0}^{[p]}$.

Let us now distinguish two different cases for the solution $\mathfrak{u}$ of (5.51):

(1) $\mathfrak{u} \in C^k(\Omega)$;
(2) $\mathfrak{u}$ satisfies (2.89).

Note that case (1) is only of theoretical interest because, in practice, this level of continuity of $\mathfrak{u}$ will seldom occur.

Theorem 5.4 *Let $u \in C^k(\Omega)$, $k = 1, 2, 3$ then*

$$\left\| e_{\Delta}^{[p]} \right\|_{\mathcal{U}}^2 \leqq C \left(\sum_{\tau_q \in \Delta} \left(h(\tau_q)^{2\min(k-1,p)+2} + h(\tau_q)^{2\min(k,p+1)+2} \right) \right) |\mathfrak{u}|^2_{C^{\min(k,p+1)}(\tau_q)}, \tag{5.57}$$

where C is independent of the mesh Δ, but depends on the mimumum angle of the mesh.

Proof The proof is completely analogous to that of one dimension. The only difference is that in one dimension we had $h(\tau)^{2\min(k-1,p)+1}$, whereas now we have $h(\tau)^{2\min(k-1,p)+2}$, and in one dimension we had C/h elements, whereas now we have C/h^2. Note that instead of $|\mathfrak{u}|_{C^{\min(k,p+1)}}$ we could have $\|\mathfrak{u}\|_{C^k}$. □

Exercise 5.9 Prove Theorem 5.4.

We can simplify Theorem 5.4 if we use $h =^{\max}_{\tau \in \Delta} h(\tau)$.

Theorem 5.5 *Let $\mathfrak{u} \in C^k(\Omega), k = 1, 2, 3$, and let Δ be quasiuniform. Then,*

$$\left\| e_{\Delta}^{[p]} \right\|_{\mathcal{U}} \leqq \left\| \mathfrak{u} - \mathcal{L}_{\Delta}^{[p]}(\mathfrak{u}) \right\|_{\mathcal{U}} \leqq C\, h^{\min(k-1,p)} \|\mathfrak{u}\|_{C^k(\Omega)}, \tag{5.58}$$

where the constant C is independent of the mesh Δ, but depends on the minimum angle of the mesh.

When the mesh is quasiuniform, the right-hand side of (5.57) has all summands of the same order; which indicates that the quasiuniform mesh is effective. Note that for non-quasiuniform meshes Theorem 5.5 can be very pessimistic.

Exercise 5.10 Prove Theorem 5.5

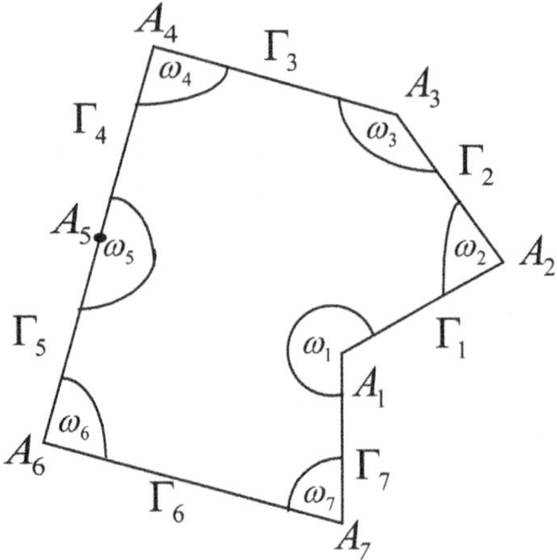

Fig. 5.4 Example of polygonal domain and boundary conditions, $Q = 7$.

We now address case (2), where u satisfies (2.89). Let us consider the specific case of problem (5.51) defined in the polygonal domain of Fig. 5.4, with vertices A_j, $j = 1, 2, \ldots, 7$, neighbourhoods ω_j of the vertices, and boundary conditions:

$$\begin{aligned} u(\mathbf{x}) &= 0, \ \mathbf{x} \in \Gamma_1 \cup \Gamma_4 \cup \Gamma_6 \cup \Gamma_7, \\ \frac{\partial u}{\partial n_c}(\mathbf{x}) &= G, \ \mathbf{x} \in \Gamma_2 \cup \Gamma_3 \cup \Gamma_5. \end{aligned}$$

Note that the vertex A_5 is present because the boundary condition changes at that point.

Following Section 2.2.2 we define $\hat{\Omega} = \Omega - \cup_{j=1}^{7} \bar{\omega}_j$ and recall from (2.89) that, since we are only taking $g(\mathbf{x}) = 0, x \in \Gamma_{\mathrm{D}}$, and $\alpha(\mathbf{x}) = 0, x \in \Gamma_{\mathrm{N}}$,

$$\|u\|_{C^k(\hat{\Omega})} \leqq C \left(\|f\|_{C^{k-1}(\Omega)} + \sum_{\Gamma_j \in \Gamma_N} \|G\|_{C^{k-1}(\Gamma_j)} \right) \leqq CQ, 1 \leqq l \leqq k, \tag{5.59a}$$

and in the neighbourhoods ω_j we have $u = u_1 + u_2$, with

$$|u_1|_{C^l(\omega_j)} \leqq CQ, \left| \frac{\partial^l u_2}{\partial x_1^{l_1} \partial x_2^{l_2}} \right| \leqq C(\varepsilon) r_j^{\beta_j - \varepsilon - l}. \tag{5.59b}$$

We now have the following theorem:

Theorem 5.6 *Let* $u(\mathbf{x})$ *be the solution of (5.51) with* Ω *and the boundary conditions of Fig. 5.4, let* $u(\mathbf{x})$ *satisfy (5.59a and b), and assume that* Δ *is quasiuniform, then*

$$\left\|e_{\Delta}^{[p]}\right\|_{\mathcal{U}} \leqq \left\|u - \mathcal{L}_{\Delta}^{[p]}(u)\right\|_{\mathcal{U}} \leqq C_1(\varepsilon)h^{\min(\beta-\varepsilon,k-1,p)} + C_2h^{\min(k-1,p)},$$

$$\text{where } \beta = \min \beta_j,\ j = 1,\ldots,7. \tag{5.60}$$

Proof For the quasiuniform mesh Δ we divide the elements $\tau_q \in \Delta$ into three groups $T_i,\ i = 1,2,3$, where, see Fig. 5.4,

$T_1 \equiv \left\{\tau_q \in \Delta \mid \tau_q \in \hat{\Omega}, \hat{\Omega} \equiv \Omega - \cup\bar{\omega}_j\right\};$

$T_2 \equiv \{\tau_q \in \Delta \mid \bar{\tau}_q \cap (\cup A_j) \neq 0,\ j = 1,\ldots,7\};$

$T_3 \equiv \{\tau_q \in \Delta \mid \tau_q \notin T_1, \tau_q \notin T_2\}.$

There are not more than $\mathcal{C}/h^2$ elements in each of groups T_1 and T_3. Group T_2 has only a finite number of elements, independent of h, and these are called *key elements.* We shall now estimate the errors separately for the individual groups.

Group T_1 : For elements $\tau_q \in T_1$ the solution u is smooth, so that $\|u\|_{C^k(\tau_q)} \leqq Q$ and hence as in Theorem 5.5 we have

$$\sum_{\tau_q\in T_1} \left\|u - \mathcal{L}_{\Delta}^{[p]}u\right\|_{\mathcal{U}(\tau_q)}^2 \leqq \mathcal{C}h^{2\min(k-1,p)}Q. \tag{5.61}$$

Elements of groups T_2 and T_3 are in ω_j, where $u = u_1 + u_2$ and we have

$$\sum_{\tau_q\in T_2\cup T_3} \left\|u_1 - \mathcal{L}_{\Delta}^{[p]}u_1\right\|_{\mathcal{U}(\tau_q)}^2 \leqq \mathcal{C}h^{2\min(k-1,p)}Q.$$

Group T_2 : For elements $\tau_q \in T_2$, with $\tau_q \subset \omega_j$, we have

$$\left\|u_2 - \mathcal{L}_{\Delta}^{[p]}u_2\right\|_{\mathcal{U}(\tau_q)}^2 \leqq \mathcal{C}h^{2(\beta_j-\varepsilon)}, \tau_q \in T_2,$$

because of (2.89c) and hence, because τ_q has only a finite number of elements, we have

$$\sum_{\tau_q\in T_2} \left\|u_2 - \mathcal{L}_{\Delta}^{[p]}u_2\right\|_{\mathcal{U}(\tau_q)}^2 \leqq \mathcal{C}h^{2\min(\beta_j-\varepsilon)}. \tag{5.62}$$

Group T_3. For the jth boundary vertex A_j we construct the layer $A^{j,i} = \{x \in \Omega | \gamma_{i+1} = ih > \| x - A_j \| > (i-1)h = \gamma_i\}, i = 1,2,..,l$ as in Fig. 5.5, where l is such that $\omega_j \subset A^{j,l}$ and we define the groups of elements $T_3^{j,i} = \{\tau_q | \tau_q \cap A^{j,i} \neq 0\}$

(Note that the i starts at 1 even though the $T_3^{j,i}$ include $\tau_q \in T_2$.) The value C_0 is selected so that the elements cover the entire domain ω_j. Because the elements satisfy the minimum angle condition, we have that in each $T_3^{j,i}$ there are at most $\mathcal{C}i$ elements, where $\mathcal{C}$ is independent of l. Further, we have that $l \leqq \mathcal{C}/h$.

In order to estimate the interpolation error in each layer we use (2.79) in order to bound $\|u\|_{C^k(\tau_q)}$ and then use the estimate of Theorem 4.11. Summing the estimates over all the elements in all the layers in T_3 we have

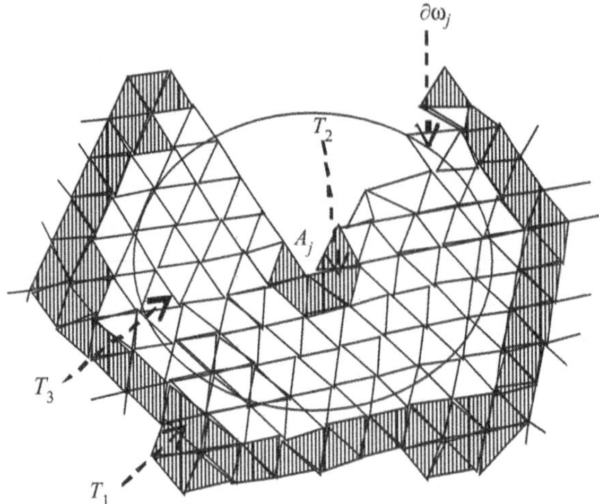

Fig. 5.5 Detail of neighbourhood ω_j of vertex A_j, with different groups of elements.

$$\sum_{\tau_q \in T_2} \left\| \mathfrak{u}_2 - \mathcal{L}_{\Delta}^{[p]} \mathfrak{u}_2 \right\|_{\mathcal{U}(\tau_q)}^2 \leqq C h_j^{\min 2(\beta_j - \varepsilon, k-1, p)}. \tag{5.63}$$

Combining the errors (5.61), (5.62) and (5.63) for the three groups T_1, T_2, T_3 we obtain the result of the theorem. □

We see from Theorem 5.6 that the sizes of the angles at the vertices A_i of the boundary Γ of Ω (which of course determine whether singularities are present) determine the rate of convergence of $\mathcal{L}_{\Delta}^{[p]}(\mathfrak{u})$. We also see from the proof that the mesh is not equilibrated, because the interpolation errors differ markedly between the groups T_1, T_2 and T_3. This means that the quasiuniform mesh does not perform well, and hence there is a need for the mesh to be refined in the neighbourhoods of the vertices. In this case the rate of convergence has to be expressed in terms of the number of elements M of the mesh.

Clearly, for a quasiuniform mesh we have that $M \approx h^{-2}$. Hence, if $\mathfrak{u} \in C^k(\Omega)$ then, from Theorem 5.6,

$$\left\| e_{\Delta}^{[p]} \right\|_{\mathcal{U}} \leqq C\, M^{-\frac{1}{2}\min(\beta - \varepsilon, k-1, p)}. \tag{5.64}$$

Hence, the question arises as to whether in terms of M we can obtain the same rate of convergence for the general case when $\mathfrak{u}$ satisfies (2.89) with a properly refined mesh. This we now address.

To do this we consider only the mesh in the neighbourhood of the vertex which we assume is located at the origin. In the proof of Theorem 5.6, the layers of elements in ω_j all had thicknesses h. We shall now use layers of differing thicknesses, and in

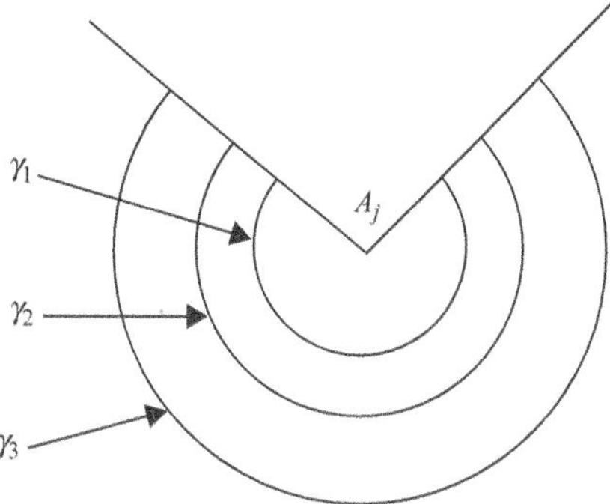

Fig. 5.6 Radically graded layers at A_j with radii $\gamma_s = (sh)^\gamma$.

particular will grade them in the manner of the radical mesh of one dimension. In order to do this at the vertex A_j in ω_j we define the layers using concentric circles with radii γ_s centred on the vertex, where $\gamma_s = (sh)^\gamma, \gamma > 1, s = 1, \ldots, c/h$, see Fig. 5.6.

Let the sth layer be

$$\Phi s = \left\{ \mathbf{x} \in \omega_j \mid \gamma_s \leqq (x_1^2 + x_2^2)^{\frac{1}{2}} \leqq \gamma_{s+1} \right\}, \; s = 1, 2, \ldots, c/h,$$

and note that

$$\Phi(0) = \{\mathbf{x} \mid |\mathbf{x}| < \gamma_1\}.$$

Now, we shall assume that the mesh Δ is such that if $\tau_q \cap \Phi(s) \neq 0$, then for $s = 1, 2, \ldots$

(1) $D_1\gamma_s \leqq \operatorname{diam} \tau_q \leqq D_2\gamma_s$;
(2) $\operatorname{dist}(\tau_q) \geqq D_3\gamma_s$,;
(3) for $\tau_q \cap \Phi(0) \neq 0, D_4\gamma_1 \leqq \operatorname{diam} \tau_q \leqq D_2\gamma_1$,

where $\operatorname{dist}(\tau_q)$ is the distance of τ_q from the vertex A_j and the D_i are independent of Δ; and of course the minimum angle condition is satisfied.

Such meshes are called *radical meshes* with index γ. We now state a theorem for such meshes.

Theorem 5.7 *Let* u *satisfy the conditions of Theorem 5.7, then there exists a family of meshes Δ such that*

$$\left\| e_\Delta^{[p]} \right\|_{\mathcal{U}} \leqq CM^{-\frac{1}{2}\min(k-1,p)}Q, \tag{5.65}$$

where C is independent of Δ and M is the number of degrees of freedom.

Proof The proof of this theorem is analogous to that of Theorem 5.6. □

Remark 5.11 *We see that we have selected γ so that for the majority of the elements the errors in each were of approximately the same magnitude; i.e. the mesh is nearly equilibrated.*

In a manner similar to the one-dimensional case, a '*good*' mesh will lead to equilibration of the error over the elements. The task of generating a suitable mesh in two dimensions is not so simple to implement, and one has to rely on available mesh generators.

In Example 5.5 we discussed the effect of pollution, which was caused by very non-uniform meshes. In two dimensions the causes of pollution are more complicated. These may be the presence of a very non-uniform mesh or, even with a quasiuniform mesh, the singular behaviour of the solution in the neighbourhoods of boundary vertices. The detailed analysis of these effects is beyond the scope of this book. Nevertheless, we illustrate these effects on our two benchmark problems.

5.2.2 Error analysis for Benchmark Problems 1 and 2

Benchmark Problems 1 and 2 were introduced in Section 3.4.3, where finite element results for each problem were presented. We now discuss those results in the light of the theory of this chapter.

Benchmark Problem 1

For this problem the solution $u \in C^3(\Omega)$, and the finite element results were produced with the meshes of Fig. 3.22 for which all the elements satisfy the minimum angle condition. Hence, using Theorem 5.5 we have that

$$\left\| e_{\Delta}^{[p]} \right\|_{\mathcal{U}} \leqq \left\| \mathfrak{u} - \mathcal{L}_{\Delta}^{[p]} \mathfrak{u} \right\|_{\mathcal{U}} \leqq C h^{p} \left\| \mathfrak{u} \right\|_{C^3}, \tag{5.66}$$

where the constant C depends on the character of the mesh, but the rate of convergence does not. This behaviour is exactly illustrated in Fig. 3.23. We shall recall that the error can also be expressed as a function of $M \approx 1/h^2$, the number of elements, so that

$$\left\| e_{\Delta}^{[1]} \right\|_{\mathcal{U}} \leqq C M^{-\frac{1}{2}}, \left\| e_{\Delta}^{[1]} \right\|_{\mathcal{U}} \leqq C M^{-1}.$$

From Fig. 3.23 we see that M is twice as large for the criss-cross mesh as for the other patterns, so as is expected the error with this mesh is the smallest of the four. If the errors were recomputed in terms of M, the constants C for all four patterns would be essentially the same. For every mesh there is a specific constant but the rate of convergence is independent of the pattern of the mesh.

Benchmark Problem 2

This problem in the L-shaped domain of Fig. 3.24 has a singularity at A so that as stated $\beta = 1/3$ and the regularity of $\mathfrak{u}$ is as characterized in (2.89). Hence, $\mathfrak{u} \in C^3(\hat{\Omega})$ and on ω_A we have $\mathfrak{u} = \mathfrak{u}_1 + \mathfrak{u}_2$, where $\mathfrak{u}_1 \in C^3(\Omega)$ and $\mathfrak{u}_2$ satisfies (2.89c). Hence, using Theorems (4.13), (4.14) and (3.7) we have, for arbitrary ε,

$$\left\| e_{\Delta}^{[1]} \right\|_{\mathcal{U}} \leqq C\left(h + h^{1/3-\varepsilon}\right) \approx Ch^{1/3-\varepsilon},$$

$$\left\| e_{\Delta}^{[2]} \right\|_{\mathcal{U}} \leqq C\left(h^2 + h^{1/3-\varepsilon}\right) \approx Ch^{1/3-\varepsilon}.$$

The values of rel, and hence of $\left\| e_{\Delta}^{[1]} \right\|_{\mathcal{U}}$, are shown as a function of h in Fig. 3.26, and once again we see that there is excellent agreement with the theoretical rate of convergence independent of p. Of course we have that $\left\| e_{\Delta}^{[2]} \right\|_{\mathcal{U}} < \left\| e_{\Delta}^{[1]} \right\|_{\mathcal{U}}$ so that, as illustrated, the constant C is smaller for $p = 2$ than for $p = 1$.

Analogously to Benchmark Problem 1, we could evaluate the estimates in terms of M.

Moral 1: If a mesh is uniform and the minimal angle condition is uniformly satisfied, then accuracy is governed by element size but is virtually independent of the mesh structure.

Moral 2: For problems in polygonal domains and with uniform meshes, accuracy is governed by the behaviour of the solutions near the boundary corners. If boundary singularaties occur, characterized by the coefficient β, then the rate of convergence is the same for $p = 1$ and 2 for $\beta < 1$.

Moral 3: For two-dimensional problems in polygonal domains the effects of boundary singularaties cannot be avoided when using uniform meshes. They can, however, be mitigated using suitable local mesh refinement.

Moral 4: The behaviour of finite element solutions for one- and two-dimensional problems is analogous with respect to interfaces in the sense that when in two dimensions the interfaces are smooth, the effects of the interfaces can be avoided by using meshes that coincide with the interfaces.

5.2.3 Error analysis for the two-dimensional heat-transfer problem; 2D Eng Problem

We consider again the 2D Eng Problem as defined in Section 2.2.3 consisting of the pipe with fins, see Fig. 2.13, which we solved after (2.90) and the region of Fig. 2.14 were mapped onto the polygonal region of Fig. 2.15. The differential equations and the boundary conditions in the transformed domain are as given in (2.91). The coefficients are smooth in the subdomains occupied by the stainless and carbon steels, and the data in the boundary conditions are also smooth.

The solution regularity was addressed in Section 2.2.2. Because the domain (see Figs. 2.15 and 3.27) has corners, the leading singularity in the solution is located at the point (5) and depends on the internal angle $\omega_5 = 1.4916\pi$. The singularity is characterized by the coefficient β_5 given in the Table 2.3, which has the value $\beta_5 = 0.6742$.

We will analyze the numerical results for the computation of Q_3, the total energy of the problem, see Section 2.23. The accuracy, as in the one-dimensional case (see (5.47)), relates to the error $e_{\Delta}^{[p]}, p = 1, 2$ and its energy norm. The meshes, which are quasiuniform, (see Fig. 3.27) and calculations are as in Section 3.4.4. Note that the meshes coincide with the interface between the stainless and carbon steels. Hence, the accuracy of the solution is not influenced by the jump of the coefficients along the interface.

We will characterize the accuracy of the calculations in terms of the degrees of freedom N. Using Theorem 5.6, and (5.60), we obtain $\text{rel}\left(Q_3^{\Delta}\right) = C \left\| e_{\Delta}^{[p]} \right\|^2 \leq CN^{-\beta_5}$ because $N \simeq h^{-2}$.

The computed values Q_3^{Δ} of Q_3 were given in Tables 3.18 and 3.19 for all mesh patterns and refinement levels for $p = 1$ and 2, whilst the 'exact' value is computed by extrapolation from the last three refinements, as explained in Chapter 3. Comparison on the extrapolated values for different mesh patterns and values of p indicates that they are accurate.

The convergence rate is close to the theoretical rate β_5, for both $p = 1$ and 2, but is slightly closer for $p = 1$. This is because we are not fully in the asymptotic range. In Fig. 3.36 we show graphically the accuracy for $p = 1$ on a ln/ln scale, and this graph takes the form of a straight line with slope 2/3. We see that the errors lie very accurately on this line and the local rate of convergence is slightly higher for low mesh density; this indicates that as $N \to \infty$ we will see the theoretical rate accurately. Results for $p = 2$ are given in Fig. 3.37 and the (asymptotic) rates of the convergence are the same for $p = 1$ and 2.

As in the one-dimensional case we conclude:

Moral 1: The numerical results agree very well with the theoretical prediction.

Moral 2: The accuracy of the numerical solution using quasiuniform finite element meshes is governed by the singularity β_5, and because $\beta_5 < 1$ the rate of convergence is independent of p. However, the constant C in front of $N^{-\beta_6}$ is slightly smaller for $p = 2$; i.e. the accuracy for $p = 2$ is better than for $p = 1$. In order to get the rate of convergence that is governed by p the mesh has to be properly refined in the neighbourhood of the corner.

Moral 3: The accuracy of the numerical solutions is essentially not governed by the mesh pattern.

Moral 4: The mesh has to be (and is) aligned with the interface between the two steels, or the convergence rate with drop to $N^{-\frac{1}{2}}$.

Moral 5: The accuracy of the solution measured in the energy norm is very acceptable for practical purposes and for the meshes with low refinement.

Before this accuracy could be accepted for any design decision, one would have to be sure of this error. This can be done by *a posteriori* error estimation, as discussed in Chapter 7, or by comparing the results for different refinements.

6

Functionals and superconvergence

Summary

- **The error in the energy norm of the finite element solution to a problem is often not the quantity of main interest. In this chapter finite element errors are estimated in terms of functionals.**
- **Points in an element where the derivatives of the finite element solution have greater accuracy than that found generally in the element are identified.**
- **Interpolation to these derivative values at the points of *superconvergence* is used to produce functions that are *superconvergent* over the whole element.**

In Chapter 5 we considered the error in the finite element method when it was measured in the energy norm. Recalling Remark 2.1 of Chapter 2, we note that, in practice, the quantity of main interest in a problem may not be the solution itself, but certain functionals of the solution; for example the values of the solution at specific points, or in linear elasticity the stress at a specific point, or the average of the displacement or the stresses over a subdomain of the region of the problem. All these are examples of functionals of the solutions. Correspondingly, we need to estimate the finite element errors in terms of such functionals, and we shall see that these computed values can be smaller than we might at first sight expect. For example, we may be interested in the error of the finite element solutions at points at which it is smaller than that found generally. We thus need to locate such points.

6.1 One-dimensional problems

6.1.1 Error in the functionals in one dimension

Let us consider our model problem (2.14 a–c), its weak formulation as in (2.19), and the finite element problem (3.18) that has solution $u_\Delta^{[p]} \in S_{\Delta,g_1}^{[p]}$, $p = 1, 2$, for the mesh defined in (3.17b). Let Φ be a linear functional defined on $\mathcal{U}_0$; i.e. Φ satisfies (2.31). We are interested in finding the value $\Phi(\mathrm{u})$, and the simplest way to obtain the approximate value is to compute $\Phi(u_\Delta^{[p]})$. The error will then be

$$e_\Phi^{[p]} = \Phi(\mathrm{u}) - \Phi(u_\Delta^{[p]}) = \Phi(e_\Delta^{[p]}), \tag{6.1}$$

because Φ is linear.

We now want to estimate the error $\left|e_\Phi^{[p]}\right|$, and for this we have the following theorem:

Theorem 6.1 *If* $\mathfrak{u} \in \mathcal{U}_{g_1}$ *is the solution of (3.2),* $u_\Delta^{[p]} \in S_{\Delta,g_1}^{[p]}$ *is the solution of the finite element problem (3.18),* $e_\Delta^{[p]} \equiv \mathfrak{u} - u_\Delta^{[p]}$, $p = 1, 2$ *and* Φ *is a bounded linear functional on* $\mathcal{U}_0$*, then*

$$\left|e_\Phi^{[p]}\right| \leqq \|\Psi_\Phi - w\|_{\mathcal{U}} \left\|e_\Delta^{[p]}\right\|_{\mathcal{U}}, \tag{6.2}$$

for any function $w \in S_{\Delta,0}^{[p]}$*, where* $\Psi_\Phi \in \mathcal{U}_0$ *satisfies*

$$B(\Psi_\Phi, v) = \Phi(v), \quad v \in \mathcal{U}_0. \tag{6.3a}$$

If, further, $\Psi_{\Phi,\Delta}^{[p]} \in S_{\Delta,0}^{[p]}$ *is the solution of the finite element problem*

$$B\left(\Psi_{\Phi,\Delta}^{[p]}, v\right) = \Phi(v) \ \forall v \in S_{\Delta,0}^{[p]}, \tag{6.3b}$$

then, defining $e_{\Phi,\Delta}^{[p]} \equiv \Psi_\Phi - \Psi_{\Phi,\Delta}^{[p]}$*, we find that* $\left|e_\Phi^{[p]}\right| \leqq \left\|e_{\Phi,\Delta}^{[p]}\right\|_{\mathcal{U}} \left\|e_\Delta^{[p]}\right\|_{\mathcal{U}}$.

Proof Because Φ is a bounded linear functional on U_0, Ψ_Φ the solution of (6.3a) exists and

$$e_\Phi^{[p]} = \Phi(\mathfrak{u}) - \Phi\left(u_\Delta^{[p]}\right) = B\left(\Psi_\Phi, \mathfrak{u} - u_\Delta^{[p]}\right). \tag{6.4}$$

Because of the Galerkin orthogonality condition, and symmetry where $B(u, v) = B(v, u)$, we have that

$$e_\Phi^{[p]} = B\left(\Psi_\Phi, e_\Delta^{[p]}\right) = B\left(\Psi_\Phi - v, e_\Delta^{[p]}\right),$$

from which result (6.2) follows on using the Schwarz inequality. Selecting $v = \Psi_{\Phi,\Delta}^{[p]}$ we obtain (6.3c). □

If we now apply Theorem 5.3 we obtain the result of Theorem 6.2.

Theorem 6.2 *If, in addition to the conditions of Theorem 6.1, we have that* Δ *is a quasiuniform mesh, then*

$$\left|e_\Phi^{[p]}\right| \leqq C\, h^{y_1+y_2} \left|\mathfrak{u}\right|_{C^{\min(k_1,p+1)}(I)} \left|\Psi_\Phi\right|_{C^{\min(k_2,p+1)}(I)}, \tag{6.5}$$

where $y_1 = \min(k_1 - 1, p)$, $y_2 = \min(k_2 - 1, p)$, *and* C *is independent of* h, u *and* Φ. *Note that* k_1 *and* k_2 *characterize the regularity of* $\mathfrak{u}$ *and* Ψ_Φ. *Usually,* $k_1 = k_2 \geqq p$ *and hence the functional error is of order* h^{2p}.

Theorem 6.2 shows that the functional error $\Phi(e_\Delta^{[p]})$ is of order h^{2p}, whereas the error in the energy norm of the finite element solution $u_\Delta^{[p]}$ is of order h^p; under suitable assumptions on the smoothness of $\mathfrak{u}$ and Ψ_Φ.

Example 6.1 Let us consider again the problem in which $\mathfrak{u}(x)$ satisfies

$$-\frac{d^2\mathfrak{u}}{dx^2}(x) = f(x), \quad x \in I = (0,l),$$
$$\mathfrak{u}(0) = 0, \tag{6.6}$$
$$\frac{d\mathfrak{u}}{dx}(l) = 0.$$

The finite element solution is $u_\Delta^{[1]}$, with Δ a quasiuniform partition of I, and $f \in C^0(I)$. Then, by (2.41), $\mathfrak{u} \in C^2(I)$ and from Theorem 5.3 we have that

$$\left\| e_\Delta^{[1]} \right\|_{\mathcal{U}} \leqq Ch \left|\mathfrak{u}\right|_{C^2(I)}.$$

Suppose now that we are interested in the functional $\Phi(\mathfrak{u}) \equiv \mathfrak{u}(x_o)$, $x_o \in I$. By (2.38) we have that $|\Phi(v)| \leqq \mathcal{C} \|v\|_{\mathcal{U}_0}$ $\forall v \in \mathcal{U}_0$, and hence there exists a function $\Psi_\Phi \in \mathcal{U}$ that satisfies

$$B(\Psi_\Phi, v) \equiv \int_I \frac{d\Psi_\Phi}{dx}\frac{dv}{dx}dx = v(x_0) \;\; \forall v \in \mathcal{U}_0. \tag{6.7}$$

It is straightforward to determine the function Ψ_Φ. Integrating by parts we see that

$$\frac{d\Psi_\Phi}{dx} = 0, \quad 0 < x < x_0 \text{ and } x_o < x < l,$$

and hence

$$\Psi_\Phi(x) = \begin{cases} ax & \text{for} \quad 0 \leqq x_0, \\ ax_0 & \text{for} \quad x_0 < x < l, \text{ because } \frac{d\Psi_\Phi}{dx}(l) = 0. \end{cases}$$

Selecting $a = 1$ we can easily verify that $B(\Psi_\Phi, v) = v(x_0)$. We distinguish between two cases:

(1) $x_0 = x_i^\Delta$ for some i; i.e. x_0 is a vertex;
(2) $x_0 \neq x_i^\Delta$ for any i; i.e. x_0 is not a vertex.

Assume first that $x_0 = x_i^\Delta$, so that $\Psi_\Phi \in S_\Delta^{[1]}$ and hence by Theorem 6.1 and (2.40) we have that $e_\Phi^{[p]} = 0$. This means that for problem (6.6) the finite element solution is exact at nodal points; note that this holds for any mesh (not necessarily quasiuniform).

Secondly, assume that $x_0 \neq x_i^\Delta$ and that $x_0 \in \left(x_q^\Delta, x_{q+1}^\Delta\right) = \tau_q$. Then, in τ_q the error in the linear interpolant $\mathcal{L}_\Delta^{[1]}(\Psi_\Phi)_{\tau_q}$ measured in the energy norm (see Fig. 6.1) is

$$\left\| \frac{d\Psi_\Phi}{dx} - \frac{d\mathcal{L}_\Delta^{[1]}(\Psi_\Phi)}{dx} \right\|^2_{L^2(I)}$$
$$\leqq \left(\frac{x_{q+1} - x_q}{h_q}\right)^2 (x_{q+1} - x_0) + \left(1 - \frac{x_{q+1} - x_q}{h_q}\right)^2 (x_0 - x_q),$$

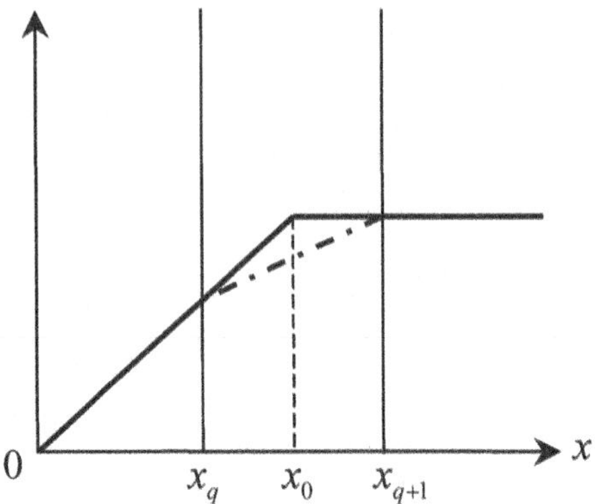

Fig. 6.1 The function Ψ_Φ, solid line, and its linear interpolant $\mathcal{L}_\Delta^{[1]}$, dotted line, in the element τ_q.

so that using Theorems 5.2 and 6.1 we obtain

$$\left|e_\Phi^{[p]}\right| \leqq Ch^{\frac{3}{2}}, \tag{6.8}$$

where C depends on x_0.

As the finite element solution $u_\Delta^{[1]}$ is exact at the nodal points, it is the linear interpolant to u. From the assumption that $f \in C^0(I)$ we have that $u \in C^2(I)$, and hence from standard interpolation theory, see Rektorys (1994), we have that $\left|e_\Phi^{[p]}\right| \leqq Ch^2$, which indicates that estimate (6.8) is pessimistic for the case that $f \in C^0(I)$.

In Example 6.1 we analyzed the case $p = 1$. Similar results hold for $p = 2$.

Exercise 6.1 Consider problem (2.17)–(2.19) with $a(x) = 1$, $c(x) = 0, \alpha_2 = 1, G_2 = 0$, and $f \in C^0[I]$. Show that $e_\Delta^{[p]}(x_q) = 0$, $q = 1, \ldots, M+1$, $p = 1, 2$.

Example 6.2 Let us consider the problem

$$-\frac{d}{dx}\left(a(x)\frac{du}{dx}(x)\right) + cu(x) = f(x), \quad x \in I = (0, l),$$
$$u(0) = 0, \tag{6.9}$$
$$\frac{du}{dx}(l) = 0.$$

We assume that a and c satisfy (2.20 b and c), that $a \in C^1(I)$, $c \in C^0(I)$, $f \in C^0(I)$, and that Δ is a quasiuniform mesh. Then, we have for $p = 1, 2$ that

$$\left\|e_\Delta^{[1]}\right\|_{\mathcal{U}} \leqq C\,h, \tag{6.10}$$

where C is independent of Δ.

If $a \in C^2(I)$, $c \in C^1(I)$ and $f \in C^1(I)$, then

$$\left\| e_{\Delta}^{[2]} \right\|_{\mathcal{U}} \leqq \mathcal{C}\, h^2. \tag{6.11}$$

Let $\Phi(v) = u(x_q^{\Delta})$, where x_q^{Δ} is a vertex of the mesh. Then, $\Psi_\Phi \in C^2(0, x_q) \cup C^2(x_q, l)$ in the first case ($p = 1$) and $\Psi_\Phi \in C^3(0, x_q) \cup C^3(x_q, l)$ in the second case ($p = 2$). (Note that $\Psi_\Phi \notin C^1(I)$).

We then have that

$$\left\| e_{\Phi,\Delta}^{[1]} \right\|_{\mathcal{U}} \leqq \mathcal{C}\, h, \ \left(\text{respectively, } \left\| e_{\Phi,\Delta}^{[2]} \right\|_{\mathcal{U}} \leqq \mathcal{C}\, h^2 \right), \tag{6.12}$$

so that we obtain $\left| e_{\Delta}^{[1]}(x_q) \right| \leqq \mathcal{C}\, h^2$ in the $p = 1$ case and $\left| e_{\Delta}^{[2]}(x_q) \right| \leqq \mathcal{C}\, h^4$ in the $p = 2$ case.

Example 6.3 Consider again problem (6.9), again with a quasiuniform mesh and with $p = 1$. We are interested in the error functional

$$\Phi\left(e_{\Delta}^{[1]}\right) = \left(e^{[1]}(x_{q+1}^{\Delta}) - e^{[1]}(x_q^{\Delta})\right) / h_q,$$

where x_q^{Δ} and x_{q+1}^{Δ} are vertices of the mesh. From Example 6.2 it is clear that

$$\left| \Phi(e_{\Delta}^{[1]}) \right| \leqq \mathcal{C} h,$$

where $\mathcal{C}$ depends on $a \in C^1(I)$ and $c \in C^0(I)$ but is independent of Δ and f.

We seek a better estimate and to this end we consider $\Psi_\Phi \in \mathcal{U}$ that satisfies

$$B(\Psi_\Phi, v) = (v(x_{q+1}^{\Delta}) - v(x_q^{\Delta})) / h_q \ \forall v \in \mathcal{U}.$$

By analysing the function Ψ_Φ we can easily see that

$$|\Psi_\Phi|_{C^2(0,x_q)} \leqq K_1,$$

$$|\Psi_\Phi|_{C^2(x_{q+1},l)} \leqq K_2,$$

$$|\Psi_\Phi|_{C^2(x_q,x_{q+1.})} \leqq \frac{1}{h_q} K_3,$$

for constants K_1, K_2, K_3. Using (6.12) we have an $0(h)$ estimate for $\left| e_{\Phi}^{[1]} \right|$, and we now seek an $0(h^2)$ estimate. Let us consider the element τ_k, where we have

$$\left\| \Psi_\Phi - \mathcal{L}^{[1]} \Psi_\Phi \right\|_{\mathcal{U}(\tau_k)} \leqq \mathcal{C}\, h^{\frac{3}{2}} \text{ for } k \neq q,$$

and

$$\left\| \Psi_\Phi - \mathcal{L}^{[1]} \Psi_\Phi \right\|_{\mathcal{U}(\tau_k)} \leqq \mathcal{C}\, h^{\frac{1}{2}} \text{ for } k = q.$$

Because

$$\|u\|_{C^2(I)} \leqq \mathcal{C}\, \|f\|_{C^0(I)},$$

we find that

$$\left\| e_{\Delta}^{[1]} \right\|_{\mathcal{U}(\tau_k)} \leqq C h^{\frac{3}{2}} \|f\|, \, k \neq q,$$

because $\left| e_{\Delta}^{[1]}(x_q) \right| \leqq C h^2$ by Example 6.2. Using Theorem 6.1 we obtain

$$\begin{aligned}
\Phi\left(e_{\Delta}^{[1]}\right) &\leqq B\left(\Psi_{\Phi} - \mathcal{L}^{[1]}\Psi_{\Phi}, e_{\Delta}^{[1]}\right) \\
&\leqq \left(\sum_{k\neq q} \left\| \Psi_{\Phi} - \mathcal{L}^{[1]}\Psi_{\Phi} \right\|_{\mathcal{U}(\tau_k)} \left\| e_{\Delta}^{[1]} \right\|_{\mathcal{U}(\tau_k)}\right) \\
&\quad + \left\| \Psi_{\Phi} - \mathcal{L}^{[1]}\Psi_{\Phi} \right\|_{\mathcal{U}(\tau_q)} \left\| e_{\Delta}^{[1]} \right\|_{\mathcal{U}(\tau_q)} \\
&\leqq C\left(\left(\sum_{k\neq q} h^{\frac{3}{2}} h^{\frac{3}{2}}\right) + h^{\frac{1}{2}} h^{\frac{3}{2}}\right) \\
&\leqq C h^2.
\end{aligned}$$

Note that if $a = 1$ and $c = 0$, then $\Phi\left(e_{\Delta}^{[1]}\right) = 0$.

Remark 6.1 *The approach that we have used in Theorem 6.1 and in the above examples is called the* ***duality approach****. It can be used for a priori error estimates provided the functional Φ is bounded on $\mathcal{U}$. The result appears to be surprising. We know that $e^{[1]}(x_q^{\Delta}) = 0(h^2)$ so that one would expect in Example 6.3 that $\Phi\left(e_{\Delta}^{[1]}\right) \approx 0(h)$, rather than $0(h^2)$. The reason is that $e_{\Delta}^{[1]}(x_q^{\Delta}) = Ch^2$ and $e_{\Delta}^{[1]}(x_{q+1}^{\Delta}) = e_{\Delta}^{[1]}(x_q^{\Delta})\,(1 + 0(h))$.*

Exercise 6.2 Consider the problem

$$-\frac{d}{dx}\left(a(x)\frac{du}{dx}(x)\right) + c(x)u(x) = f(x), \;\; x \in (0, l) = I,$$

$$u(0) = 0,$$

$$a(x)\frac{du}{dx}(x)\mid_{x=l} + \alpha_2 u(l) = G_2,$$

and the functionals

$$\Phi_1(u) \equiv \int_0^l u(x)\sin\frac{\pi x}{l}dx,$$

$$\Phi_2(u) \equiv \frac{1}{(\beta-\alpha)}\int_{\alpha}^{\beta} u(x)dx,$$

$$\Phi_3(u) \equiv \frac{1}{(\beta-\alpha)}\int_{\alpha}^{\beta} a(x)\frac{du}{dx}(x)dx,$$

where $0 \leqq \alpha < \beta \leqq l$. Estimate the error in these functionals using a finite element method with $p = 1$ and 2 and a quasiuniform mesh, and under the following smoothness assumptions:

$$a \in C^1(I), \quad c \in C^0(I), \quad f \in C^0(I),$$
$$a \in C^2(I), \quad c \in C^1(I), \quad f \in C^1(I),$$
$$a \in C^2(I), \quad c \in C^0(I), \quad f \in C^1(I),$$
$$a \in C^2(I), \quad c \in C^1(I), \quad f \in C^1(I).$$

We now use the duality principle again to estimate the finite element error in the L^2-norm.

Theorem 6.3 *Let $a \in C^{k+1}(I), c \in C^k(I), f \in C^k(I), k = 0, 1$ and consider again problem (2.14). For a piecewise pth-order finite element solution to (2.31) on a quasiuniform mesh we have that*

$$\left\| e_\Delta^{[p]} \right\|_{L^2(I)} \leqq Ch \left\| e_\Delta^{[p]} \right\|_{\mathcal{U}}, \tag{6.13a}$$

where C is independent of h and $\mathfrak{u}$. Specifically,

$$\left\| e_\Delta^{[p]} \right\|_{L^2(I)} \leqq Ch^2 \text{ for } k = 0, 1, \ p = 1, 2, \tag{6.13b}$$

$$\left\| e_\Delta^{[p]} \right\|_{L^2(I)} \leqq Ch^3 \text{ for } k = 1, \quad p = 2. \tag{6.13c}$$

Proof Let $\Phi(v) \equiv \int_0^l e_\Delta^{[p]} v \, dx$ and

$$B(\Psi_\Phi, v) = \Phi(v) \ \forall v \in \mathcal{U},$$

then, using the regularity result (2.41b), we have that

$$\|\Psi_\Phi\|_{H^2(I)} \leqq C \left\| e_\Delta^{[p]} \right\|_{L^2(I)}.$$

From Theorem 5.3, recalling Remark 4.4 that this theorem holds not only for $u \in C^{k+1}[I]$ but also for $u \in H^{k+1}[I]$, it follows that

$$\|\Psi_\Phi - \Psi_{\Phi,\Delta}\|_{\mathcal{U}} \leqq Ch \|\Psi_\Phi\|_{H^2(I)} \leqq Ch \left\| e_\Delta^{[p]} \right\|_{L^2(I)}.$$

Hence, by Theorem 6.1 we have that

$$\begin{aligned} \left\| e_\Delta^{[p]} \right\|_{L^2(I)}^2 &= \Phi(e_\Delta^{[p]}) = B(\Psi_\Phi, e_\Delta^{[p]}) \\ &\leqq \left\| e_\Delta^{[p]} \right\|_{\mathcal{U}} \|\Psi_\Phi - \Psi_{\Phi,\Delta}\|_U \\ &\leqq Ch \left\| e_\Delta^{[p]} \right\|_{\mathcal{U}} \left\| e_\Delta^{[p]} \right\|_{L^2(I)}, \end{aligned}$$

and (6.13a) follows. Then, (6.13b and c) follow by the estimation of $\left\| e_\Delta^{[p]} \right\|_{\mathcal{U}}$. □

Thus far, we have assumed that Φ is a bounded functional on $\mathcal{U}$; i.e. that $|\Phi(v)| \leqq C \|v\|_{\mathcal{U}}$, $\forall v \in \mathcal{U}$. This does of course preclude us from using the duality principle when Φ is not bounded, for example when $\Phi(v) = \frac{dv}{dx}(0)$.

Exercise 6.3 Show that the functional $\Phi(v) = \frac{dv}{dx}(0)$ is not bounded on $\mathcal{U}_0$.

In general, for $\mathrm{u} \in \mathcal{U}$ we are unable to say anything about $\Phi(\mathrm{u})$ if the functional is not bounded. However, we do have additional information about u; for example if u is the solution of (2.14) we may have the additional information that $f \in L^2(I)$.

Example 6.4 Consider problem (2.14) with $a \in C^1(I)$, $c \in C^0(I)$, $f \in C^0(I)$ and $g1 = \alpha_2 = G_2 = 0$, and suppose that we are interested in $\frac{d\mathrm{u}}{dx}(0)$. Because $u \in C^2[I]$, $\frac{d\mathrm{u}}{dx}(0)$ is well defined. Let us define $v_0(x) \equiv 1 - \frac{x}{l}$, so that $v_0(0) = 1$, and $v_0 \in \mathcal{U} - \mathcal{U}_0$ but $v_0 \notin \mathcal{U}_0$. Nevertheless,

$$B(\mathrm{u}, v_0) = \int_0^l \left(a \frac{d\mathrm{u}}{dx} \frac{dv_0}{dx} + c\mathrm{u}v_0 \right) dx$$

is well defined and by integration by parts we find that

$$\begin{aligned} B(\mathrm{u}, v_0) &= \int_0^l \left(-\frac{d}{dx}(a\frac{d\mathrm{u}}{dx}) + c\mathrm{u} \right) v_0 \, dx - (a\frac{d\mathrm{u}}{dx})(0) \\ &= \int_0^l f v_0 dx - (a\frac{d\mathrm{u}}{dx})(0). \end{aligned}$$

Hence,

$$(a\frac{d\mathrm{u}}{dx})(0) = \int_0^l f v_0 dx - Q(\mathrm{u}),$$

where

$$Q(\mathrm{u}) = \int_0^l \left(a \frac{d\mathrm{u}}{dx} \frac{dv_0}{dx} + c\mathrm{u}v_0 \right) dx.$$

Clearly, with C depending on v_0,

$$|Q(\mathrm{u})| \leqq C \|\mathrm{u}\|_{\mathcal{U}}.$$

Now, although Φ is not a bounded functional on $\mathcal{U}_0$, Q is. Hence, we should compute Ψ_Q and proceed as before. In this way, we are able to obtain, for the case of a quasiuniform mesh and with $p = 1$, that

$$\left| (a\frac{d\mathrm{u}}{dx})(0) - \int_0^l f v_0 dx - Q(u_\Delta^{[1]}) \right| \leqq Ch^2. \tag{6.14}$$

We emphasize that we have included the term $\int_0^l f v_0 dx$ in (6.14), which obviously can be computed. This example also shows that there are many functionals that lead to the same value when applied to the exact solution u, but that yield differing levels of accuracy when applied to the finite element solution $u_\Delta^{[1]}$.

Exercise 6.4 Consider the problem

$$-\frac{d^2\mathrm{u}}{dx^2}(x) + c\mathrm{u}(x) = f(x),\ x \in I,$$
$$\mathrm{u}(0) = 0,$$
$$\left(\frac{d\mathrm{u}}{dx} + u\right)(l) = 1,$$

where $f \in C^1[I], a \in C^1[I]$, and its finite element solution on a uniform mesh using elements of degree $p = 1, 2$. Show that

$$e^{[p]}(x_j^\Delta) \leqq C\,h^p \left\|e^{[p]}\right\|_{\mathcal{U}},\ p = 1, 2.$$

Explain why for $c = 1$ you cannot expect that $e^{[p]}(x_j^\Delta) = 0$, while for $c = 0$ this condition holds.

Exercise 6.5 Consider again the problem of Exercise 6.4. Propose a way for computing $\dfrac{du}{dx}(l)$ and analyze the accuracy of your method.

Exercise 6.6 Consider the problem

$$-\frac{d^2\mathrm{u}}{dx^2}(x) = 1,\ x \in I,$$
$$\mathrm{u}(0) = 1,$$
$$\left(\frac{d\mathrm{u}}{dx}\right)(l) = 5.$$

Is it true that $e^{[p]}(x_j) = 0$ for this problem?

Exercise 6.7 Consider the problem

$$-\frac{d}{dx}\left(\frac{1}{1+x}\frac{d\mathrm{u}}{dx}(x)\right) = f(x),\ x \in I,$$
$$\mathrm{u}(0) = 0,$$
$$\frac{d\mathrm{u}}{dx}(l) = 0,$$

and its finite elements solution using elements of degree $p = 2$. Is it true that $e^{[2]}(x_j^\Delta) = 0$?

Exercise 6.8 Consider the problem

$$-\frac{d}{dx}\left(a(x)\frac{du}{dx}(x)\right) = f(x),\ x \in I,$$

$$u(0) = 0,$$

$$\left(\frac{du}{dx}\right)(l) = 0,$$

and its finite solution on a uniform mesh using elements of degree $p = 1$. Find the analytical formulae for $u(x_j^\Delta)$, $u^{[1]}(x_j^\Delta)$ and $e^{[1]}(x_j^\Delta)$. From these formulae discuss the necessary and sufficient conditions for $e^{[p]}(x_j^\Delta) = 0$.

6.1.2 Local character of the error and pollution

In this section we shall analyze the relation between the interpolation error and $e_\Delta^{[p]}$. We first address the case $p = 1$, and the situation is illustrated in Fig. 6.2, for the elements τ_q, τ_{q+1}. This shows the exact solution u (top curve), the interpolant $\mathcal{L}^{[1]}(u)$ (top piecewise linear) and the finite element solution (bottom piecewise linear). We define the error in the interpolant

$$e_{\text{loc}}^{[1]} = u - \mathcal{L}^{[1]}(u),$$

which is coarse hatched in Fig. 6.2, and clearly $e_{\text{loc}}^{[1]}(x_q) = e_{\text{loc}}^{[1]}(x_{q+1}) = 0$.

The finite element solution arises from global computation over the whole of I. We thus define

$$e_{\text{glob}}^{[1]} = \mathcal{L}^{[1]}(u) - u_\Delta^{[1]},$$

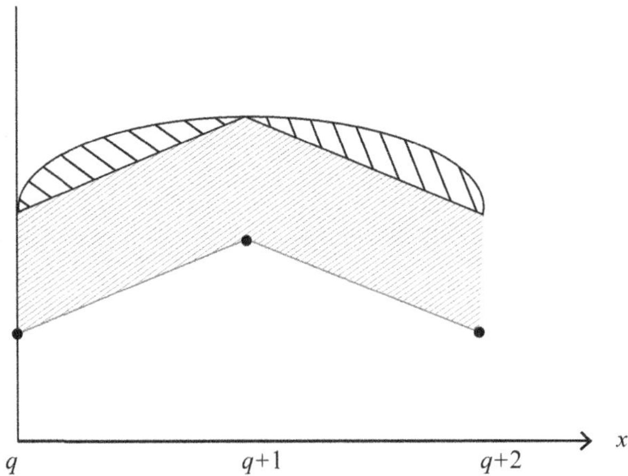

Fig. 6.2 Resolution of the error situation; error $e_{\text{glob}}^{[1]}$ fine hatched in grey, interpolation error hatched coarse, exact solution, u (top curve), finite element solution (bottom piecewise linear), interpolant $\mathcal{L}^{[1]}(u)$ (top piecewise linear).

so that $e_{\text{glob}}^{[1]}$ is linear in τ_q and

$$e_{\text{glob}}^{[1]}(x_q) = (\mathfrak{u} - u_\Delta^{[1]})(x_q) = e_\Delta^{[1]}(x_q).$$

In Fig. 6.2 $e_{\text{glob}}^{[1]}$ is hatched in red. Now, we have

$$\begin{aligned} e_{\text{glob}}^{[1]}(x) &= \frac{e_\Delta^{[1]}(x_{q+1}) + e_\Delta^{[1]}(x_q)}{2} \\ &\quad + \frac{e_\Delta^{[1]}(x_{q+1}) - e_\Delta^{[1]}(x_q)}{h_q}\left(x - \frac{x_{q+1}+x_q}{2}\right) \\ &\equiv {}^{(1)}\rho_q + {}^{(2)}\rho_q\left(x - \frac{x_{q+1}+x_q}{2}\right). \end{aligned}$$

From Fig. 6.2 we see that $e_\Delta^{[1]} = e_{\text{loc}}^{[1]} + e_{\text{glob}}^{[1]}$ and the pollution, *which is the difference between* $e_\Delta^{[1]}$ *and the interpolant,* is characterized by the values ${}^{(1)}\rho_q$ and ${}^{(2)}\rho_q$. We shall in general say that the pollution is negligible on τ_q if

$$\|e_{\text{glob}}\|_{\mathcal{U}(\tau_q)} \ll \|e_{\text{loc}}\|_{\mathcal{U}(\tau_q)};$$

more precisely that

$$\left\|e_{\text{glob}}^{[1]}\right\|_{\mathcal{U}(\tau_q)} \Big/ \left\|e_{\text{loc}}^{[1]}\right\|_{\mathcal{U}(\tau_q)} \to 0 \text{ as } h \to 0.$$

Note that $e_{\text{glob}}^{[1]}$ is linear on τ_q, whilst $e_{\text{loc}}^{[1]}$ is not. We see that the error on τ_q is close to the error of the interpolant, only if the pollution is negligible.

We also see that

$$\left\|e_{\text{glob}}^{[1]}\right\|_{\mathcal{U}(\tau_q)}^2 \leqq \mathcal{C}\, h_q \left(\left|{}^{(1)}\rho_q\right|^2 + h_q^2\left|{}^{(2)}\rho_q\right|^2\right).$$

On the other hand, for solutions that are locally smooth with

$$\left|\frac{d\mathfrak{u}^2}{dx^2}\right| > \gamma > 0,$$

we have

$$\left\|e_{\text{loc}}^{[1]}\right\|_{\mathcal{U}(\tau_q)}^2 \geqq \mathcal{C}\, h^3,$$

and we see that pollution is completely characterized by the errors at the nodal points.

Example 6.5 Under the assumptions of Example 6.2 we have that ${}^{(1)}\rho_q \leqq \mathcal{C}\, h^2$. Further, from Example 6.3 ${}^{(2)}\rho_q \leqq \mathcal{C}\, h^2$ and the pollution is negligible.

We now consider the case $p = 2$. In order to do this in the element τ_q we define the generalized interpolant to $\mathfrak{u}$, $\mathcal{P}^{[p]}(\tau)\mathfrak{u} \equiv \mathcal{P}^{[p]}(\mathfrak{u})$, $p = 1, 2$, where $\mathcal{P}^{[1]}(\mathfrak{u}) = \mathcal{L}^{[1]}(\mathfrak{u})$

and

$$\mathcal{P}^{[2]}(\mathrm{u}) = \mathcal{P}^{[1]}(\mathrm{u}) + \xi_q(\mathrm{u})\, z_q(x),$$

with $z_q(x) \equiv (x - x_q^{\Delta})(x_{q+1}^{\Delta} - x)$, and $\xi_q(\mathrm{u}) \in \mathbb{R}$ is such that

$$B_{\tau_q}(\xi_q(\mathrm{u}) z_q(x), z_q(x)) = B_{\tau_q}\left(\mathrm{u} - \mathcal{L}^{[1]}(\mathrm{u}), z_q(x)\right).$$

Hence, $\mathcal{P}^{[2]}(\mathrm{u})$ is the best approximation in $\mathcal{U}(\tau_q)$ to $\mathrm{u}(x), x \in \tau_q$ by quadratic polynomials that interpolate to the values $\mathrm{u}(x_q)$ and $\mathrm{u}(x_{q+1})$. This of course means that $\mathcal{P}^{[2]}(\mathrm{u}) \neq \mathcal{L}^{[2]}(\mathrm{u})$ in τ_q. Nevertheless, $\mathcal{P}^{[2]}(\mathrm{u})$ is still a local approximation because it is based only on u in τ_q. We have immediately that

$$\left\|\mathrm{u} - \mathcal{P}^{[2]}(\mathrm{u})\right\|_{\mathcal{U}(\tau_q)} \leq \left\|\mathrm{u} - \mathcal{L}^{[2]}(\mathrm{u})\right\|_{\mathcal{U}(\tau_q)}.$$

We now consider the finite element error $e_{\Delta}^{[2]}$. Because $u_{\Delta}^{[2]}$ arises from a global computation over the whole of Ω, on the element τ_q we form the splitting

$$e_{\Delta}^{[2]} = e_{\mathrm{loc,q}}^{[2]} + e_{\mathrm{glob,q}}^{[2]}, \tag{6.15a}$$

where

$$e_{\mathrm{loc}}^{[2]} \equiv \mathrm{u} - \mathcal{P}^{[2]}(\mathrm{u}), \tag{6.15b}$$

with

$$e_{\mathrm{loc},q}^{[2]}(x_{q+1}) = e_{\mathrm{loc},q}^{[2]}(x_q) = 0,$$

and

$$e_{\mathrm{glob},q}^{[2]} = \mathcal{P}^{[2]}(\mathrm{u}) - u_{\Delta}^{[2]}, \tag{6.15c}$$

with

$$e_{\mathrm{glob},q}^{[2]}(x_q) = (\mathrm{u} - u_{\Delta}^{[2]})(x_q) = e_{\Delta}^{[2]}(x_q),$$

and

$$e_{\mathrm{glob},q}^{[2]}(x_{q+1}) = e_{\Delta}^{[2]}(x_{q+1}).$$

Further, we define

$${}^{(1)}e_{\mathrm{glob},q}^{[2]} = \mathcal{L}^{(1)} e_{\mathrm{glob},q}^{[2]}, \tag{6.15d}$$

with

$${}^{(1)}e_{\mathrm{glob,q}}^{[2]}(x_{q+1}) = e_{\mathrm{glob,q}}^{[2]}(x_{q+1}) \text{ and } {}^{(1)}e_{\mathrm{glob},q}^{[2]}(x_q) = e_{\mathrm{glob},q}^{[2]}(x_q),$$

and

$${}^{(2)}e_{\mathrm{glob},q}^{[2]} = e_{\mathrm{glob},q} - \mathcal{L}^{(1)} e_{\mathrm{glob},q}^{[2]}, \tag{6.15e}$$

with

$${}^{(2)}e_{\mathrm{glob},q}^{[2]}(x_q) = {}^{(2)}e_{\mathrm{glob},q}^{[2]}(x_{q+1}) = 0.$$

Then, we have that

$$\begin{aligned} {}^{(1)}e^{[2]}_{\mathrm{glob},q} &= \frac{e^{[2]}_{\Delta}(x_q) + e^{[2]}_{\Delta}(x_{q+1})}{2} + \frac{e^{[2]}_{\Delta}(x_{q+1}) - e^{[2]}_{\Delta}(x_q)}{h_q}\left(x - \frac{x_q + x_{q+1}}{2}\right) \\ &\equiv {}^{(1)}\rho_q + {}^{(2)}\rho_q\left(x - \frac{x_q + x_{q+1}}{2}\right), \end{aligned}$$

and the ${}^{(1)}e^{[2]}_{\mathrm{glob},q}$ is called *pollution.*

Note that $e^{[2]}_{\mathrm{glob},q}(x)$ is a polynomial of degree 2 and the pollution is a linear function on $\tau_q, p = 1, 2$. As previously we speak about pollution being negligible.

Figure 6.3 illustrates the major notions. In Fig. 6.3a for τ_q we see the exact solution u solid curve with turning points and the generalized interpolant $\mathcal{P}^{[2]}(\mathrm{u})$ solid curve without turning points. The dashed curve is the interpolant $\mathcal{L}^{[2]}$, whilst the thick straight line is the linear interpolant, and the lowest curve is $u^{[2]}_{\Delta}$. The difference between $\mathcal{P}^{[2]}(\mathrm{u})$ and $u^{[2]}_{\Delta}$ is $e^{[2]}_{\mathrm{glob},q}$ (hatched vertically). In Fig. 6.3b $e^{[2]}_{\mathrm{glob},q}$ is split into $\mathcal{L}^{[1]}(e^{[2]}_{\mathrm{glob},q})$, the linear interpolant to $e^{[2]}_{\mathrm{glob},q}$, and ${}^{(2)}e^{[2]}_{\mathrm{glob},q}$, shown hatched.

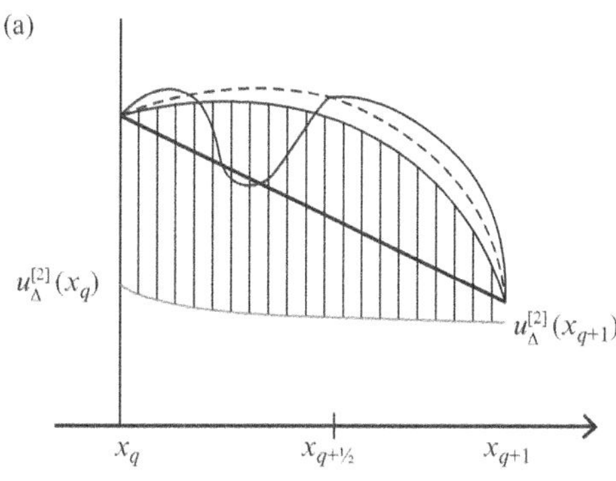

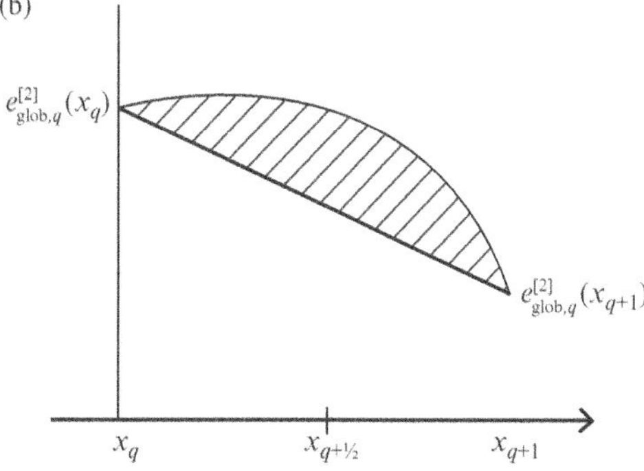

Fig. 6.3 Local and global errors.

The term *pollution* was previously mentioned in Chapter 5. We can now be more specific. Distinguishing again between $e^{[p]}_{\text{glob}}$ and $e^{[p]}_{\text{loc}}$ we say that if

$$\left\| e^{[p]}_{\text{glob},q} \right\|_{\mathcal{U}(\tau_q)} \Big/ \left\| e^{[p]}_{\text{loc},q} \right\|_{\mathcal{U}(\tau_q)}$$

tends to zero as $h \to 0$ then pollution is negligible or insignificant, whereas we have *essential pollution* if this quotient tends to ∞ as $h \to 0$.

Example 6.6 Consider the problem of Example 5.5, with $h_1 = h^{0.15}$ and with $p = 1$. Then, we have for $q > 1$ that

$$\left\| e^{[1]} \right\|_{\mathcal{U}(\tau_q)} \simeq \mathcal{C}\, h^{0.45} h^{\frac{1}{2}} = \mathcal{C}\, h^{0.95}.$$

We also have $\left\| \mathfrak{u} - \mathcal{L}^{[1]}\mathfrak{u} \right\|_{\mathcal{U}(\tau_q)} \simeq \mathcal{C}\, h^{3/2}$, and from this we see that, for $q > 1$,

$$\left\| e^{[1]}_{\text{loc},q} \right\|_{\mathcal{U}(\tau_q)} \approx \mathcal{C}\, h^{3/2},$$

$$\left\| e^{[1]}_{\text{glob},q} \right\|_{\mathcal{U}(\tau_q)} \approx \mathcal{C}\, h^{0.95}.$$

Hence,

$$\left\| e^{[1]}_{\text{glob},q} \right\|_{\mathcal{U}(\tau_q)} \Big/ \left\| e^{[1]}_{\text{loc},q} \right\|_{\mathcal{U}(\tau_q)} \to \infty$$

as $h \to 0$, and the pollution is not negligible.

Theorem 6.4 *Under the assumptions of (2.20) we have for $p = 1, 2$ that*

$$B_{\tau_q}\left({}^{(2)}e^{[p]}_{\text{glob},q}, v \right) = -B_{\tau_q}\left({}^{(1)}e^{[p]}_{\text{glob},q}, v \right)$$
$$\forall v \in S^{[p]}_{\Delta,0}(\tau_q), \tag{6.16a}$$

and

$$\left\| {}^{(2)}e^{[p]}_{\text{glob},q} \right\|_{\mathcal{U}(\tau_q)} \leqq \left\| {}^{(1)}e^{[p]}_{\text{glob},q} \right\|_{\mathcal{U}(\tau_q)}$$
$$\leqq \mathcal{C}(1 + h_q) h_q^{\frac{1}{2}} \left({}^{(1)}\rho_q^2 + h_q^2\, {}^{(2)}\rho_q^2 \right)^{\frac{1}{2}}. \tag{6.16b}$$

Proof Let $v \in S^{[p]}_{\Delta,0}(\tau_q)$, then we have

$$0 = B_{\tau_q}\left(e^{[p]}_{\Delta}, v \right) = B_{\tau_q}\left(e^{[p]}_{\text{loc},q} + e^{[p]}_{\text{glob},q}, v \right)$$
$$= B_{\tau_q}\left(e^{[p]}_{\text{loc},q}, v \right) + B_{\tau_q}\left(e^{[p]}_{\text{glob},q}, v \right).$$

But, from the definition of $e^{[p]}_{\text{loc},q}$ we have that

$$B_{\tau_q}\left(e^{[p]}_{\text{loc},q}, v \right) = 0 \;\; \forall v \in S^{[p]}_{\Delta,0}(\tau_q),$$

and hence that

$$0 = B_{\tau_q}\left(e^{[p]}_{\text{glob},q}, v\right) = B_{\tau_q}\left({}^{(1)}e^{[p]}_{\text{glob},q}, v\right) + B_{\tau_q}\left({}^{(2)}e^{[p]}_{\text{glob},q}, v\right),$$

from which (6.16a) follows. From this, we also obtain

$$\left\| {}^{(2)}e^{[p]}_{\text{glob},q} \right\|_{\mathcal{U}(\tau_q)} \leqq \left\| {}^{(1)}e^{[p]}_{\text{glob},q} \right\|_{\mathcal{U}(\tau_q)},$$

and

$$\left\| {}^{(1)}e^{[p]}_{\text{glob},q} \right\|^2_{\mathcal{U}(\tau_q)} \leqq a_{\max} \left\| \rho_q \right\|^2_{\mathcal{U}(\tau_q)} + c_{\max} \left\| {}^{(1)}e^{[p]}_{\text{glob},q} \right\|_{L^2(\tau_q)},$$

which leads to (6.16b). □

6.1.3 Superconvergence in one dimension

We now address the local error of the finite element solution and discuss how we may recover values of the derivatives at specific points, using only local data from $u^{[p]}_{\Delta}$. This is equivalent to the analysis of the error when pollution is negligible. For this, we start with an example.

Example 6.7 Consider the problem

$$\begin{aligned} -\frac{d^2u}{dx^2}(x) &= 1, \ x \in (0, l), \\ u(0) &= 0, \\ \frac{du}{dx}(l) &= 0, \end{aligned} \tag{6.17}$$

and its finite element solution with $p = 1$.

The exact solution of (6.17) is $\mathfrak{u}(x) = -\frac{1}{2}x^2 + lx$. Thus, $e^{[1]}_{\Delta}(x^{\Delta}_q) = 0$, and hence on $\tau_q \equiv (x^{\Delta}_q, x^{\Delta}_{q+1})$ with $x_q < x < x_{q+1},$ we have

$$e^{[1]}_{\Delta}(x) = \mathfrak{u} - u^{[1]}_{\Delta}(x) = -\frac{1}{2}(x - x^{\Delta}_q)(x - x^{\Delta}_{q+1}),$$

so that

$$\frac{de^{[1]}_{\Delta}}{dx} = -\left(x - \left(\frac{x^{\Delta}_q + x^{\Delta}_{q+1}}{2}\right)\right),$$

from which it follows that

$$\frac{de^{[1]}_{\Delta}}{dx} = 0 \quad \text{at} \quad \frac{x^{\Delta}_q + x^{\Delta}_{q+1}}{2},$$

and

$$\max_{x \in \tau_q} \left| \frac{de_\Delta^{[1]}}{dx} \right| = \frac{1}{2} h_q.$$

Example 6.6 shows that $de_\Delta^{[1]}/dx$ is much smaller in the centre of the element than its maximum over the element. This special point, $(x_q^\Delta + x_{q+1}^\Delta)\,/\,2$, is called a point of *superconvergence* for the derivatives of the finite element solution. We shall now prove the existence in general of points of superconvergence.

Theorem 6.5 *Let* $\mathrm{u}(x)$ *be the solution of problem, (2.14) and (2.33), where* $a \in C^{k+1}(I)$, $c \in C^k(I)$, $f \in C^k(I)$, *and* $u_\Delta^{[p]}$ *be the piecewise pth-order finite element solution on the quasiuniform mesh* Δ, *and let the pollution be negligible. Then,*

(1) *for* $p = 1$ *and* $k = 1$ *and some* $\eta > 0$

$$\left| \frac{de_\Delta^{[1]}}{dx} \left(\frac{x_q^\Delta + x_{q+1}^\Delta}{2} \right) \right| \leqq C\, h^2, \tag{6.18a}$$

whilst for sufficiently small h *and* $\left| \dfrac{d^2\mathrm{u}}{dx^2}(x) \right| \geqq \eta > 0$ *on* τ_q

$$\max_{x \in \tau_q} \left| \frac{de_\Delta^{[1]}}{dx}(x) \right| \geqq C\, h\eta, \tag{6.18b}$$

(2) *for* $p = 2$ *and* $k = 2$

$$\left| \frac{de_\Delta^{[2]}}{dx} \left(\frac{x_q^\Delta + x_{q+1}^\Delta}{2} \pm \sqrt{\frac{1}{12}} h_q \right) \right| \leqq C\, h^3, \tag{6.19a}$$

whilst for sufficiently small h *and* $\left| \dfrac{d^3\mathrm{u}}{dx^3} \right| \geqq \eta > 0$ *on* τ_q

$$\max_{x \in \tau_q} \left| \frac{de_\Delta^{[2]}}{dx}(x) \right| \geqq C\, h^2\eta. \tag{6.19b}$$

Proof We first prove the theorem under the assumption that in (2.14) a is constant and $c = 0$. We then have that $e_\Delta^{[p]}(x_j^\Delta) = 0$, $p = 1$ and 2. Let $p = 1$, then from the assumption that $f \in C^1(\tau_q)$ and $a \in C^2(\tau_q)$ it follows that $e_\Delta^{[1]} \in C^3(\tau_q)$. Hence, we have

$$\frac{de_\Delta^{[1]}}{dx} = A + B\left(x - \frac{x_q^\Delta + x_{q+1}^\Delta}{2} \right) + Z(x),$$

where $|Z(x)| \leqq C h^2$. Because $e^{[1]}(x_q) = e^{[1]}(x_{q+1}) = 0$, we have that $\int_{\tau_q} \frac{de_\Delta^{[1]}}{dx} dx = 0$, which yields $A \leqq C h^2$, $B = \frac{d^2 u}{dx^2}\left(\frac{x_q^\Delta + x_{q+1}^\Delta}{2}\right)(1 + 0(h))$, and hence that

$$\left| \frac{de_\Delta^{[1]}}{dx}\left(\frac{x_q^\Delta + x_{q+1}^\Delta}{2}\right)\right| \leqq C h^2,$$

which proves (6.18a). If $\left|\frac{d^2 u}{dx^2}(x)\right| \geqq \eta > 0$, then $B \neq 0$ and hence for sufficiently small h we get (6.18b).

Suppose now that $p = 2$, then again $e_\Delta^{[2]}(x_j^\Delta) = 0$ and additionally we have

$$\int_{\tau_q} \frac{de_\Delta^{[2]}}{dx} \frac{d}{dx}\left((x - x_q^\Delta)(x - x_{q+1}^\Delta)\right) dx = 0. \tag{6.20}$$

Since it is assumed that $f \in C^2(I)$, $e_\Delta^{[2]} \in C^4(\tau_q)$ and

$$\begin{aligned} \frac{de_\Delta^{[2]}}{dx} &= A + B\left(x - \frac{x_q^\Delta + x_{q+1}^\Delta}{2}\right) \\ &\quad + C\left(\left(x - \frac{x_q^\Delta + x_{q+1}^\Delta}{2}\right)^2 - \beta\right) + Z(x), \end{aligned}$$

with $|Z(x)| \leqq C h^3$ and

$$\beta = \int_{\tau_q}\left(x - \frac{x_q^\Delta + x_{q+1}^\Delta}{2}\right)^2 dx = \frac{1}{12} h_q^2,$$

condition $\int_{\tau_q} \frac{de_\Delta^{[2]}}{dx} dx = 0$ yields $|A| \leqq C h^3$, whilst condition (6.20) yields $|B| \leqq C h^2, C = \frac{1}{2}\frac{d^3 u}{dx^3}\left(\frac{x_j + x_{j+1}}{2}\right)(1 + 0(h))$. Now $\left(x - \frac{x_q^\Delta + x_{q+1}^\Delta}{2}\right)^2 - \beta = 0$ for $x = \frac{x_q^\Delta + x_{q+1}^\Delta}{2} \pm \sqrt{\frac{1}{12}} h_q$ and (6.19a) follows. If $\left|\frac{d^3 u}{dx^3}\right| \geqq \eta$ on τ_q, then $C \neq 0$ and (6.19b) holds.

Continuing the proof, we change the assumptions and now assume that $a \in C^{k+1}(I)$, $c \in C^k(I)$ $f \in C^k(I)$, but still assume that $e_\Delta^{[p]}(x_q) = e_\Delta^{[p]}(x_{q+1}) = 0$. In this case (6.18a) and (6.18b) follow, because in the first part of the proof with $p = 1$, the a

and c are not used. For $p = 2$ instead of (6.20) we have, writing $\varphi = (x - x_q)(x - x_{q+1})$,

$$0 = \int_{\tau_q} \left(a(x) \frac{de_\Delta^{[2]}}{dx} \frac{d\varphi}{dx} \right) dx$$

$$+ \int_{\tau_q} a \left(\frac{x_q^\Delta + x_{q+1}^\Delta}{2} \right) \frac{de_\Delta^{[2]}}{dx} \frac{d\varphi}{dx} dx + \int_{\tau_q} c(x) e_\Delta^{[2]} \varphi dx$$

$$- \int_{\tau_q} a \left(\frac{x_q^\Delta + x_{q+1}^\Delta}{2} \right) \frac{de_\Delta^{[2]}}{dx} \frac{d\varphi}{dx} dx.$$

Using the fact that $\left| a(x) - \left(a \dfrac{x_q^\Delta + x_{q+1}^\Delta}{2} \right) \right| \leqq C\, h$, we get as before that $B \leqq C\, h^2$, which yields (6.19a) and (6.19b).

We must now address the influence of $e_{\text{glob},q}^{[p]}$ on $e_\Delta^{[p]}$; i.e. the influence of pollution. Clearly we have

$$\left| \int_{\tau_q} \frac{de^{[p]}}{dx} dx \right| = \left| \int_{\tau_q} \frac{de_{\text{loc},q}^{[p]}}{dx} dx \right| + \left| \frac{e^{[p]}(x_{q+1}) - e^{[p]}(x_q)}{h_q} \cdot h_q \right| \leqq C\, h^{p+1}.$$

This yeilds $A \leqq C\, h_q^{p+1}$, and for $p = 1$ the result follows. Using (6.12) we get $|B| \leqq C\, h^{p+1}$ from which for $p = 2$ (6.19a) follows. □

We have seen that there are in an element points of superconvergence where, in the absence of pollution, the derivatives of the finite element solution have greater accuracy than that found generally in the element. By using linear or quadratic interpolation to the derivative values at these points of superconvergence, we can produce functions that are superconvergent over the whole element, see, e.g., Goodsell and Whiteman (1989) and Wheeler and Whiteman (1987). Note that in order to have sufficient points to do this we shall have to interpolate to superconvergent values at points outside the element, i.e. in a *patch* of elements. Alternatively, we could use a least squares approach to fit a linear or quadratic polynomial to specific point superconvergent values. Clearly, there are many ways in which these superconvergent derivative point values can be exploited.

Theorem 6.5 addressed superconvergence for the derivatives of the solution, we now consider superconvergence for finite element approximations to the solution itself.

Theorem 6.6 *Let* $u(x)$ *be the solution of problem (2.14) where* $a \in C^3(I)$, $c \in C^2(I)$, $f \in C^2(I)$, *let* $p = 2$, Δ *be a quasiuniform mesh and the pollution be negligible, then*

$$\left| e_\Delta^{[2]} \left(\frac{x_q^\Delta + x_{q+1}^\Delta}{2} \right) \right| \leqq C\, h^4, \tag{6.21a}$$

$$\max \left(\left| e_\Delta^{[2]}(x_q^\Delta) \right|, \left| e_\Delta^{[2]}(x_{q+1}^\Delta) \right| \right) \leqq C\, h^4, \tag{6.21b}$$

whilst, for sufficiently small h and $\left|\dfrac{d^3\mathrm{u}}{dx^3}\right| \geqq \eta > 0$ *on* τ_q *we have* $\max_{x\in\tau_q}\left|e_\Delta^{[2]}(x)\right| \geqq C\,h^3$.

Proof The proof follows exactly the same steps as those of the proof of Theorem 6.5. □

Note that Theorem 6.6 does not address the case $p = 1$, because under the assumption $\left|\dfrac{d^2\mathrm{u}}{dx^2}\right| \geqq \eta > 0$ on τ_q we have that $\max_{x\in\tau_q}\left|e_\Delta^{[1]}(x)\right| \geqq C\,\eta h^2$, and hence the superconvergence will occur only at x_q^Δ and x_{q+1}^Δ when $\left|e_\Delta^{[1]}(x_q)\right|, \left|e_\Delta^{[1]}(x_{q+1})\right| < C\,h^3$, which in general does not occur.

Remark 6.2 *We have analyzed the case for a quasiuniform mesh, because this allowed us to use estimates of the pollution error* $e_{glob,q}^{[p]}$. *For a general mesh we shall have superconvergence under the assumptions that for* $\varepsilon > 0$

$$\left\|\frac{de_{\mathrm{glob},q}^{[p]}}{dx}\right\|_{L^\infty(\tau_q)} \leqq C\,h_q^\varepsilon \left\|\frac{de_{\mathrm{loc},q}^{[p]}}{dx}\right\|_{L^\infty(\tau_q)},$$

and

$$\left\|e_{\mathrm{glob},q}^{[p]}\right\|_{L^\infty(\tau_q)} \leqq C\,h_q^\varepsilon \left\|e_{\mathrm{loc},q}^{[p]}\right\|_{L^\infty(\tau_q)},$$

where the norm on the left of each inequality is very much smaller than that on the right in each case; 'very much smaller' being such that we may neglect the h^ε *terms.*

Remark 6.3 *For general p the superconvergence points in an element occur at the Gaussian points for* $\dfrac{d\mathrm{u}}{dx}$ *and at the Lobbatto points for* u. *For the definitions of Gauss and Lobbatto points see Abramowitz and Stegun (1972), Sections 25.4.29 and 25.4.32, respectively.*

Example 6.8 Consider again the problem of the type (2.14) in which $\mathrm{u}(x)$ satisfies

$$-\frac{d}{dx}\left((1+x)\frac{d\mathrm{u}}{dx}(x)\right) = 0,\ x \in (0,1) = I,$$

$$\mathrm{u}(0) = 0,$$

$$2\frac{d\mathrm{u}}{dx}(1) + \mathrm{u}(1) = 1.$$

This is the same problem as was considered in Example 5.5. We shall now consider the general case for which $h_1 = h^\beta$. As a detailed analysis given in Babuška and Strouboulis (2001), p.263 shows,

$$|\mathrm{u}(h_1) - \hat{u}(h_1)| \approx Ch_1^3,$$

$$\left|\widehat{\mathrm{u}}(h_1) - u_\Delta^{[1]}(h_1)\right| \leqq Ch_1^2,$$

where $\widehat{\mathrm{u}}(h)$ is the solution of the modified problem as in Example 5.5. Hence, $\mathrm{u}(x) - \widehat{\mathrm{u}}(x) \equiv w(x)$ satisfies

$$-\frac{d}{dx}\left((1+x)\frac{dw}{dx}(x)\right) = 0,\ x \in \hat{I} \equiv (0, h_1),$$

$$2\frac{dw}{dx}(1) + w(1) = 0,\ |w(x)| \approx Ch_1^3.$$

From this we see that

$$\|w\|_{C^3(\hat{I})} \approx Ch_1^3$$
$$\|w\|_{\mathcal{U}(\hat{I})} \approx Ch^3$$
$$\|w\|_{L^3(\hat{I})} \approx Ch^3,$$

and also that

$$\left\|\widehat{\mathrm{u}} - u_\Delta^{[1]}\right\|_{\mathcal{U}(\hat{I})} \approx Ch$$
$$\left\|\widehat{\mathrm{u}} - u_\Delta^{[1]}\right\|_{L^2(\hat{I})} \approx Ch^2.$$

Hence,

$$\left\|\widehat{\mathrm{u}} - u_\Delta^{[1]}\right\|_{\mathcal{U}(\hat{I})} \leqq \|u - \widehat{\mathrm{u}}\|_{\mathcal{U}(\hat{I})} + \left\|\widehat{\mathrm{u}} - u_\Delta^{[1]}\right\|_{\mathcal{U}(\hat{I})}$$
$$\approx Ch_1^3 + h,$$

and the global error in $u_\Delta^{[1]}$ is characterized by $(u - \widehat{\mathrm{u}})$. Thus,

$$\left\|\frac{de_{\mathrm{glob}}^{[1]}}{dx}\right\|_{L^\infty(\tau_q)} \approx \left\|\frac{dw}{dx}\right\|_{L^\infty(\tau_q)} \approx Ch_1^3$$
$$\left\|e_{\mathrm{glob}}^{[1]}\right\|_{L^\infty(\tau_q)} \approx \|w\|_{L^\infty(\tau_q)} \approx Ch_1^3,$$

and

$$\left\|\frac{de_{\mathrm{loc}}^{[1]}}{dx}\right\|_{L^\infty(\tau_q)} \approx Ch$$
$$\left\|e_{\mathrm{loc}}^{[1]}\right\|_{L^\infty(\tau_q)} \approx Ch^2,$$

so that we have

$$\left\|\mathrm{u} - u_\Delta^{[1]}\right\|_{\mathcal{U}(\hat{I})} \approx Ch_1^3 + Ch$$
$$\left\|\mathrm{u} - u_\Delta^{[1]}\right\|_{L^2(\hat{I})} \approx Ch_1^3 + Ch^2,$$

and the size of h_1 will not influence

$$\left\| \mathrm{u} - u_{\Delta}^{[1]} \right\|_{\mathcal{U}(\hat{I})} \quad \text{if } h_1^3 \leqq h^{1+\varepsilon},$$

nor

$$\left\| \mathrm{u} - u_{\Delta}^{[1]} \right\|_{L^2(\hat{I})} \quad if\ h_1^3 \leqq h^{2+\varepsilon}.$$

Hence, in order to see the superconvergence in $\left\| \mathrm{u} - u_{\Delta}^{[1]} \right\|_{L^2(\hat{I})}$ we need $\beta > \frac{1}{3}(2+\varepsilon)$, whilst the error in $\left\| \mathrm{u} - u_{\Delta}^{[1]} \right\|_{\mathcal{U}(\hat{I})}$ is not influenced if $\beta > \frac{1}{3}$.

Let us now address superconvergence using another approach that leads to the same results for a uniform mesh. This latter approach can be generalized to two dimensions, whilst the previous one cannot.

Let ϕ_{p+1} be a monomial of degree $p+1$ and, considering again the equation of problem (6.6), with $\frac{d\mathrm{u}}{dx}(l) = \mathcal{C}$ assume that the exact solution is $\mathrm{u} = \phi_{p+1} \equiv x^{p+1}$. Let the elements of the uniform mesh on I be of size h and define $\bar{w}_p$ to be the error in the interpolant $\mathcal{L}^{[p]}\phi_{p+1}$ to ϕ_{p+1} on this mesh; i.e. $\bar{w}_p \equiv \phi_{p+1} - \mathcal{L}^{[p]}\phi_{p+1}$. Then, we have:

Lemma 6.1 *The error in the interpolant $\bar{w}_p$ is a periodic function with period h.*

Proof Clearly $\phi_{p+1}(x+h) = \phi_{p+1}(x) + q_p(x)$, where $q_p(x)$ is a polynomial of degree p. Then, we have

$$\begin{aligned}
\bar{w}_p(x+h) &= \phi_{p+1}(x+h) - \left(\mathcal{L}^{[p]}\phi_{p+1}\right)(x+h) \\
&= \phi_{p+1}(x) + q_p(x) - \left(\mathcal{L}^{[p]}\phi_{p+1} + \mathcal{L}^{[p]}q_p\right)(x) \\
&= \phi_{p+1}(x) - \mathcal{L}^{[p]}\phi_{p+1}(x) = \bar{w}_p(x).
\end{aligned}$$

In the above we have used the fact that $\mathcal{L}^{[p]}\phi = \phi$ for any polynomial ϕ of degree p. □

Let $u_{\Delta}^{[p]}$ be the finite element solution of the problem that is associated with the solution ϕ_{p+1}, which is the generalized interpolant $\mathcal{P}_{\Delta}^{[p]}$ (as introduced in Example 6.5). Then,

$$e_{\Delta}^{[p]} = \phi_{p+1} - u_{\Delta}^{[p]} = \phi_{p+1} - \mathcal{P}_{\Delta}^{[p]},$$

because there is no pollution. In the same vein as before we can now prove the following lemma.

Lemma 6.2 *The error $e_{\Delta}^{[p]}$ defined above is a periodic function with period h.*

Consider an element $\tau \in \Delta$ and let

$$S_{\Delta,\mathrm{Per}}^{[p]} = \left\{ v \mid v \in S_{\Delta}^{[p]},\ v \text{ is periodic with period } h \right\},$$

and let $\Psi_\Delta^{[p]} \equiv \bar{w}_p + w_p$, where $w_p \in S_{\Delta,\mathrm{Per}}^{[p]}$, is determined up to an additive constant from the condition

$$B_\tau\left(\bar{w}_p + w_p, v\right) = 0 \ \ \forall v \in S_{\Delta,\mathrm{Per}}^{[p]},$$

and the additive constant is determined from

$$\int_\tau \left(\bar{w}_p + w_p\right) dx = \int_\tau \Psi_\Delta^{[p]} dx = 0.$$

Using Lemma 6.1 $e_\Delta^{[p]}$ is periodic and we can construct $\bar{\bar{w}}_p \in S_{\Delta,\mathrm{Per}}^{[p]}$ such that

$$B_\tau\left(e_\Delta^{[p]} + \bar{\bar{w}}_p, v\right) = 0 \ \ \forall v \in S_{\Delta,\mathrm{Per}}^{[p]},$$

and

$$\int_\tau \left(e_\Delta^{[p]} + \bar{\bar{w}}_p, v\right) = 0.$$

Because there is no pollution $u_\Delta^{[p]} = \mathcal{P}_\Delta^{[p]}(\phi_{p+1})$, where $\mathcal{P}_\Delta^{[p]}(\phi_{p+1})$ is the generalized interpolant used in (6.15), and we find that $\bar{\bar{w}}_p = \mathcal{C}$. Hence,

$$B_\tau\left(\Psi_\Delta^{[p]}, v\right) = B_\tau\left(e_\Delta^{[p]}, v\right) = 0,$$

so that $\Psi_\Delta^{[p]} - e_\Delta^{[p]} =$ constant, because $\Psi_\Delta^{[p]}$ and $e_\Delta^{[p]}$ are periodic, and hence

$$\frac{d\Psi_\Delta^{[p]}}{dx} = \frac{de_\Delta^{[p]}}{dx}, \ \text{on } \tau.$$

If $\mathrm{u} = \phi_{p+1} + \xi_p + 0\left(h^{(p+2)}\right), \xi_p \in \mathcal{P}_\Delta^{[p]}$, then asymptotically (for small h) we have

$$\frac{de_\Delta^{[p]}}{dx} = \frac{d\Psi_\Delta^{[p]}}{dx}\left(1 + 0(h^{p+1})\right), \ \text{on } \tau.$$

The function $\Psi_\Delta^{[p]}$ and its derivative $\dfrac{d\Psi_\Delta^{[p]}}{dx}$ give the asymptotic form of the error, which can be utilized in various ways. One of these is to obtain the positions of the points x_{sup}^Δ of superconvergence at which $\dfrac{d\Psi_\Delta^{[p]}}{dx}(x_{\mathrm{sup}}^\Delta) = 0$.

Once again, we reiterate the one essential assumption that pollution is negligible.

As has already been stated, this last approach to superconvergence can be generalized to higher space dimensions when the meshes are uniform, see Babuška *et al.* (1996), (more precisely translation invariant), when the solutions are smooth and when pollution is negligible. In view of this, we now summarize the approach in a series of steps, as follows:

Step 1: For the monomial $\phi_{p+1}(x) \in \mathcal{P}^{[p+1]}(\tau)$ compute the interpolation error $\bar{w}_p$.

Step 2: Compute the correction $w_p \in S^{[p]}_{\Delta,Per}$, using the principal part of the bilinear form of the weak problem. Compute

$$\Psi^{[p]}_{\Delta} = \bar{w}_p + w_p + \mathcal{C},$$

where $\mathcal{C}$ is determined from condition $\int_\tau \Psi^{[p]}_{\Delta} dx = 0$. The function $\Psi^{[p]}_{\Delta}$ is the asymptotic form of the error $e^{[p]}_{\Delta}$ on τ for the exact solution $u = \phi_{p+1} + \xi_p + 0(h^{p+2}), \xi_p \in \mathcal{P}^{[p]}(\tau)$

Step 3: Determine the zeros of the derivatives of $\Psi^{[p]}_{\Delta}$ that are the points of superconvergence.

Returning to the one-dimensional case we note that it is possible to use either $\tau = (0, h)$ or, by scaling, $\tilde{\tau} = (0, 1)$. Note also that in the above procedure we did not see any difference between elements internal to I and the first or last elements that involve the boundary points. Thus, no boundary conditions were involved. In higher dimensions the situation is more complicated, as it is necessary to distinguish between elements interior to the domain and those involving the boundary.

We now present some examples in order to illustrate the approach.

Example 6.9 Consider the differential equation

$$-\frac{d}{dx}\left(a(x)\frac{du}{dx}(x)\right) + u(x) = f(x),$$

and let us take $p = 1$, $h = 1$ and $\tau = (0, 1)$.

For this we work through steps 1–3.

Step 1: We take $\phi_{p+1} = x^2$ and have

$$\bar{w}_p = x^2 - x = x(x - 1).$$

Step 2: Consider the principal part of the bilinear form arising from the weak form of the differential equations. This is

$$B_\tau(u, v) = \bar{a} \int_\tau \frac{du}{dx}\frac{dv}{dx} dx,$$

where $\bar{a}$ is the average of a on τ. Because $S^{[1]}_{\Delta,\mathrm{Per}}$ consists only of constant functions, we find that $w_p = \mathcal{C}$ and hence that

$$\Psi^{[1]}_{\Delta} = \bar{w}_p + \mathcal{C} = x(1 - x) + \mathcal{C},$$

where $\mathcal{C}$ is determined from the condition

$$\int_\tau \Psi^{[1]}_{\Delta}(x) dx = 0.$$

Step 3: Determining the zeros of $\dfrac{d\Psi_\Delta^{[1]}}{dx} = 2x - 1$ we find that $x_{Sup}^\Delta = \frac{1}{2}$ is the only point of superconvergence.

Example 6.10 Let us consider again the same equation as in the previous example, but now take $p = 2$.

Step 1: We have $\phi_{p+1} = x^3$ and hence that

$$\bar{w}_p = \phi_{p+1} - \mathcal{L}^2\phi_{p+1} = x^3 - x + \frac{3}{2}x(1-x).$$

Step 2: In this case $S_{\Delta,\text{Per}}^{[2]}$ is the span of constants and $x(1-x)$, so that by a simple computation we find that $w_p = C$.

Step 3: Now $\dfrac{d\Psi_\Delta^{[2]}}{dx} = 3x^2 - 3x + \frac{1}{2}$, and the roots of this polynomial are $x = \frac{1}{2} \pm \frac{1}{\sqrt{12}}$. Hence, with scaling we put $x_{\text{Sup}}^\Delta = \frac{1}{2} \pm \frac{1}{\sqrt{12}}$.

Let us now show how we can further utilize the asymptotic form of the error.

Example 6.11 Consider again the differential equation

$$-\frac{d^2u}{dx^2}(x) = f(x), \quad x \in I \equiv (-1, 1),$$

take $p = 1$, and consider a mesh with elements

$$\tau_q \equiv (-1, 0), \quad \tau_{q+1} \equiv (0, 1).$$

We are interested in determining $\dfrac{du_\Delta^{[1]}}{dx}(0)$. Note that $x = 0$ is not a point of superconvergence.

Let $y_1 \in \tau_q$, $y_2 = \tau_{q+1}$,

$$\rho = \alpha\frac{du_\Delta^{[1]}}{dx}(y_1) + \beta\frac{du_\Delta^{[1]}}{dx}(y_2),$$

and ask whether we can determine α and β so that

$$\rho(0) = \frac{du}{dx}(0) + Ch^2,$$

i.e. to get a superconvergent value of $\dfrac{du_\Delta^{[1]}}{dx}(0)$. We have for (unknown) coefficients $a,\ b,\ c$ that

$$u = a + bx + cx^2 + 0(h^3),$$

$$\frac{du}{dx} = b + 2cx + 0(h^2).$$

Hence, we take $\phi_{p+1} = x^2$ and then

$$\text{on } \tau_q: \quad \Psi_\Delta^{[1]} = x(x+1) + C, \frac{d\Psi^{[1]}}{dx} = (2x+1),$$

$$\text{on } \tau_{q+1}: \quad \Psi_\Delta^{[1]} = x(x-1) + C, \frac{d\Psi^{[1]}}{dx} = (2x-1).$$

Now, because $u - u_\Delta^{[1]}$ is asymptotically equal to $\Psi_\Delta^{[1]}$ we have that

$$u_\Delta^{[1]} = c(x^2 - x(x+1)) + a + bx, \qquad \text{on } \tau_q,$$

$$u_\Delta^{[1]} = c(x^2 - x(x-1)) + a + bx, \qquad \text{on } \tau_{q+1},$$

and

$$\rho = (\alpha + \beta)b + (\beta - \alpha)c.$$

Hence, for $\alpha + \beta = 1, \alpha - \beta = 0$, so that $\alpha = \beta = \frac{1}{2}$, we obtain the superconvergent value independent of $y_1 \in \tau_q$ and $y_2 \in \tau_{q+1}$.

Exercise 6.9 Consider the case where $p = 2$, and where we have $y_{1,1}, y_{2,2} \in \tau_q$, $y_{2,1}, y_{2,2} \in \tau_{q+1}$. Is it possible to find the values of $\alpha_1, \alpha_2, \beta_1, \beta_2$ so that

$$\alpha_1 \frac{du_\Delta^{[2]}}{dx}(y_{1,1}) + \alpha_2 \frac{du_\Delta^{[2]}}{dx}(y_{1,2})$$

$$+\beta_1 \frac{du_\Delta^{[2]}}{dx}(y_{2,1}) + \beta_2 \frac{du_\Delta^{[2]}}{dx}(y_{2,2})$$

will be the superconvergent values of $\frac{du}{dx}(0)$?

6.1.4 Engineering application; one-dimensional heat-transfer problem

We return again to the one-dimensional engineering problems 1D Eng Problem 1 and 1D Eng Problem 2 that were introduced in Section 2.1.5, for which finite element solutions were computed in Section 3.3.5, and for which the accuracy in the energy norm, the quantity Q_3 was considered in Section 5.1.2.

We now address the computation of approximations Q_1^Δ and Q_2^Δ to the data of interest $Q_1 = u(c)$ and $Q_2 = \left|\frac{du}{dn_c}(c)\right|$ as defined in Section 3.3.5.

For 1D Eng Problem 1 when the interface coincides with a vertex of the mesh we have, using (2.40) $u \in C^3(3.3.5) \cup C^3(3.5, 6.5)$. The same is valid for Ψ_Φ as in (6.3a). Hence, using (5.14b) we have that $\left\|e_\Delta^{[p]}\right\|_{\mathcal{U}}$ and $\left\|e_{\Phi,\Delta}^{[p]}\right\|_{\mathcal{U}}$ are $0(h^p), p = 1, 2$. Hence, using Theorem 6.1, $\left|Q_1 - Q_1^\Delta\right| \approx 0(h^{2p}), p = 1, 2$. Figure 3.8 illustrates exactly this behaviour with convergence rates $2p$.

In 1D Eng Problem 1 if the interface does not coincide with a vertex then again u and Ψ_Φ have the same character. Now, however, $u \in C^3(\tau_q)$ and $\Psi_\Phi \in C^3(\tau_q)$ if τ_q

does not contain the interface, whereas if τ_q contains the interface, then, from (5.49) and (5.50)

$$\left\| \mathrm{u} - \mathcal{L}^{[p]}(\tau_q)\mathrm{u} \right\|_{\mathcal{U}(\tau_q)} \approx C(s)h^{\frac{1}{2}},$$

and

$$\left\| \Psi_\Phi - \mathcal{L}^{[p]}(\tau_q)\Psi_\Phi \right\|_{\mathcal{U}(\tau_q)} \approx C(s)h^{\frac{1}{2}}.$$

Hence, using Theorem 6.1

$$\left| e_{Q_1}^{[p]} \right| \equiv \left| Q_1 - Q_1^\Delta \right| \approx 0(h), p = 1, 2.$$

In 1D Eng Problem 2 the interface never coincides with a vertex, and so in this case we always have that $\left| e_{Q_1}^{[p]} \right| \approx 0(h), p = 1, 2$. Figure 3.9 illustrates this behaviour.

The situation for the functional Q_2 is more complicated because Q_2 involves a derivative and is therefore not a continuous functional. Two approaches for calculating Q_2^Δ were introduced in Section 3.3.5.

1) $Q_2^{\Delta,1} = 2\pi c\gamma cs \dfrac{du_\Delta^{[p]}}{dx}(c) \equiv \left(-a\dfrac{du_\Delta^{[p]}}{dx}(c) \right)$, and again we address separately the cases where the interface coincides with a vertex and where it does not.

We define

$$^1Q_2^{\Delta,1} = -\left(a\frac{du_\Delta^{[p]}}{dx} \right)(c) + a\left(c - \frac{h}{2} \right)\left(\frac{\mathrm{u}(c) - \mathrm{u}(c-h)}{h} \right).$$

Then, applying Example 6.3 with $p = 1$ to this we see that the first term $\approx 0(h^2)$. The second term in $^1Q_2^{\Delta,1}$ is the flux $F(c - \frac{h}{2}) + 0(h)$ and hence

$$\left| e_{Q_2^{\Delta,1}}^{[1]} \right| \approx 0(h^2) + 0(h) = 0(h).$$

When the interface does not coincide with a vertex in 1D Eng Problem 1, or in 1D Eng Problem 2, with a slight modification of the argument of Example 6.3 using (5.49) with $p = 1$ we also get that $^1Q_2^{\Delta,1} \approx 0(h)$ and hence $\left| e_{Q_2^{\Delta,1}}^{[1]} \right| \approx 0(h)$.

When $p = 2$ the corresponding analysis is analogous to when Example 6.3 is adapted for quadratic elements. When the interface coincides with the vertex we have $\left| e_{Q_2^{\Delta,1}}^{[2]} \right| \approx 0(h^2)$, and when it does not $\left| e_{Q_{2,1}}^{[2]} \right| \approx 0(h)$. Figure 3.10 shows exactly this behaviour for problem 1, whilst Fig. 3.11 shows 'chaotic' behaviour for Problem 2.

2) In the second approach introduced in Section 3.3.5

$$Q_2^{\Delta,2} = 2\pi c\beta \left(u_\Delta^{[p]}(c) - T_1 \right) \equiv \alpha \left(u_\Delta^{[p]}(c) - T_1 \right),$$

and the error $\left| e_{Q_{2,2}}^{[p]} \right| \approx \left| e_{Q_1}^{[p]} \right| \approx 0(h^{2p})$, respectively $0(h^p)$, as in Figs. 3.12 and 3.13.

Moral: avoid computation of bounded functionals.

6.2 Two-dimensional problems

6.2.1 The error in the functional

The analysis of the error in the functional in two dimensions is very similar to that for one-dimensional problems. Let us consider again problem (2.63) and the weak formulation (2.65) with associated energy space $\mathcal{U}$. As in the one-dimensional case we have Φ, a bounded linear functional on $\mathcal{U}$. As before we define

$$e_\Phi^{[p]} \equiv \Phi(\mathrm{u}) - \Phi(u_\Delta^{[p]}) = \Phi(e_\Delta^{[p]}),$$

and we have the complete two-dimensional analogue of Theorem 6.1, which is

Theorem 6.7 *Let* $\mathrm{u} \in \mathcal{U}$ *be the solution of (2.72),* $u_\Delta^{[p]} \in S_{\Delta,g_1}^{[p]}$ *be the solution of the finite element problem,* $e_\Delta^{[p]} \equiv \mathrm{u} - u_\Delta^{[p]}$ *and* Φ *be a bounded linear functional on* $\mathcal{U}_0$*, then*

$$\left| e_\Phi^{[p]} \right| \leqq \| \Psi_\Phi - w \|_{\mathcal{U}} \left\| e_\Delta^{[p]} \right\|_{\mathcal{U}}, \tag{6.22}$$

for any $w \in S_{\Delta,0}^{[p]}$*, where* $\Psi_\Phi \in \mathcal{U}_0$ *satisfies*

$$B\left(\Psi_\Phi, v\right) = \Phi(v) \;\; \forall v \in \mathcal{U}_0. \tag{6.23a}$$

If, further, $\Psi_{\Phi,\Delta}^{[p]} \in S_{\Delta,0}^{[p]}$ *is the solution of the finite element problem*

$$B\left(\Psi_{\Phi,\Delta}, v\right) = \Phi(v) \;\; \forall v \in S_{\Delta,0}^{[p]}, \tag{6.23b}$$

then defining $e_{\Phi,\Delta}^{[p]} \equiv \Psi_\Phi - \Psi_{\Phi,\Delta}$*, we find that*

$$\left| e_\Phi^{[p]} \right| \leqq \left\| e_{\Phi,\Delta}^{[p]} \right\|_{\mathcal{U}} \left\| e_\Delta^{[p]} \right\|_{\mathcal{U}}. \tag{6.24}$$

Proof The proof is identical to that of Theorem 6.1 □

We note that the major differences between the one-and two-dimensional cases are

(1) the functional $\Phi(u) = u(x_0), x_0 \in \Omega$ is not a bounded linear functional,
(2) in practice, as we saw in Chapter 2, the boundary Γ of Ω is not smooth so that, even with smooth coefficients in the governing differential equation and right-hand side, the solution $\mathrm{u}(x)$ and Ψ_Φ are not smooth.

We now show with examples how this affects the application of Theorem 6.7.

Example 6.12 Let us again consider a simple example of problem (2.63) in which

$$-\Delta \mathrm{u}(\mathbf{x}) = 1, \;\; \mathbf{x} \in \Omega, \tag{6.25a}$$

$$\mathrm{u}(\mathbf{x}) = 0, \;\; \mathbf{x} \in \Gamma, \tag{6.25b}$$

where $\Omega \equiv (-1,1) \times (-1,1)$ is a square with vertices $A_i,\, i = 1, \ldots, 4$. We also let $\Phi(u) = \int_\Omega u dx$. We have that

$$\Phi(\mathrm{u}) = B(\Psi_\Phi, \mathrm{u}),$$

where Ψ_Φ satisfies (6.23a) and $\mathrm{u} \in \mathcal{U}$ is the solution of the weak form of (6.25).

We must now address the regularity of the solution u. In this case, we have $\beta = 2$ in (2.89) and so, using (5.60) we get for a uniform mesh that

$$\left\| e_\Delta^{[1]} \right\|_{\mathcal{U}} \leqq C\,h, \quad \left\| e_\Delta^{[2]} \right\|_{\mathcal{U}} \leqq C\,h^{2-\varepsilon},$$

and

$$\left\| e_{\Phi,\Delta}^{[1]} \right\|_{\mathcal{U}} \leqq C\,h, \quad \left\| e_{\Phi,\Delta}^{[2]} \right\|_{\mathcal{U}} \leqq C\,h^{2-\varepsilon}.$$

Hence,

$$\left| e_\Phi^{[p]} \right| = \left| \Phi(\mathrm{u}) - \Phi\left(u_\Delta^{[p]}\right) \right| \leqq C\,h^2 \leqq \begin{cases} C\,h^2 & p = 1 \\ C\,h^{4-\varepsilon} & p = 2, \end{cases}$$

where $\varepsilon > 0$ is arbitrarily small and the constant C depends on ε.

Example 6.13 We now consider the problem of (6.25), but with Ω taken as the L-shaped domain of Fig. 6.4 and again $\Phi(\mathrm{u}) = \int_\Omega u d\mathbf{x}$. In this case, the solution of (6.25) has a low level of regularity due to the presence of a singularity at A_1 and we have that $\beta = 2/3$.

Hence,

$$\left| e_\Phi^{[p]} \right| \leqq C\,h^{4/3-\varepsilon}, p = 1, 2,$$

because

$$\left\| e_\Delta^{[p]} \right\|_{\mathcal{U}} \leqq C\,h^{2/3-\varepsilon},$$

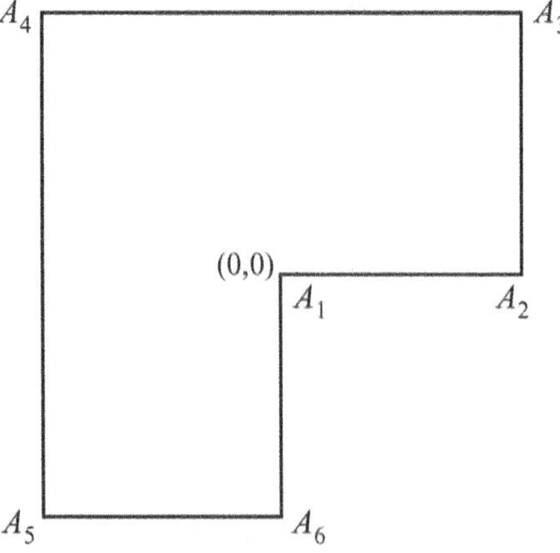

Fig. 6.4 L-shaped domain.

and

$$\left\| e_{\Phi,\Delta}^{[p]} \right\|_{\mathcal{U}} \leqq C\, h^{2/3-\varepsilon}.$$

We therefore see that the convergence of $\left| e_{\Phi}^{[p]} \right| \equiv O(h^{4/3-\epsilon})$ is the same for $p = 1, 2$, on account of the low regularity of $\mathfrak{u}$.

Theorem 6.8 *Consider again problem (2.63) with solution $\mathfrak{u}(x)$, where Ω is a polygonal domain. Let $u_{\Delta}^{[p]}$ be the finite element solution on a quasiuniform mesh, then*

$$\left\| e_{\Delta}^{[p]} \right\|_{L^2(\Omega)} \leqq Ch^{\mu} \left\| e_{\Delta}^{[p]} \right\|_{\mathcal{U}},$$

where $\mu = \min(\beta - \varepsilon, p)$, assuming that a_{mn}, c and α are sufficiently smooth.

Proof The proof is analogous to that of Theorem 6.3 in one dimension, but of course with proper use of the regularity of the solution $\mathfrak{u}(\mathbf{x})$ and of Φ. □

In most cases, we have

$$\left\| e_{\Delta}^{[p]} \right\|_{\mathcal{U}} \leqq Ch^{\mu},$$

so that

$$\left\| e_{\Delta}^{[p]} \right\|_{L^2(\Omega)} \leqq Ch^{2\mu}.$$

Example 6.14 In Example 6.4 we showed how to treat unbounded functionals for one-dimensional problems. The idea was to transform the unbounded functional into one that is both bounded and that also gives the same result. This technique is now applied in the two-dimensional context.

Let $\Omega \equiv (-1, 1) \times (-1, 1)$ and consider the problem

$$-\Delta \mathfrak{u}(\mathbf{x}) = 0, \ \mathbf{x} \in \Omega$$

$$\frac{\partial \mathfrak{u}}{\partial n_c}(\mathbf{x}) + \alpha \mathfrak{u}(\mathbf{x}) = g(\mathbf{x}), \ \mathbf{x} \in \Gamma,$$

and define the quantity of interest Q, where

$$Q \equiv \frac{\partial^3 \mathfrak{u}}{\partial x^3}(\mathbf{x}) \mid_{\mathbf{x}=0} .$$

We note first that $\mathfrak{u}$ is analytic in Ω, but not necessarily on $\bar{\Omega}$, and so all its derivatives exist at $\mathbf{x} = 0$.

Let $\Psi \equiv \dfrac{1}{r^3} \cos 3\theta$, where (r, θ) are polar coordinates centred at the origin. It is easy to see that $\Delta \Psi = 0$ in $\Omega - \{0\}$. Let $\bar{S}_e$ be a closed ball of radius ε centred at the origin and define $\Omega_\varepsilon = \Omega - \bar{S}_e$. Then,

$$0 = -\int_{\Omega_\varepsilon} \Delta u \Psi d\mathbf{x} = -\int_{\partial\Omega_\varepsilon} \frac{\partial u}{\partial n_c} \Psi ds + \int_{\partial\Omega_\varepsilon} u \frac{\partial \Psi}{\partial n_c} ds$$

$$= -\int_\Gamma \frac{\partial u}{\partial n_c} \Psi ds - \int_{\partial S_\varepsilon} \frac{\partial u}{\partial n_c} \Psi ds + \int_\Gamma u \frac{\partial \Psi}{\partial n_c} ds + \int_{\partial S_\varepsilon} u \frac{\partial \Psi}{\partial n_c} ds.$$

Now, on ∂S_ε

$$\frac{\partial \Psi}{\partial n_c} = -\frac{\partial \Psi}{\partial r} = \frac{3}{r^4} \cos 3\theta.$$

Also, we can write the solution u in the form

$$u(r,\theta) = u(0) + \frac{\partial u}{\partial x_1} r \cos\theta + \frac{\partial u}{\partial x_2} r \sin\theta$$
$$+ \frac{1}{2} \frac{\partial^2 u}{\partial x_1^2} r^2 \cos 2\theta - \frac{\partial^2 u}{\partial x_1 \partial x_2} r^2 \sin 2\theta$$
$$+ \frac{1}{6} \frac{\partial^3 u}{\partial x_1^3} r^3 \cos 3\theta + \ldots\ldots\ldots$$

Then, we see that

$$\int_{\partial S_\varepsilon} \frac{\partial \Psi}{\partial r} u \, ds = -\int_{\partial S_\varepsilon} \frac{\partial u}{\partial n_c} \Psi ds = -\frac{1}{2} \frac{\partial^3 u}{\partial x_1^3}(0) \int_0^{2\pi} \cos^2 3\theta d\theta$$
$$= -\frac{\pi}{2} \frac{\partial^3 u}{\partial x_1^3}(0).$$

Hence, as $\varepsilon \to 0$ we see that

$$\pi \frac{\partial^3 u}{\partial x_1^3}(0) = \int_\Gamma \frac{\partial u}{\partial n_c} \Psi ds - \int_\Gamma u \frac{\partial \Psi}{\partial n_c} ds.$$

But,

$$\frac{\partial u}{\partial n_c} = g - \alpha u,$$

and so

$$\pi \frac{\partial^3 u}{\partial x_1^3}(0) = \int_\Gamma g \Psi ds - \int_\Gamma \alpha u \Psi ds - \int_\Gamma u \frac{\partial \Psi}{d n_c} ds.$$

We compute the first term explicitly; the other terms are bounded functionals.

6.2.2 Local character of the error

As might be expected the analysis of the *local* character of the error is more complicated than in one dimension. Once more, we have to distinguish between local error and global error. In order to do the analysis, we make use of interior error estimates.

Theorem 6.9 *Let $\mathrm{u}(x) \in \mathcal{U}$ be the solution of problem (2.65) with $\alpha_{i,j}$ sufficiently smooth defined on a polygonal region Ω and $u_\Delta^{[p]}$ be the finite element solution using a quasiuniform mesh Δ. Let the subdomains Ω^0 and Ω^1 be such that $\overline{\Omega^0} \subset \Omega^1 \subset \overline{\Omega^1} \subset \Omega$, where*

$$0 < d = \operatorname{dist}(\Omega - \Omega^1, \Omega^0) = \inf_{x,y}(|\mathbf{x} - \mathbf{y}|, \mathbf{x} \in \Omega^0, \mathbf{y} \in \Omega - \overline{\Omega^1}), d > \mathcal{C}h.$$

Then, we have

$$\left\|\nabla e_\Delta^{[p]}\right\|_{L^2(\Omega^0)} \leq \mathcal{C} \min_{\chi \in S_\Delta^{[p]}(\Omega^1)} \left(\|\nabla(\mathrm{u} - \chi)\|_{L^2(\Omega^1)}\right) + \frac{1}{d}\|\mathrm{u} - \chi\|_{L^2(\Omega^1)} + \mathcal{C}\frac{1}{d}\left\|e_\Delta^{[p]}\right\|_{L^2(\Omega^1)}, \tag{6.26}$$

$$\left\|\nabla e_\Delta^{[p]}\right\|_{L^\infty(\Omega^0)} + \frac{1}{d}\left\|e_\Delta^{[p]}\right\|_{L^\infty(\Omega^0)} \leq \mathcal{C} \min_{\chi \in S_\Delta^{[p]}(\Omega^0)} \begin{pmatrix} \|\nabla(\mathrm{u} - \chi)\|_{L^\infty(\Omega^1)} \\ +\frac{1}{d}\|\mathrm{u} - \chi\|_{L^\infty(\Omega^1)} \end{pmatrix} + \mathcal{C}\frac{1}{d}\left\|e_\Delta^{[p]}\right\|_{L^2(\Omega^1)}, \tag{6.27a}$$

$$\left\|e_\Delta^{[p]}\right\|_{L^\infty(\Omega^0)} \leq \mathcal{C}(p) \min_{\chi \in S_\Delta^{[p]}(\Omega^0)} \left(\|\mathrm{u} - \chi\|_{L^\infty(\Omega^1)}\right) + \mathcal{C}\frac{1}{d}\left\|e_\Delta^{[p]}\right\|_{L^2(\Omega^1)},$$

where

$$\mathcal{C}(1) = \mathcal{C}\ln\left(\frac{d}{h}\right) \quad \textit{and } \mathcal{C}(2) = \mathcal{C}. \tag{6.27b}$$

Proof The proof of this theorem is very technical and we refer to Nitsche and Schatz (1974), Schatz and Wahlbin (1977), and, if $f \neq 0$, to Schatz and Wahlbin (1995). □

This theorem is essentially the theorem on pollution for $d > 0$ and h sufficiently small. The theorem states that the error of the finite element solution on Ω^0 is the error of the best approximation of u on the larger domain Ω^1 – this is the local error – together with the error $e_\Delta^{[p]}$ on the entire domain Ω measured in the $L_2 - norm$ – this is the global error. If this second term is of higher order than the first, then there is no pollution.

We now illustrate the behaviour of this theorem in the context of two Laplacian problems in the L-shaped domain of Fig. 6.5. The first of these is Benchmark Problem 2 of Section 3.4.2.

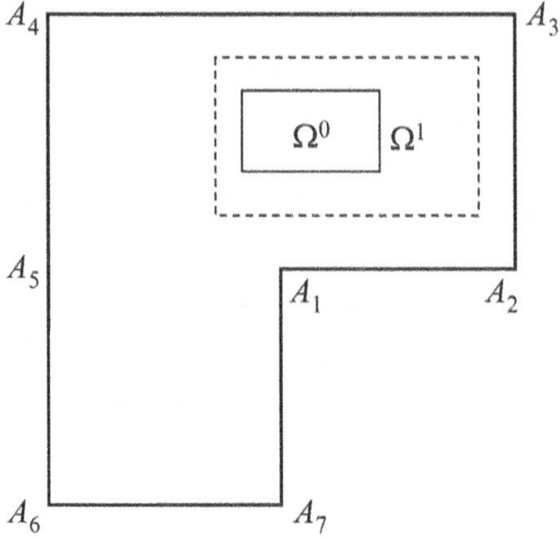

Fig. 6.5 L-shaped domain Ω with subdomains Ω^0 and Ω^1.

Benchmark Problem 2. This has, see Section 3.4.3 and Fig. 3.24, an exact solution $u(\mathbf{x}) = r^{\frac{1}{3}} \sin \theta/3, 0 < \theta < \frac{3\pi}{2}$ and boundary conditions

$$u(\mathbf{x}) = 0, \mathbf{x} \in \overline{A_1, A_2},$$
$$\frac{\partial u}{\partial n}(\mathbf{x}) = g(\mathbf{x}), \mathbf{x} \in \overline{A_2, A_3, A_4, A_5, A_6, A_7, A_1}.$$

Consider the case where Ω^0 is a square of side $\frac{1}{4}$ and centre $(\frac{1}{2}, \frac{1}{2})$ and Ω^1 is a square of side $\frac{1}{2}$ also with centre $(\frac{1}{2}, \frac{1}{2})$. We compute finite element solutions to this problem using uniform meshes with a regular pattern and we are interested in $\left\| e_{\Delta}^{[p]} \right\|_{\mathcal{U}(\Omega^0)}$. We saw in Section 3.4.3 that $\left\| e_{\Delta}^{[p]} \right\|_{\mathcal{U}(\Omega)} \approx Ch^{1/3-\varepsilon}, p = 1, 2$, and we have that $\left\| e_{\Delta}^{[p]} \right\|_{L^2(\Omega)} \approx Ch^{2/3-\varepsilon}$. We now apply Theorem 6.7 to estimate $\left\| e_{\Delta}^{[p]} \right\|_{\mathcal{U}(\Omega^0)}$. We have from (6.26) that

$$\left\| e_{\Delta}^{[p]} \right\|_{\mathcal{U}(\Omega^0)} \approx C\left(h^p + h^{2/3}\right) = C\left(h^{2/3}\right), \quad p = 1, 2.$$

Figure 6.6 shows results for $p = 2$, where $e_I^{[2]}$ is interpolation error.

Benchmark Problem 3. This has the same form as Benchmark Problem 2, except that the exact solution is $u(\mathbf{x}) = r^{\frac{2}{3}} \sin 2\theta/3, 0 < \theta < \frac{2\pi}{3}$, which leads to the boundary conditions

$$u(\mathbf{x}) = 0, \mathbf{x} \in \overline{A_6, A_1, A_2},$$
$$\frac{\partial u}{\partial n}(\mathbf{x}) = g(\mathbf{x}), \mathbf{x} \in \overline{A_2, A_3, A_4, A_5, A_6}.$$

For this problem, we have

$$\left\| \nabla e_{\Delta}^{[p]} \right\|_{L^\infty(\Omega^0)} \approx C\left(h^p + h^{4/3}\right) = Ch^{\min(p,4/3)}, p = 1, 2.$$

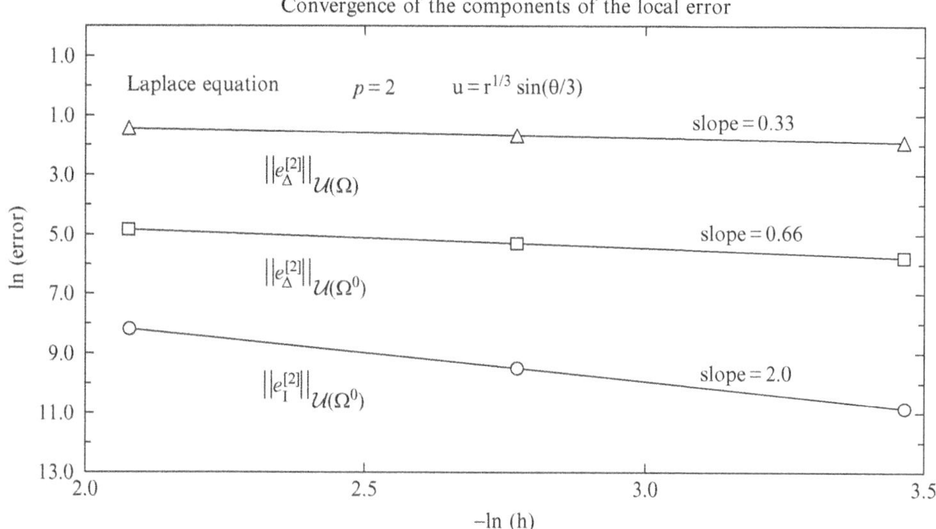

Fig. 6.6 $\ln \left\| e_{\Delta}^{[p]} \right\|_{\mathcal{U}(\Omega^0)}$ v $- \ln(h)$ for Benchmark Problem 2.

We observe that pollution is present for $p = 1, 2$ for Benchmark Problem 2, but that there is pollution for Benchmark Problem 3 only for $p = 2$.

Moral: The presence of pollution depends on the degree of the elements.

6.2.3 Superconvergence in two dimensions

The question of superconvergence in two dimensions is complex, due to the fact that finite element meshes can be very general. Again, as for one dimension, superconvergence can occur only if pollution is negligible, and the solution of a problem is sufficiently smooth. Additionally, now it is important that the mesh be structured.

We shall therefore consider here meshes that are locally uniform, i.e. uniform on Ω, and composed of *cells* that are repeated; such meshes will be called *locally periodic*. Examples of such meshes are given in Fig. 6.7, where the area of interest is the square of size 1, and the cells are of size h.

We now study superconvergence on these locally uniform meshes, expressing asymptotically the error of the finite element solution in a single cell.

Assumption 6.1 The mesh is locally periodic on the rectangle

$$\mathfrak{S}\left(\bar{\mathbf{x}}, H\right) \equiv \{\mathbf{x} = (x_1, x_2) \mid |x_i - \bar{x}_i| < H, \ i = 1, 2\} \subset \Omega$$

of size H and with centre at $\bar{\mathbf{x}}$. The cells $\mathfrak{c}(y, h)$ are also rectangular of size h and centred at y; $\mathfrak{c}(y, h) \subset \mathfrak{S}(\bar{\mathbf{x}}, H)$.

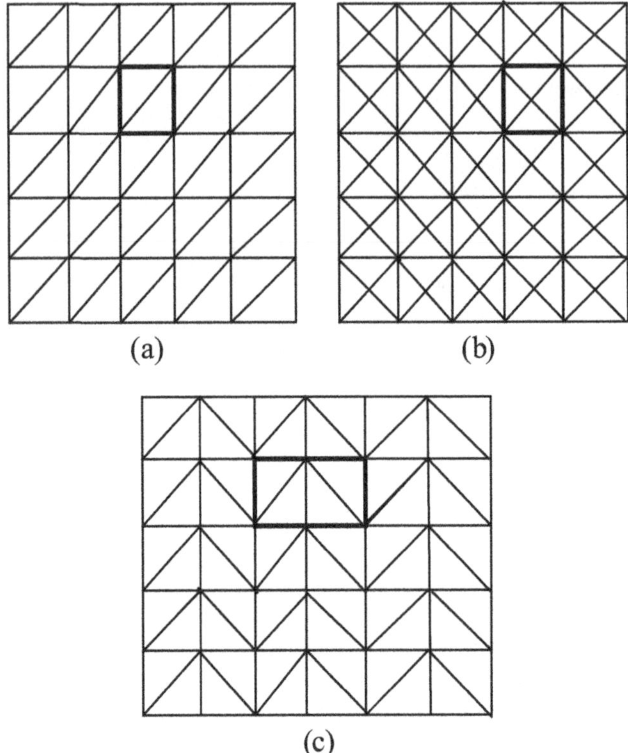

Fig. 6.7 Meshes with cells.

Assumption 6.2 We have

$$\mathfrak{u}(\mathbf{x}) \in C^{p+2}\left(\mathfrak{S}\left(\bar{\mathbf{x}}, H\right)\right), \quad p = 1 \text{ and } 2, \tag{6.28a}$$

$$\sum_{\substack{0 \leqq i,j \leqq p+1 \\ i+j=p+1}} \left| \frac{\partial^{p+1}\mathfrak{u}}{\partial x_1^i \, \partial x_2^j}(\bar{x}) \right| > 0, \tag{6.28b}$$

and the coefficients a_{ij} of the differential equations are constants.
Assumption 6.3 The mesh Δ is such that

$$\left\| e_\Delta^{[p]} \right\|_{L^2(\mathfrak{S}(\bar{x},H))} \leqq \mathcal{C}\, h^\beta H, \tag{6.29}$$

with $p+1-\varepsilon < \beta$, $0 < \varepsilon < 1 - p/(p+1)$, $p = 1, 2$.

Clearly, (6.28a) is an assumption on the smoothness of the solution $\mathfrak{u}$, whilst (6.28b) relates to the assumption that $\left\| e_\Delta^{[p]} \right\|_{\mathcal{U}(\mathfrak{S}(\bar{x},H))} \geqq \mathcal{C}\, h^{p+1}$. Assumption 6.3 indicates that pollution is negligible.

Under these assumptions we can express the error $e_{\Delta}^{[p]}$ of the finite element solution asymptotically by analysing a single cell with centre at $\bar{\mathbf{x}}$ of a given pattern.

Realizing that error $e_{\Delta}^{[p]}$ depends only on the $(p+1)$th (and higher) derivatives of $\mathfrak{u}$, we shall write

$$u(\mathbf{x}) = \sum_{|k|=0}^{p} \phi^{(k)}(\mathbf{x}) + \sum_{|k|=p+1} \phi^{(k)}(\mathbf{x}) + R^{p+1}(\mathbf{x}),$$

with $k = (k_1, k_2), \bar{\mathbf{x}} = (\bar{x}_1, \bar{x}_2)$ and

$$\phi^{(k)}(\mathbf{x}) = \frac{1}{k_1!k_2!} \frac{\partial^{|k|}\mathfrak{u}}{\partial x^{k_1} \partial x^{k_2}}(\bar{\mathbf{x}})(x_1 - \bar{x}_1)^{k_1} (x_2 - \bar{x}_2)^{k_2},$$

$|k| \leqq p+1$ and $R^{(p+1)}$ the remainder.

We now proceed in a manner analogous to that of the one-dimensional case.

Step 1: On $\mathfrak{S}(\bar{\mathbf{x}}, h)$ we define the interpolant $\mathfrak{L}^{[p]}\phi^{(k)}(\mathbf{x})$ and its error

$$\bar{w}_k(\mathbf{x}) = \phi^{(k)}(\mathbf{x}) - \mathfrak{L}^{[p]}\phi^{(k)}(\mathbf{x})$$

For $|k| \leqq p$ we obviously have $\bar{w}_k = 0$. Similarly to the one-dimensional case $\bar{w}_k$ is a periodic function. Of course, the interpolant $\mathfrak{L}^{[p]}\phi^{(k)}$ is not the finite element solution and $\bar{w}_k$ is not its error.

Step 2: In order to estimate this error we need to construct the (periodic) *correction* $\bar{w}_k(\mathbf{x})$ on every cell $\mathfrak{c}(y, h)$, which is a periodic function on $\mathfrak{c}(y, h)$, satisfying

$$\sum_{i,j} \int_{\mathfrak{c}(y,h)} \bar{a}_{ij} \frac{\partial w_k}{\partial x_i} \frac{\partial v}{\partial x_j} d\mathbf{x} = -\sum_{i,j} \int_{\mathfrak{c}(y,h)} \bar{a}_{ij} \frac{\partial \bar{w}_k}{\partial x_i} \frac{\partial v}{\partial x_j} d\mathbf{x}, \tag{6.30a}$$

$$\int_{\mathfrak{c}(y,h)} w_k dx + \int_{\mathfrak{c}(y,h)} \bar{w}_k d\mathbf{x} = 0, \quad \forall \mathbf{v} \in S_{\Delta,per}^{[p]} \subset \mathfrak{c}(y, h). \tag{6.30b}$$

Here, by $S_{\Delta,per}^{[p]} \subset S_{\Delta}^{[p]}(\mathfrak{c}(y, h))$ we mean the subspace of periodic functions in $S_{\Delta}^{[p]}(\mathfrak{c}(y, h))$. The function w_k exists because the right-hand side of (6.29a) is a bounded functional on $S_{\Delta,per}^{[p]}$. From (6.29b) it is also unique. We define

$$\Psi_k = \bar{w}_k - w_k.$$

Step 3: We define

$$\Psi(\mathbf{x}) \equiv \sum_{|k|=p+1} \frac{1}{k_1!k_2!} \frac{\partial^{p+1}\mathfrak{u}}{\partial x_1^{k_1} \partial x_2^{k_2}}(\bar{x})\Psi_k(\mathbf{x}). \tag{6.31}$$

With all this structure we are now able to prove:

Theorem 6.10 *Let* $u(\mathbf{x})$ *be the solution of (5.51) and* $u_\Delta^{[p]}$ *be the finite element solution on* $\mathfrak{S}(\bar{\mathbf{x}}, H)$, *then, under assumptions 6.1–6.3, there exists an* α, $0 < \alpha < 1$ *and* C_1 *and* $C_2, 0 < C_1 \leqq C_2$, *such that, if* $C_1 h^\alpha \leqq H \leqq C_2 h^\alpha$, *we have*

$$\max_{i=1,2} \left\| \frac{\partial}{\partial x_i}(e_\Delta^{[p]}) - \frac{\partial}{\partial x_i}(\Psi) \right\|_{L^\infty(\Omega_1)} \leqq C\, h^{p+\nu}, \nu > 0, \tag{6.32}$$

and

$$\max \left\| \frac{\partial}{\partial x_i}(e_\Delta^{[p]}) \right\|_{L^\infty(\Omega_1)} \geqq C\, h^p. \tag{6.33}$$

Proof We have seen that the procedure in two dimensions is analogous to the one-dimensional case. The proof in two dimensions is technical. □

Theorem 6.10 shows that asymptotically $\frac{\partial}{\partial x_i}(e_\Delta^{[p]}) = \frac{\partial}{\partial x_i}(\Psi)$. It is essential that we are able to obtain both Ψ_k and Ψ locally. The superconvergent points are obtained as follows:

Step 1: Select monomials $\phi_{p+1}^{(j)}(x)$ for the cell. In contrast to the one-dimensional case, we see that for $p = 1$ there are three monomials $\phi_2^{(j)}, j = 1, 2, 3$, and for $p = 2$ there are four monomials $\phi_3^{(j)}, j = 1, 2, 3, 4$. For every $\phi_{p+1}^{(j)}$ we create the interpolation error $\tilde{w}_p^{\{j\}}$ on the single cell.

Step 2: For every $\tilde{w}_p^{(j)}$ we compute the corrections $w_p^{(j)}$ that are periodic in the cell. Then, we compute

$$\Psi_\Delta^{(j)} = \tilde{w}_p^{(j)} + w_p^{(j)} + C^{(j)},$$

i.e. $B_{\mathfrak{c}}(w_p^{(j)}, v) = B_{\mathfrak{c}}(\tilde{w}_p^{(j)}, v)\ \forall v \in S_{\Delta,Per}^{[p]}$, where $\mathfrak{c}$ is the cell, $C^{(j)}$ is such that $\int_{\mathfrak{c}} \Psi_\Delta^{(j)} d(\mathbf{x}) = 0$ and then $\Psi_\Delta^{(j)}$ is asymptotically equal to the error $e_\Delta^{[p]}$, when the exact solution $u = \phi_{p+1}^{(j)}$.

Step 3: Find the sets of points (which lie on lines) in the cell at which $\frac{\partial \Psi_\Delta^{(j)}}{\partial x_1}$ and $\frac{\partial \Psi_\Delta^{(j)}}{\partial x_2} = 0$. If there is a point of superconvergence for $\frac{\partial u}{\partial x_1}$, then all the lines for which $\frac{\partial \Psi_\Delta^{(j)}}{\partial x_1} = 0$ have to intersect at this point; similarly for $\frac{\partial u}{\partial x_2} = 0$. Such points may or may not exist.

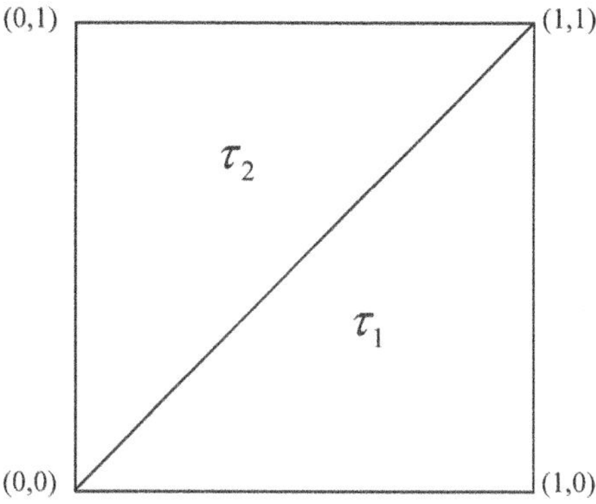

Fig. 6.8 Mesh cell.

The main assumptions in the above are the periodicity of the mesh, the smoothness of the solution u and the negligibility of pollution.

Example 6.15 Consider a mesh with a cell as in Fig 6.8. By scaling it is sufficient to consider the case $h = 1$.

Let us consider the case $p = 1$, i.e. linear shape functions on the elements. For this, we describe the steps;

Step 1: We have the monomials $\phi_2^{(1)} = x_1^2$, $\phi_2^{(2)} = x_1x_2$, $\phi_2^{(3)} = x_2^2$. From these we construct the functions $\tilde{w}_1^{(j)}$ on τ_1 and τ_2 by interpolation so that

$$\begin{aligned}
\tilde{w}_1^{(1)} &= x_1^2 - x_1 \text{ on } \tau_1 \text{ and } \tau_2,\\
\tilde{w}_1^{(2)} &= x_1x_2 - x_2 \quad \text{on } \tau_1,\\
&= x_1x_2 - x_1 \quad \text{on } \tau_2,\\
\tilde{w}_1^{(3)} &= x_2^2 - x_2 \text{ on } \tau_1 \text{ and } \tau_2.
\end{aligned}$$

Step 2: The correction $w_1^{(j)}$ is a constant, i.e. $\Psi_\Delta^{\{j\}} = \tilde{w}_1^{(j)} + C$, and hence,

$$\frac{\partial \Psi_\Delta^{(j)}}{\partial x_1} = \frac{\partial \tilde{w}_1^{(j)}}{\partial x_1}, \frac{\partial \Psi_\Delta^{(j)}}{\partial x_2} = \frac{\partial \tilde{w}_1^{(j)}}{\partial x_2}.$$

Step 3: We have to find the sets of points in the cell at which

$$\frac{\partial \Psi_\Delta^{(j)}}{\partial x_1} = 0 \text{ and } \frac{\partial \Psi_\Delta^{(j)}}{\partial x_2} = 0.$$

From step 2 we see that:

(1) $\dfrac{\partial \Psi_{\Delta}^{(1)}}{\partial x_1} = 2x_1 - 1$ on $\tau_1$and τ_2, and hence the zero in both elements lies on the line $x_1 = \frac{1}{2}$.

(2) $\dfrac{\partial \Psi_{\Delta}^{(1)}}{\partial x_2} = 0.$

(3) $\dfrac{\partial \Psi_{\Delta}^{(2)}}{\partial x_1} = x_2$ on τ_1, and $x_2 - 1$ on τ_2. Hence, the zero lines are $x_2 = 0$ on τ_1 and $x_2 = 1$ on τ_2.

(4) $\dfrac{\partial \Psi_{\Delta}^{(2)}}{\partial x_2} = x_1 - 1$ on τ_1, and $= x_1$ on τ_2. Hence, the zero lines are $x_1 = 1$ on τ_1 and $x_1 = 0$ on τ_2.

(5) $\dfrac{\partial \Psi_{\Delta}^{(3)}}{\partial x_1} = 0.$

(6) $\dfrac{\partial \Psi_{\Delta}^{(3)}}{\partial x_2} = 2x_2 - 1,$

so that the zero line is $x_2 = \frac{1}{2}$ on $\tau_1$and τ_2.

Hence, the points of superconvergence for $\dfrac{\partial u}{\partial x_1}$ are the intersections of the lines:

$$\text{for element } \tau_1; \quad x_1 = \frac{1}{2}, \; x_2 = 0,$$

$$\text{for element } \tau_2; \quad x_1 = \frac{1}{2}, \; x_2 = 1,$$

which are shown in Fig. 6.9a. In the same way we get the points of superconvergence for $\dfrac{\partial u}{\partial x_2}$ as shown in Fig. 6.9b.

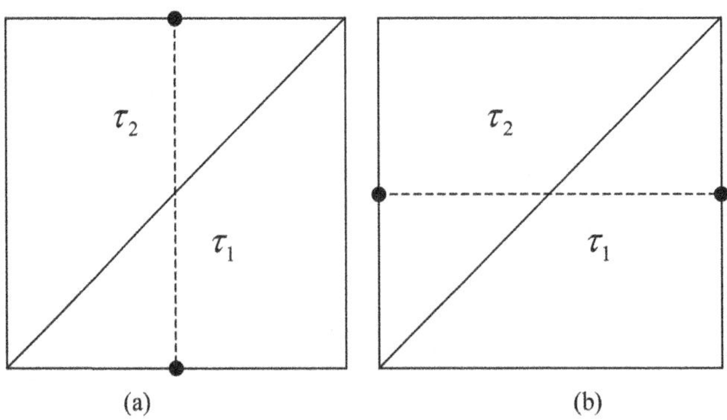

Fig. 6.9 Points of superconvergence.

In the analogous manner we can obtain the points of superconvergence for the case $p = 2$. In this case, for $\dfrac{\partial u}{\partial x_1}$ these are (as in the one-dimensional case) the Gauss points on the cell edges that are parallel to the x_1-axis, as shown in Fig. 6.10.

Exercise 6.10 Show that the points in Fig. 6.10 are superconvergence points when $p = 2$.

For the cells of the meshes in Figs. 6.8 the points of superconvergence for $\dfrac{\partial u}{\partial x_1}$ are as shown in Figs. 6.11a and 6.11b, respectively, for $p = 1$ and $p = 2$.

Example 6.16 Consider the mesh as in Fig. 6.12.

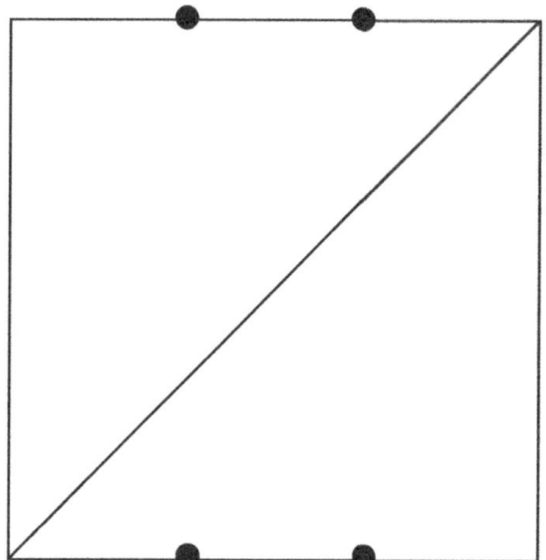

Fig. 6.10 Points of superconvergency for $\dfrac{\partial u}{\partial x_1}$.

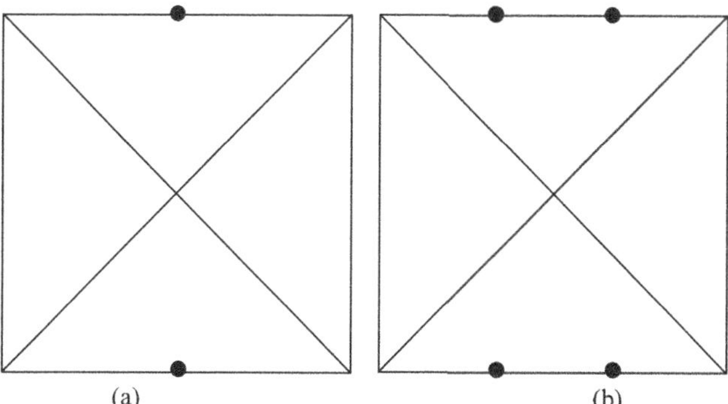

Fig. 6.11 Points of superconvergence for $\dfrac{\partial u}{\partial x_1}$ for (a) $p = 1$, (b) $p = 2$.

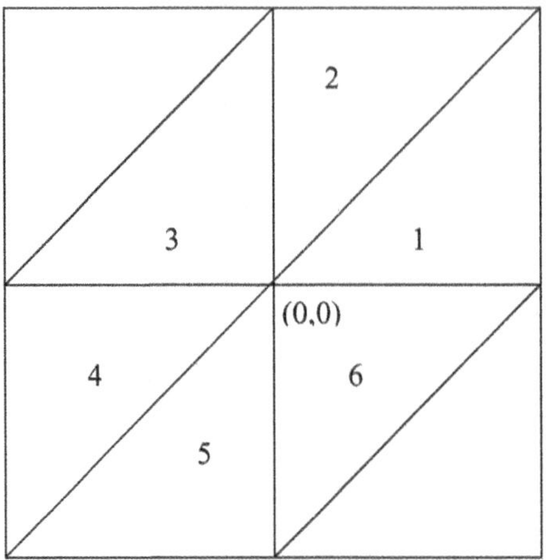

Fig. 6.12 Mesh with (0,0) at centre.

Let there be specified in the elements the points $y_i, i = 1, \ldots, 6$ where $y_1 = (\frac{2}{3}, \frac{1}{2})$, $y_2 = (\frac{1}{3}, \frac{2}{3})$, $y_3 = (-\frac{1}{3}, \frac{1}{3})$, $y_4 = (-\frac{2}{3}, \frac{1}{3})$, $y_5 = (-\frac{1}{3}, -\frac{2}{3})$, $y_6 = (\frac{1}{3}, -\frac{1}{3})$. For the case $p = 1$, these points are not points of superconvergence for $\frac{\partial u}{\partial x_1}$ or $\frac{\partial u}{\partial x_2}$. Show that the values

$$\rho_j = \frac{1}{6} \sum \frac{\partial u_{\Delta}^{[1]}}{\partial x_j}(y_1),\ j = 1, 2$$

are superconvergent values for $\dfrac{\partial u}{\partial x_j}$ at the point (0,0).

Remark 6.4 *As was said in Step 3 of the procedure for locating the points of superconvergence, such points occur where the lines on which* $\frac{\partial \Psi_{\Delta}^j}{\partial x_1} = 0, j = 1, \ldots, s(p)$ *intersect, where* $s(p)$ *is a member of the set of monomials of degree* p*; i.e.* $s(1) = 3$ *and* $s(2) = 4$*, which is the case for* $\Delta u = f$*. A similar situation exists for* $\frac{\partial \Psi_{\Delta}^j}{\partial x_2} = 0$.

For the special case of the Laplace equation $\Delta u = 0$*, the solution is of course a harmonic function so that for any* p *there are only* ***two*** *harmonic monomials. In this case, the superconvergence points always exist.*

6.2.4 Engineering application: two-dimensional heat-transfer problem

In Section 5.2.3 we addressed the finite element computation of our 2D Eng Problem, (2.90), the cooling pipe with fins. The computed results with different mesh patterns of the quantity of interest Q_3 (the total energy) were computed with the theoretical predictions. In this section, we now analyze the accuracy of the computed quantities $Q_1^{\Delta}(A)$ and $Q_1^{\Delta}(B)$, the values of the temperature at points A and B, see Figs. 2.14 and 2.15, and the quantity Q_2^{Δ}, the amount of heat going into the environment. As explained in Section 5.2.3 the exact solution of the problem has a singularity in the neighbourhood of the corner at point (5), see Fig. 2.15, with the strength of the singularity characterized by the parameter $\beta = 0.6742 \approx 2/3$.

(1) We first address the accuracy of the quantities of $Q_1^\Delta(A)$ and $Q_1^\Delta(B)$. These quantities are functionals but, in contrast to the one-dimensional case, they are not continuous. For the analysis we use here Theorem 6.9 and (6.26) to estimate $\left\| e_\Delta^{[p]} \right\|_{L^\infty(\Omega^0)}$. This theorem is also true when $\bar{\Omega}^0$ intersects a smooth part of the $L^\infty(\Omega^0)$ boundary. Using now Theorem 4.11 we get $\left\| e_I^{[p]} \right\|_{C^0(\Omega^0)} \leq Ch^{p+1}$, so that the right-hand side of (4.54a) is of $0(h^{p+1})$. By contrast, the second term $\left\| e_\Delta^{[p]} \right\|_{L^2(\Omega^0)} = 0\left(h^{\min(p+1,2\beta)}\right)$ by Theorem 6.9. Hence, because $N = 0(h^{-2})$ we expect theoretically that the error in $Q_1^\Delta(A)$ and $Q_1^\Delta(B)$ to be $0\left(h^{2\beta}\right) \approx 0\left(N^{-\beta}\right)$ for $p = 1$ and 2 for all mesh patterns. The accuracies for $Q_1^\Delta(A)$ are shown in Tables 3.10 and 3.11 and Figs. 3.28 and 3.29 for $p = 1, 2$. From these figures we see clearly the theoretical slope $\approx 2/3$ and that the accuracy is practically the same for all four mesh patterns; as theoretcially it should be.
Tables 3.12 and 3.13 show analogous results for $Q_1^\Delta(B)$, whilst Figs. 3.30 and 3.31 show the graph of the relative errors for $p = 1$ and 2. We see once more the rate $0\left(N^{-\beta}\right)$ but, interestingly and by contrast to $Q_1^\Delta(A)$, that now the pre-asymptotic range is larger.

(2) We return to the quantity of interest Q_2, the amount of heat going into the environment. In Section 3.4.4 we presented two different strategies for dealing with QoI; first, direct computation by integrating the flux $\partial u/\partial n_c$ at the boundary so that we calculated $Q_2^{\Delta,1}$, and secondly, using the boundary condition when we integrate u, where we calculated $Q_2^{\Delta,2}$. This second case clearly produces a bounded functional, while the first case does not and $\partial u/\partial n_c \notin L^\infty$. We have no theory for addressing this case.

As previously, Tables 3.16 and 3.17 and Figs. 3.34 and 3.35 give the values for $Q_2^{\Delta,2}$. Because the functional is bounded Theorem 6.5 predicts a convergence rate $0\left(N^{-\beta}\right) \approx 0\left(N^{-2/3}\right)$ for both $p = 1, 2$. This is clearly visible from Figs. 3.34 and 3.35.

Turning again to $Q_2^{\Delta,1}$, we have Tables 3.14 and 3.15 and Figs. 3.32 and 3.33 that report the errors for the direct computations using $\partial u/\partial n_c$. We see in this case much lower accuracy and a convergence rate $0\left(N^{-1/3}\right)$ and much larger influence of a particular mesh patterns. As mentioned above we cannot address this case in the light of the theory. However, we can say that in general using derivatives leads to results of low quality.

Remark 6.5 *As was said, Q_2 is not a bounded functional. However, we could use the idea of changing it into a bounded functional and there are many options for this. Let us mention one possibility. First, it is obvious that $Q_2 = Q_2^* = 16 \int_{\Gamma_1} \partial u/\partial n_c$. Let now $U(x, y) = (x - r_3)/(r_3 - r_1)$ for $r_1 < r < r_3$ and $U = 0$ elsewhere. Than, the functional of interest will be $Q_2^*(u) = B(u, U)$, which is clearly a bounded functional and function U lies in the finite element space. We have not computed this method because it is a very special case.*

7

a posteriori error estimates

Summary

- ***a posteriori*** **error estimates, which are calculable in terms of the computed finite element solution to a problem, are derived.**
- **These estimates require the use of error indicators and estimators. Various error indicators and estimators are defined, first in a one-dimensional setting, and then in two dimensions.**
- **The estimators are applied to the 1D and 2D Eng Problems and to Benchmark Problem 1, and their performance is discussed. Various morals are drawn from the results.**

In Chapters 4 and 5 we addressed the task of deriving *a priori* error estimates for the finite element method. In order to produce these estimates we assumed that we knew the norms of the higher derivatives of the solutions $\mathfrak{u}$. The estimates then had the form

$$\left\| e_{\Delta}^{[p]} \right\|_{\mathcal{U}} \leq \mathcal{C}\, h^{\mu} \left\| \frac{\partial^r \mathfrak{u}}{\partial x^r} \right\|_{\zeta},$$

where ζ was either $C(\Omega)$ or $L^2(\Omega)$. We concentrated on finding the appropriate exponent μ, and did not try to specify the constant $\mathcal{C}$, which was independent of both $\mathfrak{u}$ and the mesh size h. Estimates of this type are important for theoretical reasons; in particular because they show that a method coverges and the conditions under which the convergence will take place. However, they cannot be used for quantitative error estimation because, even if we knew $\mathcal{C}$, we do not know the values of the norms of the higher derivatives of $\mathfrak{u}$.

In order to quantify the errors in the finite element solutions we thus now seek *a posteriori* estimates. In constrast to the *a priori* case such estimates do not give any information about the rate of convergence, nor on what convergence depends. They depend on computed results, and are of major importance in the use of finite element methods in practice.

Since the original concept of *a posteriori* error estimation was made and the original estimators were proposed, Babuška and Rheinboldt (1978 a,b), many types of estimator have appeared in the literature, see Ainsworth and Oden (2000), Ladeveze and Leguillon (1983), Babuška and Strouboulis (2001), Zienkiewicz and Taylor (2005). We here describe only a few estimators that are representative of the groups of residual-based and smoothing-based estimators. In particular, we shall address only those estimators that seem to perform well in the two-dimensional context.

7.1 Error indicators and estimators in one dimension

7.1.1 The Dirichlet element-based error estimator

We shall first discuss *residual estimators*, and this we do by again considering the weak problem (2.31) and the finite element problem (3.18) so that we have

$$\text{find } \mathfrak{u} \in \mathcal{U}_0, \text{ such that } B(\mathfrak{u}, v) = F(v)\ \forall v \in \mathcal{U}_0, \tag{7.1a}$$

and, for $p = 1, 2$,

$$\text{find } u_\Delta^{[p]} \in S_{\Delta,0}^{[p]}, \text{such that } B(u_\Delta^{[p]}, v) = F(v)\ \forall v \in S_{\Delta,0}^{[p]}. \tag{7.1b}$$

As before, we define the error $e_\Delta^{[p]} \equiv \mathfrak{u} - u_\Delta^{[p]}$, $p = 1, 2$, and for any $v \in \mathcal{U}_0$ we have that

$$B(e_\Delta^{[p]}, v) = B(\mathfrak{u} - u_\Delta^{[p]}, v) = F(v) - B(u_\Delta^{[p]}, v) \equiv \Re(v).$$

The $\Re(v)$ is called the *residual* and is computable for any $v \in \mathcal{U}_0$. Note that $\Re(v) = 0, \forall v \in S_{\Delta,0}^{[p]}$; this important property will be utilized later.

Let us now write the residual $\Re(v)$ in a different form. For $a \in C^1[I]$ and $f \in L^2(I)$ we have that

$$\Re(v) = \sum_{q=1}^{M} \int_{\tau_q} r_q v dx + \sum_{q=2}^{M+1} \mathcal{J}_q v(x_q), \tag{7.2a}$$

where

$$r_q = f + \frac{d}{dx}\left(a \frac{du_\Delta^{[p]}}{dx}\right) - cu_\Delta^{[p]}, \tag{7.2b}$$

$$\mathcal{J}_q = \left(a \frac{du_\Delta^{[p]}}{dx}\right)(x_q + 0) - \left(a \frac{du_\Delta^{[p]}}{dx}\right)(x_q - 0),\ q = 2, \ldots, M, \tag{7.2c}$$

and, see (2.14c),

$$\mathcal{J}_{M+1} = G_2 - \alpha_2 u_\Delta^{[p]}(x_{M+1}) - \left(a \frac{d}{dx}(u_\Delta^{[p]})(x_{M+1} - 0)\right). \tag{7.2d}$$

In the above, we have denoted by $x_q \pm 0$, limiting values from the right and left, respectively. The $\mathcal{J}_q$ in (7.2a) are *jumps* in fluxes, and follow immediately from integration by parts.

Remark 7.1 *We have assumed here that, with a, f as in (2.14), $a \in C^1[I]$ and $f \in C^0[I]$ so that $\Re(v)$ has meaning in the classical sense. These assumptions will persist throughout the chapter.*

Exercise 7.1 Establish the forms of (7.2).

Example 7.1 Let $a = 1, c = 0, g_1 = 0, G_2 = 0, \alpha_2 = 0, l = 1$, and let $f(x) = \sin\left(\frac{\pi}{2}x\right)$. The boundary value problem (2.14) then becomes

$$-\frac{d^2\mathrm{u}}{dx}(x) = \sin\left(\frac{\pi}{2}x\right), x \in (0,1)$$

$$\mathrm{u}(0) = \frac{d\mathrm{u}}{dx}(1) = 0,$$

and has exact solution $u = \frac{4}{\pi^2}\sin\left(\frac{\pi}{2}x\right)$.

Using a finite element method with $p = 1$ and a uniform mesh Δ we have from Exercise 6.1 that $\mathrm{u}(x_q) = \mathrm{u}_\Delta^{[1]}(x_q)$, $q = 1, \ldots, M+1$, and hence $u_\Delta^{[1]}(x_q^\Delta)$ is the linear interpolant to u over Δ. Now, for this problem

$$r_q(x) = \sin\left(\frac{\pi}{2}x\right)$$

$$\mathcal{J}_q = \frac{4}{h\pi^2}\left(\sin\left(\frac{\pi}{2}x_{q+1}\right) - 2\sin\left(\frac{\pi}{2}x_q\right) + \sin\left(\frac{\pi}{2}x_{q-1}\right)\right)$$

$$= -\frac{8}{h\pi^2}\left(1 - \cos\frac{\pi}{2}h\right)\sin\frac{\pi}{2}x_q,\ q = 1, \ldots, M$$

$$\mathcal{J}_{M+1} = -\frac{4}{h\pi^2}\left(\sin\left(\frac{\pi}{2}x_{M+1}\right) - \sin\left(\frac{\pi}{2}x_M\right)\right).$$

For the problems (7.1a and b) and the quantities as defined in (7.2) we have the following theorem.

Theorem 7.1 *Let $\mathrm{u} \in \mathcal{U}_0$ be the solution of problem (7.1a) and let $u_\Delta^{[p]} \in S_{\Delta,0}^{[p]}$ be the solution of the finite element problem (7.1b), then for $\Re(v), v \in \mathcal{U}_0$, as in (7.2a)*

$$\left\|e_\Delta^{[p]}\right\|_{\mathcal{U}} = \sup_{v\in\mathcal{U}_0}\frac{\Re(v)}{\|v\|_{\mathcal{U}}},\ p = 1, 2. \tag{7.3}$$

Proof By definition, $\sup_{v\in\mathcal{U}_0}\frac{\Re(v)}{\|v\|_{\mathcal{U}}} = \sup_{v\in\mathcal{U}_0}\frac{B(e_\Delta^{[p]}, v)}{\|v\|_{\mathcal{U}}}$.

From the definition of the energy norm, see Chapter 2, this yields (7.3). □

Theorem 7.1 gives immediately a computable lower bound for the error in the energy norm; i.e.

$$\left\|e_\Delta^{[p]}\right\|_{\mathcal{U}} \geqq \frac{\Re(v)}{\|v\|_{\mathcal{U}}},\ \forall v \in \mathcal{U}_0.$$

We now introduce the concept of an *error indicator* η_q, an *error estimator* ε and the effectivity index $\varkappa$. Specifically, we define

$$\text{indicator: } \eta_q^{\text{Dir}} \equiv \left\| \hat{e}_q^{[p]} \right\|_{\mathcal{U}(\tau_q)}, \tag{7.4a}$$

$$\text{estimator: } \varepsilon^{\text{Dir}} \equiv \left(\sum_{q=1}^{M} \left(\eta_q^{\text{Dir}} \right)^2 \right)^{\frac{1}{2}}, \tag{7.4b}$$

$$\text{effectivity index: } \varkappa^{\text{Dir}} \equiv \frac{\varepsilon^{\text{Dir}}}{\left\| e_\Delta^{[p]} \right\|_{\mathcal{U}}}, \tag{7.4c}$$

where

$$\hat{e}_q^{[p]} \in \mathcal{U}_0(\tau_q) = \left\{ v \mid \int_{\tau_q} \left(\frac{dv}{dx} \right)^2 dx < \infty, v\left(x_{q+1}\right) = v\left(x_q\right) = 0 \right\} \tag{7.4d}$$

is the solution of the (local) Dirichlet problem on $\tau_q, q = 1, \ldots, M$,

$$\begin{aligned} B_{\tau_q}\left(\hat{e}_q^{[p]}, v\right) &= \int_{\tau_q} \left(a \frac{d\hat{e}_q^{[p]}}{dx} \frac{dv}{dx} + c\hat{e}_q^{[p]} v \right) dx \\ &= \int_{\tau_q} r_q v \, dx, \ \forall v \in \mathcal{U}_0(\tau_q). \end{aligned} \tag{7.5}$$

In the local context the notation $\mathcal{U}_0(\tau_q)$ will have meaning as defined in (7.4d), even though globally it has a different meaning.

The η_q^{Dir} is called the Dirichlet error indicator for the element τ_q and the Dir notation is used to emphasize that η_q^{Dir} comes from $\hat{e}_q^{[p]}$, which is found by solving a Dirichlet problem on each element. The function $\hat{e}_q^{[p]}$ exists and is unique because $\Re(v) \leqq \mathcal{C} \|v\|_{\mathcal{U}(\tau_q)} \ \forall v \in \mathcal{U}_0(\tau_q)$, where $\mathcal{C}$ depends on a, c and f. The ε^{Dir} is an *error estimator* based on the error indicator η_q^{Dir}, and the effectivity index $\varkappa^{\text{Dir}}$ characterizes the quality of the estimator. We now give a theorem that shows that ε^{Dir} is a lower bound for the error.

Theorem 7.2 *Let* $\mathrm{u} \in \mathcal{U}_0$ *and* $u_\Delta^{[p]} \in S_{\Delta,0}^{[p]}$ *be as in Theorem 7.1 and* ε^{Dir} *be as defined in (7.4b). Then*

$$\varepsilon^{Dir} \leqq \left\| e_\Delta^{[p]} \right\|_{\mathcal{U}}.$$

Proof Let $\hat{e}^{[p]}$ be defined on I such that $\hat{e}^{[p]}|_{\tau_q} = \hat{e}_q^{[p]}$. Since $\hat{e}_q^{[p]}(x_q) = \hat{e}_q^{[p]}(x_{q+1}) = 0$, $q = 1, \ldots, M$, we have that $\hat{e}^{[p]} \in \mathcal{U}_0$ and hence, using Theorem 7.1, it follows that

$$\left\| e_\Delta^{[p]} \right\|_{\mathcal{U}} \geqq \frac{\Re(\hat{e}^{[p]})}{\left\|\hat{e}^{[p]}\right\|_{\mathcal{U}}} = \frac{B(e_\Delta^{[p]}, \hat{e}^{[p]})}{\left\|\hat{e}^{[p]}\right\|_{\mathcal{U}}} = \frac{\sum_{q=1}^{M} \left(\eta_q^{\text{Dir}}\right)^2}{\left\|\hat{e}^{[p]}\right\|_{\mathcal{U}}} = \varepsilon^{\text{Dir}}.$$

□

We now seek to show that there exists a constant $\mathcal{K}$ that depends on a and c but not on u and Δ such that $\left\|e^{[p]}\right\|_{\mathcal{U}} \leqq \mathcal{K}\varepsilon^{\mathrm{Dir}}$. We start by proving a lemma.

Lemma 7.1 *If a and c satisfy assuptions (2.20b,c) and if τ_q is an element of the mesh Δ, then*

$$\left\|v - \mathcal{L}^{[1]}v\right\|_{\mathcal{U}(\tau_q)} \leqq \mathcal{K}\left\|v\right\|_{\mathcal{U}(\tau_q)} \ \forall v \in \mathcal{U}_0, \tag{7.6a}$$

where

$$\mathcal{K} \leqq \left(\frac{a_{\max}}{a_{\min}}\right)^{\frac{1}{2}}\left(1 + \frac{c_{\max}}{a_{\max}}h_q^2\right)^{\frac{1}{2}}. \tag{7.6b}$$

Proof Let us first consider the case where $a = 1, c = 0$. Noting that for $v \in \mathcal{U}_0$

$$\frac{d}{dx}\left(\mathcal{L}^{[1]}v\right)|_{\tau_q} = \frac{1}{h_q}\left(v(x_{q+1}) - v(x_q)\right) = const,$$

and

$$\int_{\tau_q} \frac{d}{dx}\left(v - \mathcal{L}^{[1]}v\right) dx = 0,$$

we now find that

$$\int_{\tau_q}\left(\frac{dv}{dx}\right)^2 dx = \int_{\tau_q}\left(\frac{d}{dx}\left(v - \mathcal{L}^{[1]}v\right)\right)^2 dx + \int_{\tau_q}\left(\frac{d}{dx}\left(\mathcal{L}^{[1]}v\right)^2\right) dx.$$

Thus,

$$\int_{\tau_q}\left(\frac{d}{dx}\left(v - \mathcal{L}^{[1]}v\right)\right)^2 \leqq \int_{\tau_q}\left(\frac{dv}{dx}\right)^2 dx,$$

which proves (7.6a); note that this last inequality is that of Remark 4.4.

In the general case, we have that

$$\begin{aligned}\int_{\tau_q} a\left(\frac{d}{dx}\left(v - \mathcal{L}^{[1]}v\right)\right)^2 dx &\leqq a_{\max}\int_{\tau_q}\left(\frac{d}{dx}\left(v - \mathcal{L}^{[1]}v\right)\right)^2 dx \\ &\leqq a_{\max}\int_{\tau_q}\left(\frac{dv}{dx}\right)^2 dx \leqq \frac{a_{\max}}{a_{\min}}\left\|v\right\|^2_{\mathcal{U}(\tau_j)}.\end{aligned}$$

Further, because $(v - \mathcal{L}^{[1]}v)(x_q^{\Delta}) = 0$, for $x \in \tau_q$ we have

$$\left|v - \mathcal{L}^{[1]}v\right|(x) \leqq h_q^{\frac{1}{2}}\left(\int_{\tau_q}\frac{d}{dx}\left(v - \mathcal{L}^{[1]}v\right)^2 dx\right)^{\frac{1}{2}} \leqq h_q^{\frac{1}{2}}\left(\int_{\tau_q}\left(\frac{dv}{dx}\right)^2 dx\right)^{\frac{1}{2}},$$

which yields

$$\int_{\tau_q} c\left(v - \mathcal{L}^{[1]}v\right)^2 dx \leqq h_q^2\frac{c_{\max}}{a_{\min}}\left\|v\right\|^2_{\mathcal{U}(\tau_q)},$$

and hence that

$$\left\| v - \mathcal{L}^{[1]} v \right\|_{\mathcal{U}(\tau_{qj})}^2 \leqq \left(\frac{a_{\max}}{a_{\min}} + \frac{c_{\max}}{a_{\min}} h_q^2 \right) \|v\|_{\mathcal{U}(\tau_q)}^2,$$

which proves (7.6). □

Remark 7.2 *So far, we have assumed that a and c satisfy only (2.20b and c). If in addition $\left\| \frac{da}{dx} \right\|_{L^\infty(I)} \leqq Q < \infty$, for some Q, then we have that*

$$\max_{x \in \tau_q} a(x) \ / \ \min_{x \in \tau_q} a(x) = 1 + \mathcal{C}h,$$

which yields

$$\mathcal{K} \leqq 1 + \mathcal{C}h. \tag{7.7}$$

We now have

Theorem 7.3 *Let a and c satisfy the assumptions as in Lemma 7.2, then*

$$\left\| e_\Delta^{[p]} \right\|_{\mathcal{U}} \leqq \mathcal{K} \varepsilon^{Dir}, \tag{7.8}$$

where $\mathcal{K}$ satisfies (7.6b). If, further, the assumptions of Remark 7.2 hold, then $\mathcal{K}$ satisfies (7.7)

Proof For $v \in \mathcal{U}_0$ we have that

$$\Re(v) = \Re(v - \mathcal{L}^{[1]}v) = \sum_{q=1}^{M} B_{\tau_q}\left(e_\Delta^{[p]}, v - \mathcal{L}^{[1]}v\right) = \sum_{q=1}^{M} B_{\tau_q}\left(\hat{e}^{[p]}, v - \mathcal{L}^{[1]}v\right)$$
$$\leqq \varepsilon^{\mathrm{Dir}} \left\| v - \mathcal{L}^{[1]}(v) \right\|_{\mathcal{U}} \leqq \varepsilon^{\mathrm{Dir}} \mathcal{K} \|v\|_{\mathcal{U}},$$

and then Theorem 7.1 yields (7.8). Here, we use the fact that $\Re(v) = 0 \ \forall v \in S_{\Delta,0}^{[1]}$; obviously $\mathcal{L}^{[1]}v \in S_{\Delta,0}^{[1]}$. □

Result (7.8) means that $\mathcal{K}\varepsilon^{\mathrm{Dir}}$ is an upper bound. We have thus from (7.5) and (7.8) shown that

$$\varepsilon^{\mathrm{Dir}} \equiv \varepsilon^{\mathrm{Dir},L} \leqq \left\| e_\Delta^{[p]} \right\|_{\mathcal{U}} \leqq \varepsilon^{\mathrm{Dir},U} \equiv \mathcal{K}\varepsilon^{\mathrm{Dir}},$$

where $\varepsilon^{\mathrm{Dir},L}$ and $\varepsilon^{\mathrm{Dir},U}$ are, respectively, upper and lower bounds, so that we have good estimation of $\left\| e_\Delta^{[p]} \right\|_{\mathcal{U}}$, even though we do not know u, and this is in practice very important. The effectivity index $\varkappa^{\mathrm{Dir}}$ then converges to 1 as $h \to 0$ when the coefficient a is smooth.

Example 7.2 Let us consider again the problem of Example 5.5, and let $M = 80$, $h_1 = h^{0.15}$ and $h_q = \frac{1-h_1}{M}$, $q = 2, \ldots, M$. Then, for $p = 1$ we have in the first element that

$$\varkappa_{\tau_1}^{\mathrm{Dir}} \equiv \frac{\eta_1^{\mathrm{Dir}}}{\left\| e_{\Delta}^{[1]} \right\|_{\mathcal{U}(\tau_1)}} = \frac{4068\mathrm{E}-5}{4069\mathrm{E}-5} \doteq 0.9998,$$

whilst for the rest of I

$$\varkappa_{I-\tau_1}^{\mathrm{Dir}} \equiv \frac{\left(\sum_{q=2}^{M} \left(\eta_q^{\mathrm{Dir}}\right)^2 \right)^{\frac{1}{2}}}{\left\| e_{\Delta}^{[1]} \right\|_{\mathcal{U}(I-\tau_1)}} = \frac{3682\mathrm{E}-7}{9863\mathrm{E}-7} \doteq 0.3733.$$

Thus, the indicators underestimate the errors in all elements $\tau_q, q > 1$. Nevertheless,

$$\left\| e_{\Delta}^{[1]} \right\|_{\mathcal{U}(I-\tau_1)} = 9863\mathrm{E}-7 \ll \left\| e_{\Delta}^{[1]} \right\|_{\mathcal{U}(\tau_1)} = 4060\mathrm{E}-5,$$

and

$$\varkappa^{\mathrm{Dir}} \equiv \frac{\varepsilon^{\mathrm{Dir}}}{\left\| e_{\Delta}^{[1]} \right\|_{\mathcal{U}_0}} = 0.9995 \approx \varkappa_{\tau_1}^{\mathrm{Dir}} = 0.9998.$$

We see that $\varepsilon^{\mathrm{Dir}} \leqq \left\| e_{\Delta}^{[1]} \right\|_{\mathcal{U}}$. On the other hand, we have $a_{\max} = 2, a_{\min} = 1, C_{\max} = 0$, and so

$$\mathcal{K} = \left(\frac{a_{\max}}{a_{\min}} \right)^{\frac{1}{2}} = \sqrt{2},$$

which gives

$$\begin{aligned} \varepsilon^{\mathrm{Dir}} &= 4069\mathrm{E}-5 \leqq \left\| e_{\Delta}^{[1]} \right\|_{\mathcal{U}_0} = 4070\mathrm{E}-5 \\ &\leqq \mathcal{K}\, \varepsilon^{\mathrm{Dir}} = 5754\mathrm{E}-5. \end{aligned}$$

We also see here that $\varepsilon^{\mathrm{Dir}}$ is a very high quality estimator on I, although it is low quality on $(I - \tau_1)$ because of the pollution in $I - \tau_1$ caused by the large element τ_1. Note that no *local* indicator is able to identify pollution.

Returning to (7.4a) we see that, if we wish to compute the indicator η_q^{Dir}, we have to be able to obtain the indicator function $\hat{e}_q^{[p]}$ and then to compute its norm. This will, in general, not be possible and we resort to computing an approximation to $\hat{e}_q^{[p]}$ by solving the Dirichlet problem on τ_q either by using a higher-order element or by using a finer mesh over τ_q, and then computing the indicator. This indicator is then computable and is lower than the theoretical one. This is because, as was stated earlier, the energy of the computed solution is always lower than the true energy because the computed solution is a projection onto a subspace. This approximating function still produces a lower bound, although it is very close to the exact error η_q^{Dir}.

There are various ways of obtaining an upper bound for η_q^{Dir}, one of which based on $\|r_q\|_{L_2(\tau_q)}$ we now describe.

We have that

$$\begin{aligned}
\left\|e_q^{[p]}\right\|_{\mathcal{U}_0(\tau_q)} &= \sup_{v\in\mathcal{U}_0(\tau_q)} \frac{\Re(v)}{\|v\|_{\mathcal{U}(\tau_q)}} \\
&= \sup_{v\in\mathcal{U}_0(\tau_q)} \frac{\int_{\tau_q} r_q v dx}{\|v\|_{\mathcal{U}(\tau_q)}} \\
&\leqq \|r_q\|_{L_2(\tau_j)} \sup_{v\in\mathcal{U}_0(\tau_q)} \frac{\left(\int_{\tau_q} v^2 dx\right)^{\frac{1}{2}}}{\|v\|_{\mathcal{U}(\tau_q)}}.
\end{aligned}$$

But, $v(x_q) = v(v_{q+1}) = 0$, so that

$$|v(x)| \leqq (x - x_q)^{\frac{1}{2}} \left(\int_{x_q}^{\frac{x_{q+1}+x_q}{2}} \left(\frac{dv}{dx}\right)^2 dx\right)^{\frac{1}{2}}, x_q \leqq x \leqq \frac{x_{q+1}+x_q}{2}$$

$$|v(x)| \leqq (x_{q+1} - x)^{\frac{1}{2}} \left(\int_{\frac{x_{q+1}+x_q}{2}}^{x_{q+1}} \left(\frac{dv}{dx}\right)^2 dx\right)^{\frac{1}{2}}, \frac{x_{q+1}+x_q}{2} < x < x_{q+1}.$$

Thus,

$$\int_{\tau_q} v^2 dx \leqq \frac{1}{4} h_q^2 \int_{\tau_j} \left(\frac{dv}{dx}\right)^2 dx, \tag{7.9a}$$

and hence

$$\sup_{v\in\mathcal{U}_0(\tau_j)} \frac{\left(\int_{\tau_j} v^2 dx\right)^{\frac{1}{2}}}{\left(\int_{\tau_j} \left(\frac{dv}{dx}\right)^2 dx\right)^{\frac{1}{2}}} \leqq \frac{1}{2} h_q,$$

giving

$$\eta_q^{\mathrm{Dir}} = \left\|\hat{e}_q^{[p]}\right\|_{\mathcal{U}(\tau_q)} \leqq \|r_q\|_{L_2(\tau_q)} \frac{1}{a_{\min}^{\frac{1}{2}}} \frac{h_q}{2}, \tag{7.9b}$$

because $\|v\|_{\mathcal{U}(\tau_q)}^2 \geqq \int_{\tau_j} a_{\min} \left(\dfrac{dv}{dx}\right)^2 dx$.

Remark 7.3 *Using eigenvalue theory we can replace the* $\frac{1}{2}$ *in (7.9b) by* $\dfrac{1}{\pi}$.

We have that

$$\varepsilon^{\text{Dir}} \leqq \left(\sum_{q=1}^{M} \left(\frac{1}{4a_{\min}} h_q^2 \|r_q\|_{L_2(\tau_q)}^2 \right) \right)^{\frac{1}{2}}, \tag{7.10a}$$

and using Theorem 7.2 we get

$$\left\| e_{\Delta}^{[p]} \right\|_{\mathcal{U}_0} \leqq C \left(\sum_{q=1}^{M} h_q^2 \|r_q\|_{L_2(\tau_q)}^2 \right)^{\frac{1}{2}}, \tag{7.10b}$$

where C depends only on $a(x)$. Note that the Dirichlet indicators are not influenced by the jumps because $\hat{e}_q^{[p]} \in \mathcal{U}_0(\tau_q)$

We now can introduce the *explicit indicator*

$$\eta_q^{\text{Dir,E}} \equiv \|r_j\|_{L_2(\tau_q)} h_q, \tag{7.11a}$$

and the explicit estimator

$$\varepsilon^{\text{Dir,E}} \equiv \left(\sum_{q=1}^{M} \left(\eta_q^{\text{Dir,E}} \right)^2 \right)^{\frac{1}{2}}, \tag{7.11b}$$

and we have that

$$\left\| e_{\Delta}^{[p]} \right\|_{\mathcal{U}_0} \leqq C \varepsilon^{\text{Dir,E}}, \tag{7.12}$$

where C is computable and depends only on $a(x)$ and l.

Remark 7.4 *The $\eta_q^{\text{Dir,E}}$ and the $\varepsilon^{\text{Dir,E}}$ are called, respectively, the explicit indicator and estimator, because we obtain $\eta_q^{\text{Dir,E}}$ without solving any auxiliary problem, as was the case for η_q^{Dir}; sometimes called an implicit estimator.*

We showed that the estimator ε^{Dir} is a lower estimate for the error. We must ask whether $\varepsilon^{\text{Dir,E}}$ is also a lower estimate; in general it is not, but if a, c, f are smooth input data, then

$$C \varepsilon^{\text{Dir,E}} \leqq \left\| e_{\Delta}^{[p]} \right\|_{\mathcal{U}_0}, \tag{7.13}$$

where C is independent of Δ and u.

Remark 7.5 *Because $\varepsilon^{\text{Dir,E}}$ is based only on an L_2 estimator of r_q the constant of which may be pessimistic, this estimate is most often not very accurate. However, the Dirichlet element estimator demonstrates how the error can be estimated, when the exact solution is not known. It is the simplest estimator to implement and is a lower estimate of the error. It has the disadvantage that for higher space dimensions with $p = 1$ it may very much underestimate the error, although for $p = 2$ it may give reasonable results. For this reason, we do not present any numerical results for this estimator; (see also Remark 7.9).*

7.1.2 The Neumann element-based error estimator

The Dirichlet estimator discussed in Section 7.1.1 uses only one part of the residual, namely the function r_q, see (7.2b), and not the jump $\mathcal{J}_q$. This is because the function $\hat{e}_q^{[p]} \subset \mathcal{U}_0(\tau_q)$; i.e. $\hat{e}_q^{[p]} = 0$ on the boundary of the element τ_q.

We now derive a second estimator, which involves both r_q and the jump $\mathcal{J}_q$, and uses the function $\hat{e}_q^{[p]} \subset \mathcal{U}(\tau_q)$, (i.e. not $\mathcal{U}_0(\tau_q)$), which satisfies in the element $\tau_q \equiv (x_q, x_{q+1}), q = 2, \ldots M$,

$$B_{\tau_q}\left(\hat{e}_q^{[p]}, v\right) = \int_{\tau_q} r_q v dx - \mathcal{J}_q^R v\left(x_q\right) + \mathcal{J}_{q+1}^L v\left(x_{q+1}\right)$$
$$\forall v \in \mathcal{U}(\tau_q), \tag{7.14a}$$

and

$$\hat{e}_1^{[p]} \in \mathcal{U}_0(\tau_q) \equiv \left\{u \in \mathcal{U}(\tau_1),\ u(x_1) = 0\right\},$$
$$B_{\tau_1}\left(\hat{e}_1^{[p]}, v\right) = \int_{\tau_1} r_1 v dx + \mathcal{J}_2^L v\left(x_2\right)\ \forall v \in \mathcal{U}_0(\tau_1), \tag{7.14b}$$

where $\mathcal{J}_q^R$ and $\mathcal{J}_q^L$ are defined in (7.16).

For simplicity we have used the same notation $\hat{e}_q^{[p]}$ here as for the Dirichlet estimator.

The term Neumann stems from the fact that now $\hat{e}_q^{[p]}$ satisfies non-homogeneous Neumann boundary conditions on the boundary of τ_q. In (7.14) we have that $\mathcal{J}_q = \mathcal{J}_q^R + \mathcal{J}_q^L$, see (7.15b), where $\mathcal{J}_q$ has been split into two parts, the right part $\mathcal{J}_q^R$ associated with the element τ_q and the left part $\mathcal{J}_q^L$ associated with the element τ_{q-1}.

Let us now address this splitting of $\mathcal{J}_q$. For this we let

$$\Phi_1^{[1]}(x) = \begin{matrix} \left(x_2 - x\right)/h_1,\ x \in \bar{\tau}_1 \\ 0,\ x \notin \bar{\tau}_1 \end{matrix},$$

$$\Phi_q^{[1]}(x) = \begin{matrix} \left(x_{q+1} - x\right)/h_q,\ x \in \bar{\tau}_q \\ \left(x - x_{q-1}\right)/h_{q-1},\ x \in \bar{\tau}_{q-1} \end{matrix},$$
$$q = 2, \ldots, M \tag{7.15a}$$

$$\Phi_{M+1}^{[1]}(x) = \begin{matrix} \left(x - x_M\right)/h_M,\ x \in \tau_M \\ 0,\ x \notin \tau_M \end{matrix}.$$

We know that for $q = 2, \ldots, M$

$$0 = B_{\tau_q \cup \tau_{q-1}}\left(e_\Delta^{[1]}, \Phi_q^{[1]}\right)$$
$$= \int_{\tau_{q-1}} r_{q-1}\Phi_q^{[1]} dx + \int_{\tau_q} r_q \Phi_q^{[1]} dx + \mathcal{J}_q \Phi_q^{[1]}(x_q), \tag{7.15b}$$

and

$$0 = B_{\tau_M}\left(e_{\Delta}^{[p]}, \Phi_{M+1}^{[1]}\right) = \int_{\tau_{M-1}} r_M \Phi_{M+1}^{[1]} dx + \mathcal{J}_{M+1}^L \Phi_{M+1}^{[1]}.$$

For $q = 2, \ldots, M-1$ we define

$$\mathcal{J}_q^L \equiv -\int_{\tau_{q-1}} r_{q-1} \Phi_q^{[1]} dx, \tag{7.16a}$$

$$\mathcal{J}_q^R \equiv -\int_{\tau_q} r_q \Phi_q^{[1]} dx, \tag{7.16b}$$

$$\mathcal{J}_{M+1}^L = -\int_{\tau_{M-1}} r_M \Phi_{M+1}^{[1]} dx = \mathcal{J}_{M+1}. \tag{7.16c}$$

Obviously, from (7.15b) we have $\mathcal{J}_q^L + \mathcal{J}_q^R = \mathcal{J}_q$, and

$$\int_{\tau_q} r_q dx + \mathcal{J}_{q+1}^L + \mathcal{J}_q^R = 0. \tag{7.17}$$

The solution $\hat{e}_q^{[p]}$ of (7.14) exists because of (7.17). For $q = 1$ it is unique and for $q = 2, \ldots, M$ it is unique if $c \neq 0$ on τ_q or $\alpha_2 \neq 0$. If $c = 0$ on $\tau_q, q = 2, \ldots, M-1$ and $c = 0, \alpha_2 = 0$ on τ_M, then the solution of (7.14) exists up to an additive constant. This will be determined so that

$$\int_{\tau_q} \hat{e}_q^{[p]} dx = 0.$$

Considering the function $\hat{e}^{[p]}(x), x \in I$, where

$$\hat{e}^{[p]}(x) \mid_{\tau_q} \equiv \hat{e}_q^{[p]}(x),$$

then $\hat{e}^{[p]} \notin \mathcal{U}_0(I)$ because it has discontinuities at x_q. We now define a function $\Psi(x)$ that is linear on every element τ_q and such that

$$\Psi(x_q) = \left(\hat{e}_q^{[p]} - \hat{e}_{q-1}^{[p]}\right)(x_q),$$

$$q = 2, \ldots, M,$$

$$\Psi(x_0) = 0, \Psi(x_{M+1}) = \hat{e}^{[p]}(x_{M+1}).$$

This is called a *gap function*, and clearly

$$\tilde{e}^{[p]} = \hat{e}^{[p]} - \Psi \in \mathcal{U}_0(I). \tag{7.18}$$

We can now create upper and lower estimators. From (7.2a) we have

$$
\begin{aligned}
\left\| e_\Delta^{[p]} \right\|_{\mathcal{U}} &= \frac{B\left(e_\Delta^{[p]}, e_\Delta^{[p]}\right)}{\left\| e_\Delta^{[p]} \right\|_{\mathcal{U}}} \\
&= \frac{1}{\left\| e_\Delta^{[p]} \right\|_{\mathcal{U}}} \left(\sum_{q=1}^{M} \int_{\tau_q} r_q e_\Delta^{[p]} dx + \sum_{q=2}^{M+1} \mathcal{J}_q e_\Delta^{[p]} (x_q) \right) \\
&= \frac{1}{\left\| e_\Delta^{[p]} \right\|_{\mathcal{U}}} \left(\sum_{q=1}^{M} \left(\int_{\tau_q} r_q e_\Delta^{[p]} dx + \mathcal{J}_q^R e_\Delta^{[p]} (x_q) + \mathcal{J}_q^L e_\Delta^{[p]} (x_{q+1}) \right) \right) \\
&= \frac{1}{\left\| e_\Delta^{[p]} \right\|_{\mathcal{U}}} \left(\sum_{q=1}^{M} B_{\tau_q} \left(\hat{e}_q^{[p]}, e_q^{[p]} \right) \right) \\
&\leq \frac{1}{\left\| e_\Delta^{[p]} \right\|_{\mathcal{U}}} \left(\sum_{q=1}^{M} B_{\tau_q} \left(e_\Delta^{[p]}, e_\Delta^{[p]} \right) \right)^{\frac{1}{2}} \left(\sum_{q=1}^{M} B_{\tau_q} \left(\hat{e}_q^{[p]}, \hat{e}_q^{[p]} \right) \right)^{\frac{1}{2}} \\
&\leq \left(\sum_{q=1}^{M} B_{\tau_q} \left(\hat{e}_q^{[p]}, \hat{e}_q^{[p]} \right) \right)^{\frac{1}{2}} .
\end{aligned} \tag{7.19}
$$

Hence, we define the upper and lower indicators

$$
\eta^{\text{Neu,U}} \equiv \left\| \hat{e}_q^{[p]} \right\|_{\mathcal{U}(\tau_q)},
$$

$$
\eta^{\text{Neu,L}} \equiv \left\| \tilde{e}_q^{[p]} \right\|_{\mathcal{U}(\tau_q)},
$$

and corresponding estimators

$$
\varepsilon^{\text{Neu,U}} \equiv \left(\sum_{q=1}^{M} \left\| \hat{e}_q^{[p]} \right\|_{\mathcal{U}(\tau_q)}^2 \right)^{\frac{1}{2}},
$$

$$
\varepsilon^{\text{Neu,L}} \equiv \left(\sum_{q=1}^{M} \left\| \tilde{e}_q^{[p]} \right\|_{\mathcal{U}(\tau_q)}^2 \right)^{\frac{1}{2}}.
$$

Theorem 7.4 *Let a and c satisfy the assumptions of (2.20b–d), then*

$$
\left\| e_\Delta^{[p]} \right\|_{\mathcal{U}} \leq \varepsilon^{\text{Neu,U}}. \tag{7.20}
$$

Proof This result follows from (7.19). □

Further, we have

Theorem 7.5 *We have*

$$\left\| e_{\Delta}^{[p]} \right\|_{\mathcal{U}} \geqq \varepsilon^{\text{Neu,L}}. \tag{7.21}$$

Proof This result follows directly from Theorem 7.1 and (7.18). □

As in the case of the Dirichlet estimator, with the Neumann estimator we have to solve local problems numerically and these underestimate the error. This can be avoided by obtaining upper bounds using complementary (Castigliano) principles; see, e.g., Ladeveze and Leguillon (1983).

We emphasize again that the Neumann estimator utilizes not only r_q, but also the jumps $\mathcal{J}_q$. It was originally introduced in Ladeveze and Leguillon, (1983).

The estimation of $\left\| \hat{e}_q^{[p]} \right\|_{\mathcal{U}(\tau_q)}^2$ leads to the *explicit* indicator η_q^{NeuE} and estimator $\varepsilon^{\text{NeuE}}$, where

$$\eta_q^{\text{NeuE}} \equiv \left(h_q^2 \int_{\tau_q} r_q^2 + \frac{h_q}{2} \left(\mathcal{J}_q^2 + \mathcal{J}_{q+1}^2 \right) \right)^{\frac{1}{2}}, \tag{7.22a}$$

$$\varepsilon^{\text{NeuE}} \equiv \left(\sum_{q=1}^{M} \left(\eta_q^{NeuE} \right)^2 \right)^{\frac{1}{2}}, \tag{7.22b}$$

for which, analogously to (7.12),

$$\left\| e_{\Delta}^{[p]} \right\|_{\mathcal{U}} \leqq C\varepsilon^{\text{NeuE}}. \tag{7.22c}$$

In the above we have assumed that $\hat{e}_q^{[p]}$ and all the relevant integrals have been evaluated exactly. In practice, $\hat{e}_q^{[p]}$ is computed numerically by using a higher-order element on τ_q or a refined mesh on τ_q.

7.1.3 The performance of the Neumann element-based error estimator

The Neumann element-based estimator will now be applied to the 1D Eng Problem introduced in Section 2.1.5. As was pointed out in Section 2.1.5 the solution of this problem can be expressed analytically, with the result that values of the quantities of interest $Q_i, i = 1, 2, 3$ can be computed for 1D Eng Problems 1 and 2.

Knowing the value of the total energy $Q_3(\mathrm{u})$ and having computed an approximation $Q_3(u_{\Delta}^{[p]})$, using the finite element solution $u_{\Delta}^{[p]}$, we can compute the energy norm of the error $e_{\Delta}^{[p]}$, where $\left\| e_{\Delta}^{[p]} \right\|_{\mathcal{U}}^2 = 2\left(Q_3(\mathrm{u}) - Q_3(u_{\Delta}^{[p]}) \right)$. Similarly, $\hat{e}_{\Delta}^{[p]}, q - 1, \ldots, M$ can be computed analytically.

We now present finite element results for both Problems 1 and 2, where, in the numerical experiments, only uniform meshes with mesh size h have been used with linear and quadratic elements. Here, the functions $\hat{e}_{\Delta}^{[p]}$ and $\tilde{e}_{\Delta}^{[p]}$ have been computed exactly, as have all the integrals.

Recall again that in Problem 1, the interface between the two types of steel is $b = 3.5\,cm$ (see (2.54)), and that this is a vertex point for the meshes with $M = 7, 14, 21$. However, for Problem 2 the coordinate $r_2 = b$ of the interface is irrational (see (2.55)) and hence never coincides with a vertex point of the mesh. As was seen in Chapter 3 the behaviour of the finite element solution is orderly for 1D Eng Problem 1, whilst it is 'chaotic' for 1D Eng Problem 2.

In Tables 7.1 to 7.4 we give values for:

the energy norm of the error: $\left\| e_{\Delta}^{[p]} \right\|_{\mathcal{U}}$,

the relative error: rel $= \left\| e_{\Delta}^{[p]} \right\|_{\mathcal{U}} \,/\, \|\mathrm{u}\|_{\mathcal{U}}$,

the upper and lower estimates $\varepsilon^{\mathrm{Neu,U}}, \varepsilon^{\mathrm{Neu,L}}$ of $\left\| e_{\Delta}^{[p]} \right\|_{\mathcal{U}}$,

based on the Neumann estimator, together with the associated effectivity indices: $\varkappa_{\mathrm{U}}, \varkappa_{\mathrm{L}}$, which illustrate the above behaviour.

Figures 7.1–7.4 show plots of the effectivity index for the upper and lower estimates based on the data of Tables 7.1 to 7.4. We see the following:

(1) The effectivity index of the upper and lower estimates depends on the mesh. Further, $\left\| e_{\Delta}^{[p]} \right\|_{\mathcal{U}} \to 0$ and $\varkappa \to 1$ as $h \to 0$.
(2) The upper estimates are better than those of the lower ones.
(3) The quality of the estimates is similar for Problems 1 and 2. In Problem 1 the estimates are better for $M = 7, 14, \ldots$, when the interface coincides with a nodal point of the mesh.
(4) If the error is large for M small, for example 30%, then the effectivity index is very good; note that for $M = 1$ the estimator recovers the error and hence in this case $\varkappa = 1$.

The indicators are the norms of $\hat{e}_q^{[p]}$on τ_q. Figure 7.5 shows the functions $e_{\Delta}^{[1]}, \hat{e}^{[1]}, \tilde{e}^{[1]}$ for Problem 1 with $M = 10$, where the interface does not coincide with a node of the mesh. We see clearly that the indicator function has not recognized the pollution. Figure 7.6 shows the derivatives of these three functions. Pollution is still present, but is not visible because of the scale. The errors are much more complex in the neighbourhood of the interface, as can be seen from Figs. 7.6 and 7.7. If the interface coincides with a node of the mesh, then no pollution is present and the behaviour is very orderly in the neighbourhood of the interface, as can be seen from Figs. 7.8 and 7.9.

The difference between the Dirichlet and Neumann error estimators is that the upper and lower Neumann estimators do not contain constants, whilst the upper Dirichlet estimate does.

7.1.4 The Dirichlet subdomain (patch) estimator

Returning to the Dirichlet estimator $\varepsilon^{\mathrm{Dir}}$, we saw that it is a lower bound for the error and Example 7.2 demonstrated that it is reasonably accurate as an upper estimate. We

Table 7.1 Results for Problem 1 using linear elements.

M	h	$\varepsilon^{\mathrm{Neu,L}}$	$\varkappa_{\mathrm{L}}$	$\left\|e_{\Delta}^{[p]}\right\|_{\mathcal{U}}$	rel	$\varepsilon^{\mathrm{Neu,U}}$	$\varkappa_{\mathrm{U}}$
1	3.50000	211.62649	1.00000	211.62649	0.36503	211.6249	1.00000
2	1.75000	170.75368	0.96415	177.10240	0.30548	185.87535	1.04954
3	1.16667	150.51578	0.95137	158.20987	0.27289	164.49418	1.03972
4	0.87500	133.32367	0.94606	140.92577	0.24308	145.38492	1.03164
5	0.70000	114.52991	0.95039	120.50846	0.20786	123.30831	1.02323
6	0.58333	87.71471	0.96482	90.91306	0.15681	92.12116	1.01329
7	0.50000	16.00964	0.99490	16.09178	0.02776	16.09853	1.00042
8	0.43750	53.44795	0.97938	54.57312	0.09413	54.83553	1.00481
9	0.38889	66.61923	0.96980	68.69404	0.11849	69.21668	1.00761
10	0.35000	72.43248	0.96597	74.98433	0.12934	75.66360	1.00906
11	0.31818	73.21045	0.96707	75.70375	0.13058	76.40271	1.00923
12	0.29167	68.71151	0.97267	70.64181	0.12185	71.21007	1.00804
13	0.26923	55.82024	0.98286	56.79369	0.09796	57.08940	1.00521
14	0.25000	8.01312	0.99379	8.06322	0.01391	8.06407	1.00011
15	0.23333	37.68394	0.98206	38.37227	0.06619	38.46361	1.00238
16	0.21875	48.49039	0.97063	49.95746	0.08617	50.15885	1.00403
17	0.20588	53.97908	0.95475	55.89342	0.09641	56.17532	1.00504
18	0.19444	55.66011	0.96644	57.59298	0.09934	57.90133	1.00535
19	0.18421	53.16786	0.97220	54.68814	0.09433	54.95221	1.00483
20	0.17500	43.86205	0.98309	44.61663	0.07696	4476014	1.00322
21	0.16667	5.34405	0.99375	5.37765	0.00928	5.37790	1.00005
22	0.15909	30.66431	0.98223	31.21918	0.05385	31.26839	1.00158
23	0.15217	39.92335	0.97031	41.14495	0.07097	41.25753	1.00274
24	0.14583	44.85077	0.96517	46.46945	0.08015	46.63157	1.00349
25	0.14000	46.62661	0.96581	48.27741	0.08327	48.45918	1.00377

Table 7.2 Results for Problem 2 using linear elements.

M	h	$\varepsilon^{\mathrm{Neu,L}}$	$\varkappa_{\mathrm{L}}$	$\left\Vert e_{\Delta}^{[p]}\right\Vert_{\mathcal{U}}$	rel	$\varepsilon^{\mathrm{Neu,U}}$	$\varkappa_{\mathrm{U}}$
1	3.50000	217.93709	1.00000	217.93700	0.37848	217.93709	1.00000
2	1.75000	174.94401	0.96205	181.84439	0.31580	191.45620	1.05286
3	1.16667	152.36560	0.94998	160.38774	0.27854	167.01998	1.04135
4	0.87500	131.60199	0.94714	138.94727	0.24130	143.28102	1.03119
5	0.70000	105.86302	0.95696	110.62465	0.19212	112.82400	1.01988
6	0.58333	57.03740	0.98340	58.00013	0.10073	58.31937	1.00550
7	0.50000	51.41596	0.98206	52.35544	0.09092	52.59037	1.00449
8	0.43750	68.52368	0.97068	70.59374	0.12260	71.16861	1.00814
9	0.38889	75.46711	0.96641	78.08998	0.13562	78.86741	1.00996
10	0.35000	75.89775	0.96811	78.39801	0.13615	79.18465	1.01003
11	0.31818	69.17392	0.97524	70.93026	0.12318	71.51337	1.00822
12	0.29167	48.81723	0.98802	49.40940	0.08581	49.60691	1.00400
13	0.26923	29.07000	0.99108	29.33163	0.05094	29.37300	1.00141
14	0.25000	46.68351	0.97479	47.89093	0.08317	48.07080	1.00376
15	0.23333	54.80654	0.96704	56.67465	0.09842	56.97254	1.00526
16	0.21875	57.74769	0.96644	59.75285	0.10377	60.10186	1.00584
17	0.20588	55.50396	0.97214	57.09434	0.09915	57.39889	1.00533
18	0.19444	44.69067	0.98410	45.41294	0.07887	45.56635	1.00338
19	0.18421	14.85669	0.99773	14.89048	0.02586	14.89589	1.00036
20	0.17500	35.14523	0.97849	35.91770	0.06238	35.99365	1.00211
21	0.16667	43.82472	0.96797	45.27496	0.07863	45.42698	1.00336
22	0.15909	47.83225	0.96526	49.55356	0.08606	49.75280	1.00402
23	0.15217	47.72099	0.96931	49.23179	0.08550	49.42718	1.00397
24	0.14583	41.61005	0.97985	42.46565	0.07375	42.59111	1.00295
25	0.14000	17.55305	0.99754	17.59636	0.03056	17.60530	1.00051

Table 7.3 Results for Problem 1 using quadratic elements.

M	h	$\varepsilon^{\mathrm{Neu,L}}$	$\varkappa_{\mathrm{L}}$	$\left\Vert e_{\Delta}^{[p]} \right\Vert_{\mathcal{U}}$	rel	$\varepsilon^{\mathrm{Neu,U}}$	$\varkappa_{\mathrm{U}}$
1	3.50000	150.07728	1.00000	150.07728	0.25886	150.07728	1.00000
2	1.75000	111.45122	0.98196	113.49891	0.19577	115.84106	1.02064
3	1.16667	80.56148	0.98470	81.81327	0.14112	82.69479	1.01077
4	0.87500	59.21579	0.98894	59.87796	0.10328	60.22442	1.00579
5	0.70000	52.25311	0.98958	52.80359	0.09108	53.04134	1.00450
6	0.58333	51.03725	0.98861	51.62551	0.08905	51.84772	1.00430
7	0.50000	0.57013	0.99999	0.57013	0.00098	0.57013	1.00000
8	0.43750	44.58849	0.99003	45.03753	0.07768	45.18514	1.00328
9	0.38889	47.62148	0.98860	48.17046	0.08309	48.35102	1.00375
10	0.35000	41.59789	0.99130	41.96278	0.07238	42.08220	1.00285
11	0.31818	35.39651	0.99278	35.65381	0.06150	35.72709	1.00206
12	0.29167	34.52483	0.99020	34.86646	0.06014	34.93499	1.00197
13	0.26923	35.26202	0.98724	35.71794	0.06161	35.79161	1.00206
14	0.25000	0.14359	1.00000	0.14359	0.00025	0.14359	1.00000
15	0.23333	32.28380	0.98876	32.65075	0.05632	32.70704	1.00172
16	0.21875	35.29553	0.98725	35.75140	0.06167	35.82528	1.00207
17	0.20588	31.57068	0.99030	31.87992	0.05499	31.93232	1.00164
18	0.19444	27.59390	0.99203	27.81545	0.04798	27.85026	1.00125
19	0.18421	27.55075	0.98958	27.84095	0.04802	27.87586	1.00125
20	0.17500	28.52968	0.98670	28.91428	0.04987	28.95338	1.00135
21	0.16667	0.06390	1.00000	0.06390	0.00011	0.06390	1.00000
22	0.15909	26.57971	0.98831	26.89422	0.04639	26.92569	1.00117
23	0.15217	29.30839	0.98673	29.70249	0.05123	29.74487	1.00143
24	0.14583	26.45970	0.98989	26.72993	0.04611	26.76082	1.00116
25	0.14000	23.38717	0.99171	23.58272	0.04068	23.60394	1.00090

Table 7.4 Results for Problem 2 using quadratic elements.

M	h	$\varepsilon^{\text{Neu,L}}$	$\varkappa_{\text{L}}$	$\left\|e_{\Delta}^{[p]}\right\|_{\mathcal{U}}$	rel	$\varepsilon^{\text{Neu,U}}$	$\varkappa_{\text{U}}$
1	3.50000	153.12224	1.00000	153.12224	0.26592	153.12224	1.00000
2	1.75000	108.87622	0.98225	110.84412	0.19250	113.05650	1.01996
3	1.16667	74.77245	0.98654	75.79268	0.13163	76.50370	1.00938
4	0.87500	56.09598	0.99001	56.66197	0.09840	56.95967	1.00525
5	0.70000	53.48341	0.98916	54.06973	0.09390	54.32847	1.00479
6	0.58333	40.44157	0.99293	40.72964	0.07073	40.84034	1.00272
7	0.50000	43.26534	0.99143	43.63929	0.07579	43.77543	1.00312
8	0.43750	49.55175	0.98874	50.11609	0.08703	50.32219	1.00411
9	0.38889	42.85928	0.99158	43.22327	0.07506	43.35556	1.00306
10	0.35000	36.07763	0.99281	36.33897	0.06311	36.41761	1.00216
11	0.31818	36.03133	0.98933	36.42009	0.06325	36.49927	1.00217
12	0.29167	34.23695	0.98821	34.64543	0.06017	34.71359	1.00197
13	0.26923	25.80399	0.99318	25.98129	0.04512	26.01005	1.00111
14	0.25000	36.48259	0.98711	36.95891	0.06418	37.04165	1.00224
15	0.23333	34.08141	0.98943	34.44553	0.05982	34.51252	1.00194
16	0.21875	28.98946	0.99213	29.21950	0.05074	29.26040	1.00140
17	0.20588	28.33765	0.98989	28.62700	0.04972	28.66547	1.00134
18	0.19444	29.34280	0.98679	29.73566	0.05164	29.77877	1.00145
19	0.18421	13.16065	0.99718	13.19781	0.02292	13.20159	1.00029
20	0.17500	29.19830	0.98712	29.57928	0.05137	29.62171	1.00143
21	0.16667	29.51248	0.98786	29.87524	0.05188	29.91895	1.00146
22	0.15909	25.42644	0.99137	25.64767	0.04454	25.67533	1.00108
23	0.15217	23.88899	0.99066	24.11422	0.04188	24.13722	1.00095
24	0.14583	25.45293	0.98681	25.79313	0.04479	25.82126	1.00109
25	0.14000	15.22770	0.99479	15.30749	0.02658	15.31338	1.00038

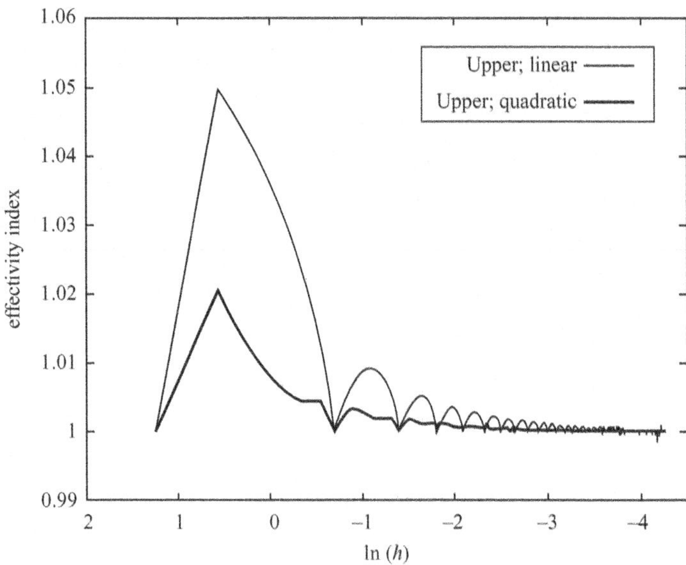

Fig. 7.1 Plot of effectivity index of the upper estimator $\varepsilon^{\mathrm{Neu,U}}$ for Problem 1.

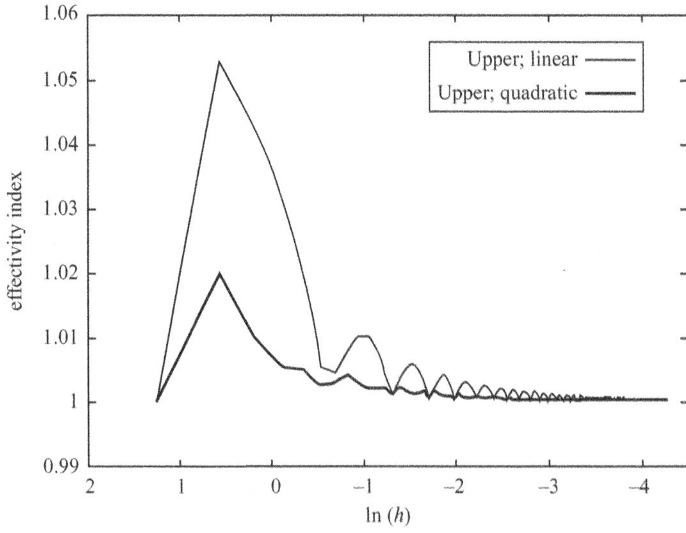

Fig. 7.2 Plot of effectivity index of the upper estimator $\varepsilon^{\mathrm{Neu,U}}$ for Problem 2.

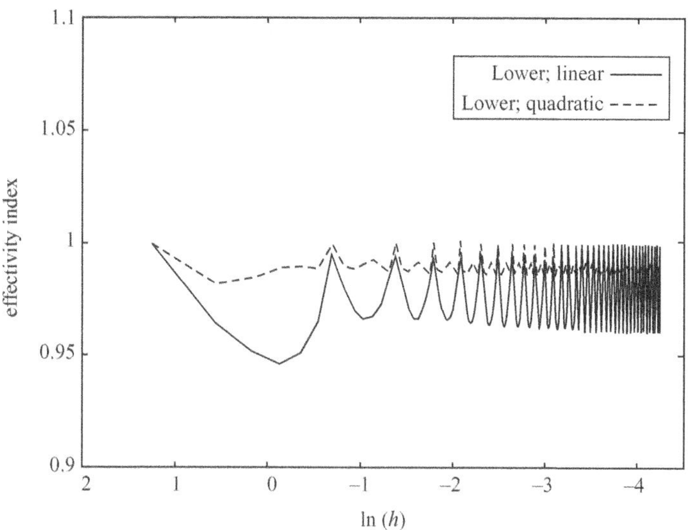

Fig. 7.3 Plot of effectivity index of the lower estimator $\varepsilon^{\mathrm{Neu,L}}$ for Problem 1.

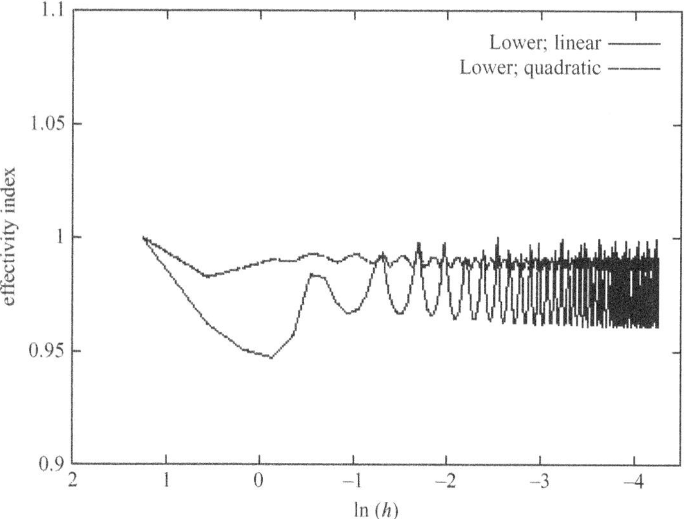

Fig. 7.4 Plot of effectivity index of the lower estimator $\varepsilon^{\mathrm{Neu,L}}$ for Problem 2.

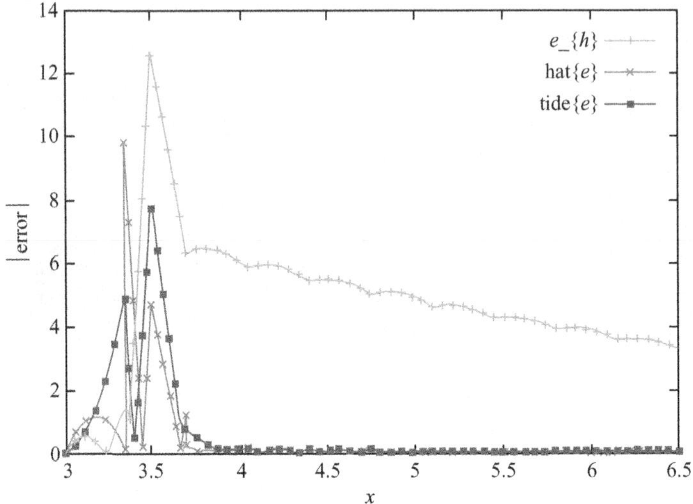

Fig. 7.5 Plot of $\left|e_{\Delta}^{[1]}(x)\right|$, $\left|\hat{e}^{[1]}(x)\right|$, and $\left|\tilde{e}^{[1]}(x)\right|$ for Problem 1 with $M = 10$. Note that the interface does not coincide with a node.

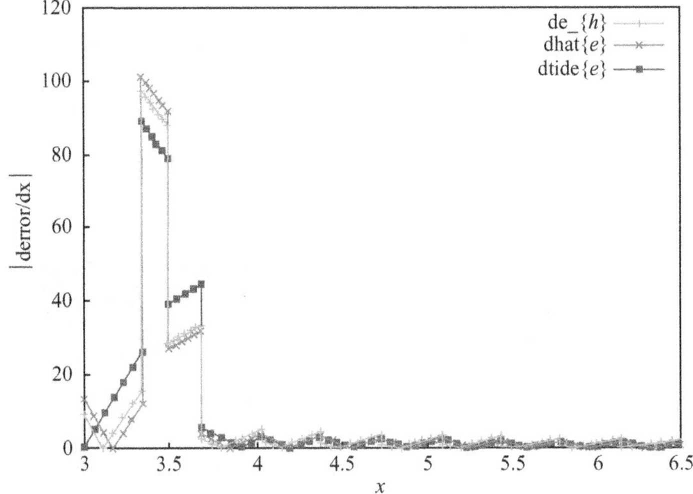

Fig. 7.6 Plot of $\left|e_{\Delta}^{[1]\prime}(x)\right|$, $\left|\hat{e}^{[1]\prime}(x)\right|$, and $\left|\tilde{e}^{[1]\prime}(x)\right|$ for Problem 1 with $M = 10$. Note that the interface does not coincide with a node.

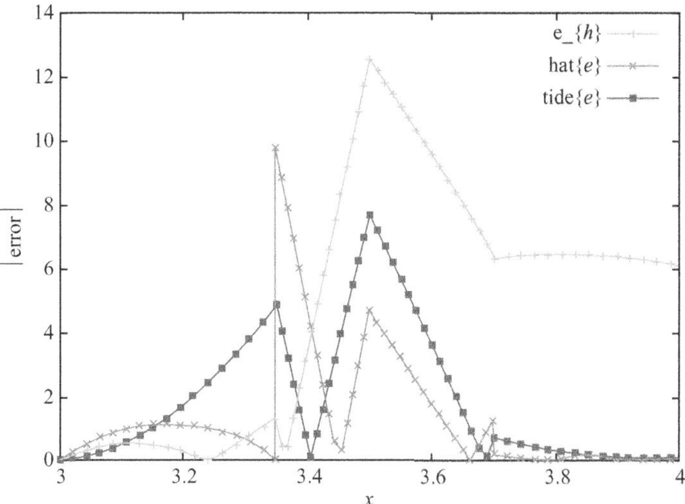

Fig. 7.7 Zoom of the plot of $\left|e_{\Delta}^{[1]}(x)\right|$, $\left|\hat{e}_{\Delta}^{[1]}(x)\right|$, and $\left|\tilde{e}_{\Delta}^{[1]}(x)\right|$ for Problem 1 with $M = 10$. Note that the interface does not coincide with a node.

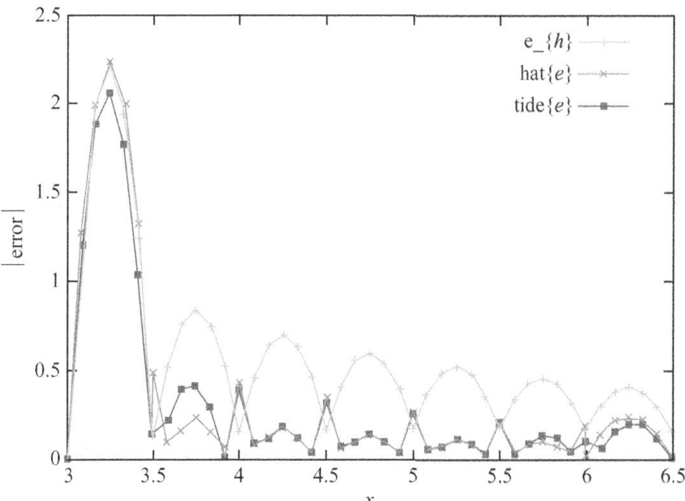

Fig. 7.8 Plot of $\left|e_{\Delta}^{[1]}(x)\right|$, $\left|\hat{e}_{\Delta}^{[1]}(x)\right|$, and $\left|\tilde{e}_{\Delta}^{[1]}(x)\right|$ for Problem 1 with $M = 7$. Note that the interface coincides with a node.

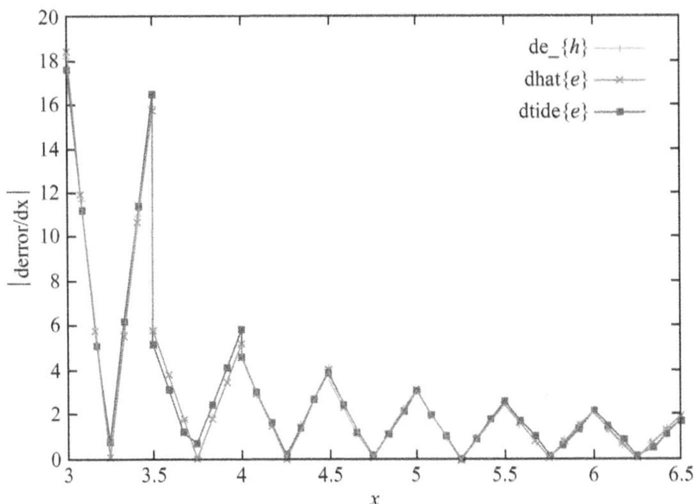

Fig. 7.9 Plot of $\left|e_\Delta^{[1]'}(x)\right|, \left|\hat{e}_\Delta^{[1]'}(x)\right|$, and $\left|\tilde{e}_\Delta^{[1]'}(x)\right|$ for Problem 1 with $M = 7$. Note that the interface coincides with a node.

note that this is only true for the one-dimensional case; as will be seen later $\varepsilon^{\mathrm{Dir}}$ can very much underestimate the error for two-dimensional problems. In order to address this we shall produce for the one-dimensional case a modified estimator for which the two-dimensional analogue performs successfully. For this, we use a patch of elements, rather than the single element as previously.

Let $\Phi_q^{[1]}(x)$ be the hat function associated with the nodal point $x_q^\Delta, q=1,\ldots,M=1$. Referring back to (3.20) we have

$$\Phi_1^{[1]}(x) = \begin{cases} (x_2 - x)/h_1 & x \in \bar{\tau}_1, \\ 0 & x \notin \bar{\tau}_1, \end{cases}$$

$$\Phi_q^{[1]}(x) = \begin{cases} (x_{q+1} - x)/h_q & x \in \bar{\tau}_q, \\ (x - x_{q-1})/h_{q-1} & x \in \bar{\tau}_{q-1}, \\ 0 & x \notin \bar{\tau}_q \cup \bar{\tau}_{q-1}, \end{cases}$$

for $q = 2, \ldots, M$

$$\Phi_{M+1}^{[1]}(x) = \begin{cases} (x - x_M)/h_M & x \in \bar{\tau}_M, \\ 0 & x \notin \bar{\tau}_M. \end{cases}$$

We denote by $\bar{\omega}_q^\Delta$ the support of $\Phi_q^{[1]}$, so that $\bar{\omega}_1^\Delta = \bar{\tau}_1; \bar{\omega}_q^\Delta = \bar{\tau}_q \cup \bar{\tau}_{q-1}$, $q = 2, \ldots, M$; $\bar{\omega}_{M+1} = \bar{\tau}_M$; the bars denote the closures of the sets.

We now introduce the subdomain residual error indicator function $\hat{e}^{[p]}_{\omega_q} \in \mathcal{U}_0(\omega_q), q = 1, \ldots, M+1$ that for $q = 1, \ldots, M$ satisfies for every $v \in \mathcal{U}_0(\omega_q) \equiv \{v \in \mathcal{U}(\omega_q) \mid v = 0 \text{ at } \partial\omega_q\}$

$$B_{\omega_q}\left(\hat{e}^{[p]}_{\omega_q}, v\right) = \Re_{\omega_q}(v) = \int_{\omega_q} r_{\omega_q} v dx + \mathcal{J}_q v(x^{\Delta}_q),$$

$$= B\left(e^{[p]}_{\Delta}, v\right) = \int_{\omega_q} f v dx - B\left(u^{[p]}_{\Delta}, v\right), \tag{7.23a}$$

where

$$r_{\omega_q} = \begin{cases} r_{q-1} & \text{on } \tau_{q-1}, \\ r_q & \text{on } \tau_q, \end{cases}$$

$\mathcal{J}_q$ is as given in (7.2c) and for $q = M+1$ satisfies

$$\hat{e}^{[p]}_{\omega_{M+1}} \in \mathcal{U}_0(\omega_{M+1}) \tag{7.23b}$$

$$B_{\omega_{M+1}}\left(\hat{e}^{[p]}_{\omega_{M+1}}, v\right) = \Re_{\omega_{M+1}} = \int_{\tau_M} r_{M+1} v dx + \mathcal{J}_{M+1} v(x^{\Delta}_{M+1}), \tag{7.23c}$$

where

$$\mathcal{U}_0(\omega_{M+1}) = \{v \in \mathcal{U}(\omega_{M+1}, v(x_M) = 0\}.$$

The function $\hat{e}^{[p]}_{\omega_q}$ exists and is unique. With this, we are now able to define both the *Dirichlet subdomain indicator*, $\eta^{\text{Dir}}_{\omega_q}$, where

$$\eta^{\text{Dir}}_{\omega_q} \equiv \left\| \hat{e}^{[p]}_{\omega_q} \right\|_{\mathcal{U}(\omega_q)}, \tag{7.24a}$$

and the *Dirichlet subdomain estimator*

$$\varepsilon^{\text{Dir}}_{\text{Subd}} \equiv \left(\sum_{q=1}^{M+1} \left(\eta^{\text{Dir}}_{\omega_q}\right)^2 \right)^{\frac{1}{2}}. \tag{7.24b}$$

Note that although both the estimators ε^{Dir} and $\varepsilon^{\text{Dir}}_{\omega_q}$ employ indicators that are the solutions of Dirichlet problems with zero boundary conditions, respectively, on τ_q and ω_q, they are markedly different. The estimator $\varepsilon^{\text{Dir}}_{\text{Subd}}$ takes the derivative jumps $\mathcal{J}_q$ on the element boundaries into account, see (7.23c), whilst the estimator ε^{Dir} does not, see (7.5).

Theorem 7.6 *Let a and c satisfy the assumptions as given in (2.20b–d) let $C_1 \leqq h_{q+1}/h_q \leqq C_2$, and $\varepsilon^{\text{Dir}}_{\text{Subd}}$ be defined as in (7.24b), then*

$$\frac{1}{2}\varepsilon^{\text{Dir}}_{\text{Subd}} \leqq \left\| e^{[p]}_{\Delta} \right\|_{\mathcal{U}} \leqq C\varepsilon^{\text{Dir}}_{\text{Subd}}, \tag{7.25}$$

where C is independent of the mesh and of the solution.

Proof Recalling that $B(e_\Delta^{[p]}, v) = 0$ for any $v \in S_{\Delta,0}^{[p]} \subset \mathcal{U}_0$ and that $\sum_{q=1}^{M+1} \Phi_q^{[1]} = 1$, we have for any $v \in S_{\Delta,0}^{[p]}$ that

$$\left\| e_\Delta^{[p]} \right\|_{\mathcal{U}_0}^2 = B\left(e_\Delta^{[p]}, e_\Delta^{[p]}\right) = B\left(e_\Delta^{[p]}, e_\Delta^{[p]} - v\right)$$
$$= B\left(e_\Delta^{[p]}, \sum_{q=1}^{M+1} \Phi_q^{[1]}\left(e_\Delta^{[p]} - v\right)\right).$$

Because $\Phi_q^{[1]}\left(e_\Delta^{[p]} - v\right) \in \mathcal{U}_0\left(\omega_q^\Delta\right), q = 1, \ldots, M$, and $\Phi_{M+1}^{[1]}\left(e_\Delta^{[p]} - v\right) \in \mathcal{U}_0\left(\omega_{M+1}^\Delta\right)$, we obtain from (7.23)

$$B\left(e_\Delta^{[p]}, \Phi_q^{[1]}\left(e_\Delta^{[p]} - v\right)\right) = \Re_{\omega_q}\left(\Phi_q\left(e_\Delta^{[p]} - v\right)\right)$$
$$= B_{\omega_q}\left(\hat{e}_{\omega_q}^{[p]}, \Phi_q\left(e_\Delta^{[p]} - v\right)\right),$$

and hence for any $v \in S_{\Delta,0}^{[p]}$,

$$\left\| e_\Delta^{[p]} \right\|_{\mathcal{U}_0}^2 = \sum_{q=1}^{M+1} B\left(e_\Delta^{[p]}, \Phi_q^{[p]}\left(e_\Delta^{[p]} - v\right)\right)$$
$$= \sum_{q=1}^{M+1} B_{\omega_q}\left(\hat{e}_{\omega_q}^{[p]}, \Phi_q\left(e_\Delta^{[p]} - v\right)\right)$$
$$\leqq \left(\sum_{q=1}^{M+1} \left\| \hat{e}_{\omega_q}^{[p]} \right\|_{\mathcal{U}(\omega_q)}^2\right)\left(\sum_{q=1}^{M+1} \left\| \Phi_q^{[1]}\left(e_\Delta^{[p]} - v\right)\right\|_{\mathcal{U}(\omega_q)}^2\right).$$

Employing Lemma 7.2, which is given below and states that

$$\min_{v \in S_{\Delta,0}^{[p]},} \sum_{q=1}^{M+1} \left\| \Phi_q^{[1]}\left(e_\Delta^{[p]} - v\right)\right\|_{\mathcal{U}(\omega_q)}^2 \leqq \mathcal{C} \left\| e_\Delta^{[p]} \right\|_{\mathcal{U}}^2,$$

we get the right-hand inequality of (7.25), the upper estimate.

We now prove the left-hand inequality (7.25), the lower estimate. We have

$$\left\|\varepsilon_{\text{Subd}}^{\text{Dir}}\right\|^2 = \sum_{q=1}^{M+1} \left\|\hat{e}_{\omega_q}^{[p]}\right\|_{\mathcal{U}(\omega_q)}^2 = \sum_{q=1}^{M+1} B_{\omega_q}\left(e_\Delta^{[p]}, \hat{e}_{\omega_q}^{[p]}\right),$$

$$= B\left(e_\Delta^{[p]}, \sum_{q=1}^{M+1} \hat{e}_{\omega_q}^{[p]}\right) \leqq \left\|e_\Delta^{[p]}\right\|_{\mathcal{U}} \left\|\sum_{q=1}^{M+1} \hat{e}_{\omega_q}^{[p]}\right\|_{\mathcal{U}}.$$

Casting ahead, Lemma 7.3 contains the result

$$\left\|\sum_{q=1}^{M+1} \hat{e}_{\omega_q}^{[p]}\right\|_{\mathcal{U}} \leqq 2\left(\sum_{q=1}^{M+1} \left\|\hat{e}_{\omega_q}^{[p]}\right\|_{\mathcal{U}(\omega_j)}^2\right)^{\frac{1}{2}} = 2\varepsilon_{\text{Subd}}^{\text{Dir}}, \qquad \square$$

which yields the left-hand inequality of (7.25).

If $\left\|e_\Delta^{[p]}\right\|_{\mathcal{U}} \approx h$, then also $\varepsilon_{\text{Subd}}^{\text{Dir}} \approx h$. However, we do not have that $\varepsilon_{\text{Subd}}^{\text{Dir}} = \left\|e_\Delta^{[p]}\right\|_{\mathcal{U}} (1 + 0(h))$.

Lemma 7.2 *Let a and c satisfy the assumptions of (2.20b–d) and let $w \in \mathcal{U}_0$, then*

$$\min_{v \in S_{\Delta,0}^{[p]},} \sum_{q=1}^{M} \left\|\Phi_q^{[1]}(w - v)\right\|_{\mathcal{U}(\omega_q)} \leqq C \|w\|_{\mathcal{U}_0}^2, \tag{7.26}$$

and C depends only on a, c and α_2.

Proof We choose $v = \mathcal{L}^{[1]}w$ and estimate $\left\|\Phi_q^{[1]}(w - v)\right\|_{\mathcal{U}(\omega_q)}$ term by term. We have

$$\left\|\frac{d}{dx}\left(\Phi_q^{[1]}(w - v)\right)\right\|_{L^2(\omega_q)}^2$$

$$\leqq 2\left(\left\|\Phi_q^{[1]}\frac{d}{dx}(w - v)\right\|_{L^2(\omega_q)}^2 + \left\|(w - v)\frac{d\Phi_q^{[1]}}{dx}\right\|_{L^2(\omega_q)}^2\right).$$

Then, using Theorem 4.5 and Remark 4.4 we obtain

$$\left\|\frac{d}{dx}\left(\Phi_q^{[1]}(w - v)\right)\right\|_{L^2(\omega_q)}^2 \leqq C \|w - v\|_{\mathcal{U}(\omega_q)}^2,$$

because $\left|\frac{d\Phi_q^{[1]}}{dx}\right| \leqq \frac{1}{h_q}$ on τ_q and $\leqq \frac{1}{h_{q-1}}$ or τ_{q-1}. Because $\left|\Phi_q^{[1]}\right| \leqq 1$ we have that

$$\left\|\Phi_q^{[1]}\frac{d(w-v)}{dx}\right\|_{L^2(\omega_q)} \leqq C\,\|w-v\|_{\mathcal{U}(\omega_q)},$$

from which after summing over q the result follows. Here again, C depends only on a, c and α_2. □

Lemma 7.3 *For $\hat{e}_{\omega_q}^{[p]} \in \mathcal{U}_0(w_q), q = 1, \ldots, M$ defined as in (7.14a) we have that*

$$\left\|\sum_{q=1}^{M+1} \hat{e}_{\omega_q}^{[p]}\right\|_{\mathcal{U}}^2 \leqq 2\sum_{q=0}^{M+1} \left\|\hat{e}_{\omega_q}^{[p]}\right\|_{\mathcal{U}_0(\omega_q)}^2. \tag{7.27}$$

Proof We first note that for $x \in \tau_q$ we can define

$$\overset{\barwedge}{e}_{\tau_q}^{[p]}(x) \equiv \hat{e}_{\omega_{q+1}}^{[p]}(x) + \hat{e}_{\omega_q}^{[p]}(x),$$

for $q = 1, \ldots, M$, and hence

$$\left\|\overset{\barwedge}{e}_{\tau_q}^{[p]}\right\|_{\mathcal{U}(\tau_q)}^2 \leqq 2\left(\left\|\hat{e}_{\omega_{q+1}}^{[p]}\right\|_{\mathcal{U}(\tau_q)}^2 + \left\|\hat{e}_{\omega_q}^{[p]}\right\|_{\mathcal{U}(\tau_q)}^2\right),$$

which when summed over all q gives (7.27). □

Remark 7.6 *The lower bound can also be computed from Theorem 7.2. For this, we realize that the function w on I with $w\,|_{\omega_q} = \overset{\barwedge}{e}_{\tau_q}^{[p]}$ belongs to $\mathcal{U}_0$, and hence we get*

$$\left\|\varepsilon_{\text{Subd}}^{\text{Dir}}\right\|_{\mathcal{U}} \geqq \frac{\Re(w)}{\|w\|_{\mathcal{U}}}.$$

Note that the functions $\hat{e}_{\omega_q}^{[p]}$ were computed, so that w can also be computed. This estimate is very accurate.

Once again, we cannot compute $e_{\omega_q}^{[p]}$, and hence $\eta_{\omega_q}^{\text{Dir}}$, exactly but we can compute it numerically using higher-order elements or a finer mesh on ω_q. We can also obtain an upper estimate analogous to (7.9b) and obtain

$$\left(\eta_{\omega_q}^{\text{Dir}}\right)^2 \leqq C\left(\left(\frac{h_q + h_{q-1}}{2}\right)^2 \frac{1}{a_{\min}}\left(\|r_q\|_{L^2(\tau_q)}^2 + \|r_{q-1}\|_{L^2(\tau_{q-1})}^2\right)\right. \tag{7.28}$$

$$\left. + \mathcal{J}_q^2 \frac{2}{a_{\min}}\left(\frac{h_q + h_{q-1}}{2}\right)\right),$$

where the constant C depends only on a and c.

Exercise 7.2 Derive the inequality (7.28).

Using (7.28) we can define $\varepsilon_{\omega}^{\mathrm{Dir,E}}$ such that

$$\varepsilon_{\omega}^{\mathrm{Dir,E}} \equiv \left(\sum_{q=1}^{M+1} \left(h_q^2 \left\| r_q \right\|_{L^2_{(\tau_q)}}^2 \right) \right)^{\frac{1}{2}}, \tag{7.29a}$$

and for a quasiuniform mesh this gives

$$\left\| e^{[p]} \right\|_{\mathcal{U}_0} \leq C\varepsilon_{\omega}^{\mathrm{Dir,E}}, \tag{7.29b}$$

where C is independent of the mesh h and the solution, but depends on a, c and α. Again, if b and c are smooth and Δ is a quasiuniform mesh, then we also have that

$$(1 + 0(h))\, C\varepsilon_{\omega}^{\mathrm{Dir,E}} \leq \left\| e^{[p]} \right\|_{\mathcal{U}}.$$

As before, the estimate includes constants that cannot easily be computed.

Let us again emphasize the main difference between $\varepsilon^{\mathrm{Dir,E}}$ and $\varepsilon_{\omega}^{\mathrm{Dir,E}}$, which is that in $\varepsilon_{\omega}^{\mathrm{Dir,E}}$ the jump $\mathcal{J}_q$ is taken into consideration, whilst in $\varepsilon^{\mathrm{Dir,E}}$ it is not. The subdomain estimator is equivalent to $\varepsilon^{\mathrm{Neu,E}}$, so that again we shall not illustrate it numerically.

7.1.5 The Neumann subdomain (patch) estimator

Let us now introduce a second subdomain indicator, the *Neumann subdomain indicator* $\eta_{w_q}^{\mathrm{Neu}}$. We know that for $q = 2, \ldots, M+1$, because the hat function $\Phi_q^{[1]} \in S_{\Delta,0}^{[1]}$ satisfies

$$\begin{aligned} \Re\left(\Phi_q^{[1]}\right) &= \int_{w_q} r_{w_q} \Phi_q^{[1]} dx + \mathcal{J}_q \Phi_q^{[1]}(x_q) \\ &= \int_{w_q} r_{w_q} \Phi_q^{[1]} dx + \mathcal{J}_q = 0. \end{aligned} \tag{7.30a}$$

Hence, on every $\omega_q, q = 1, \ldots, M+1$ there exists a solution $\overset{=[p]}{\hat{e}}_{\omega_q} \in \mathcal{U}(\omega_q)$ to a Neumann problem, such that for any $v \in \mathcal{U}(\omega_q)$

$$\begin{aligned} B_{\omega_q}\left(\overset{=[p]}{\hat{e}}_{\omega_q}, v\right) &= \int_{\omega_q} r_{\omega_q} \Phi_q^{[1]} v dx + \mathcal{J}_q v(x_q), \\ &= \int_{\omega_q} f \Phi_q^{[1]} v dx - B\left(u_{\Delta}^{[p]}, \Phi_q^{[1]} v\right), \end{aligned} \tag{7.30b}$$

and on ω_1 we have $\overset{=[p]}{\hat{e}}_{\omega_1} \in \mathcal{U}_0(\omega_1) \equiv \{v \in \mathcal{U}(\omega_1, v(0) = 0\}$ and for any $v \in \mathcal{U}_0(\omega_1)$

$$B_{\omega_1}\left(\overset{=[p]}{\hat{e}}_1, v\right) = \int_{\omega_1} r_{\omega_1} \Phi_q^{[1]} v dx. \tag{7.30c}$$

Because of (7.30a) the functions $\overset{=}{\hat{e}}{}^{[p]}_{\omega_q}, q = 2, \ldots, M+1$, exist whether or not c is zero in ω_q and $\alpha_2 = 0$. If c is zero then $\overset{=}{\hat{e}}{}^{[p]}_{\omega_q}$ is determined up to an additive constant that does not influence the energy norm. This constant will be fixed, for example so that $\overset{=}{\hat{e}}(x_q) = 0$. For $q = M+1$, if c and α_2 are zero, then $\overset{=}{\hat{e}}$ is once more determined up to a constant.

We define two Neumann subdomain indicators ${}^1\eta^{\text{Neu}}_{\omega_q}$ and ${}^2\eta^{\text{Neu}}_{\omega_q}$ and the corresponding estimators

$$ {}^1\varepsilon^{\text{Neu}}_{\text{Subd}} \equiv \left(\sum_{q=1}^{M+1} \left({}^1\eta^{\text{Neu}}_{\omega_q} \right)^2 \right)^{\frac{1}{2}}, \tag{7.31a} $$

$$ {}^2\varepsilon^{\text{Neu}}_{\text{Subd}} \equiv \left(\sum_{q=1}^{M+1} \left({}^2\eta^{\text{Neu}}_{\tau_q} \right)^2 \right)^{\frac{1}{2}}, \tag{7.31b} $$

where

$$ {}^1\eta^{\text{Neu}}_{\omega_q} \equiv \sqrt{2} \left\| \overset{=}{\hat{e}}{}^{[p]}_{\omega_q} \right\|_{\mathcal{U}(\tau_q)}, \tag{7.31c} $$

and

$$ {}^2\eta^{\text{Neu}}_{\tau_q} \equiv \left\| \overset{=}{\hat{e}}{}^{[p]}_{\omega_{q+1}} + \overset{=}{\hat{e}}{}^{[p]}_{\omega_q} \right\|_{\mathcal{U}(\tau_q)}. \tag{7.31d} $$

We then have the following theorem

Theorem 7.7 *Let a and c satisfy the assumptions as in (2.29a–c). Then,*

$$ \left\| e^{[p]}_{\Delta} \right\|_{\mathcal{U}} \leqq {}^1\varepsilon^{\text{Neu}}_{\text{Subd}}, \tag{7.32a} $$

$$ \left\| e^{[p]}_{\Delta} \right\|_{\mathcal{U}} \leqq {}^2\varepsilon^{\text{Neu}}_{\text{Subd}}. \tag{7.32b} $$

Proof We have that

$$ \left\| e^{[p]}_{\Delta} \right\|_{\mathcal{U}} = \sup_{v \in \mathcal{U}_o} \frac{\Re(v)}{\|v\|_{\mathcal{U}}}. $$

We also have, because $\Phi^{[1]}_q + \Phi^{[1]}_{q+1} = 1$ on τ_q that

$$ \begin{aligned} |\Re(v)| &= \left| \sum_{q=1}^{M} B_{\tau_q}\left(\left(\overset{=}{\hat{e}}{}^{[p]}_{\omega_{q+1}} + \overset{=}{\hat{e}}{}^{[p]}_{\omega_q} \right), v \right) \right| \\ &\leqq \left(\sum_{q=1}^{M} \left({}^2\eta^{\text{Neu}}_{\tau_q} \right)^2 \right)^{\frac{1}{2}} \|v\|_{\mathcal{U}}, \end{aligned} $$

from which (7.32b) follows. Realising that on every τ_q

$$\left(\overset{=[p]}{\hat{e}}_{\omega_{q+1}} + \overset{=[p]}{\hat{e}}_{\omega_q}\right)^2 \leq 2\left(\left(\overset{=[p]}{\hat{e}}_{\omega_{q+1}}\right)^2 + \left(\overset{=[p]}{\hat{e}}_{\omega_q}\right)^2\right),$$

we get the estimate (7.32a) from (7.32b). □

As with other indicators and estimators we need to compute local problems. This is again done using either higher-order elements or finer meshes.

The Neumann estimators were proposed by Pares *et al.* (2006). Although both the Neumann element and the Neumann subdomain estimators have good quality, the subdomain estimator is both more robust and easier to implement. In order to see this the reader should compare the results of Sections 7.1.3 and 7.1.6. They can also be generalized to the two-dimensional context. It is important to note here that no unidentified constants appear in these estimators.

Let us now address the situation regarding lower bounds. With ${}^1\Phi_q$ we let

$$w = \sum_{q=1}^{M+1} {}^1\Phi_q \overset{=}{\hat{e}}_{\omega_q}.$$

Then, $w \in \mathcal{U}_0(I)$ and we obtain the lower bound

$$\left\|e_\Delta^{[p]}\right\|_{\mathcal{U}} \geq \frac{|\Re(w)|}{\|w\|_{\mathcal{U}}} = {}^1\varepsilon^{\mathrm{Neu,L}}.$$

Contrastingly, if w is such that $w\,|_{\tau_q} = \overset{=[p]}{\hat{e}}_{\omega_{q+1}} + \overset{=[p]}{\hat{e}}_{\omega_q}$, then w is a discontinuous function so that $w \notin \mathcal{U}_0$. Nevertheless, a piecewise linear gap function ${}^2\Psi$ can be constructed as in Section 7.1.2, so that $w - {}^2\Psi \in U_0(I)$ and so we get the lower estimator $\left\|e_\Delta^{[p]}\right\|_{\mathcal{U}} \geq \frac{|\mathcal{R}(w-{}^2\Psi)|}{\|w-{}^2\Psi\|_{\mathcal{U}}} = {}^2\varepsilon^{\mathrm{Neu,L}}$.

7.1.6 The performance of the Neumann subdomain estimators for the one-dimensional engineering problems

In this section, we present results that demonstrate the accuracy of the Neumann subdomain estimators (7.31a and b) when finite element methods are used to solve 1D Eng Problems 1 and 2. Tables 7.5 and 7.6 and Tables 7.7 and 7.8 show the effectivity of the upper error estimators when, respectively, linear and quadratic elements are used for 1D Eng Problems 1 and 2, using uniform meshes with mesh size h. Values are given for:

(1) the energy norm of the error: $\left\|e_\Delta^{[p]}\right\|_{\mathcal{U}}$;
(2) the upper bounds ${}^1\varepsilon^{\mathrm{Neu,U}}$ and ${}^2\varepsilon^{\mathrm{Neu,U}}$ for Problems 1 and 2;
(3) the corresponding effectivity indexes ${}^1\varkappa_{\mathcal{U}}$ and ${}^2\varkappa_{\mathcal{U}}$.

We see that the estimator ${}^2\varepsilon^{\mathrm{Neu}}_{\mathrm{Subd}}$ has high quality, which is similar for both linear and quadratic elements. The effecivity index of the estimator ${}^1\varepsilon^{\mathrm{Neu}}_{\mathrm{Subd}}$ is for linear elements approximately 1.4, which is close to the value of the factor $\sqrt{2}$ that appears

Table 7.5 Estimator 1 and Estimator 2 for Problem 1 using linear elements.

M	h	$\left\| e_{\Delta}^{[1]} \right\|_{\mathcal{U}}$	${}^{1}\varepsilon^{\mathrm{Neu,U}}$	${}^{1}\varkappa_{\mathrm{U}}$	${}^{2}\varepsilon^{\mathrm{Neu,U}}$	${}^{2}\varkappa_{\mathrm{U}}$
1	3.50000	211.62649	270.09603	1.27629	211.62648	1.00000
2	1.75000	177.10240	271.15073	1.53104	186.20450	1.05139
3	1.16667	158.20987	237.70482	1.50247	164.49383	1.03972
4	0.87500	140.92577	209.83913	1.48900	145.38496	1.03164
5	0.70000	120.50846	178.41745	1.48054	123.30839	1.02323
6	0.58333	90.91306	134.17324	1.47584	92.12123	1.01329
7	0.50000	16.09178	24.93005	1.54924	16.09881	1.00044
8	0.43750	54.57312	78.79789	1.44390	54.83559	1.00481
9	0.38889	68.69404	98.97787	1.44085	69.21672	1.00761
10	0.35000	74.98433	108.00331	1.44035	75.66362	1.00906
11	0.31818	75.70375	109.05956	1.44061	76.40272	1.00923
12	0.29167	70.64181	101.82772	1.44147	71.21008	1.00804
13	0.26923	56.79369	81.97393	1.44336	57.08941	1.00521
14	0.25000	8.06322	12.49155	1.54920	8.06413	1.00011
15	0.23333	38.37227	54.91948	1.43123	38.46362	1.00238
16	0.21875	49.95746	71.42050	1.42963	50.15885	1.00403
17	0.20588	55.89342	79.91093	1.42970	56.17532	1.00504
18	0.19444	57.59298	82.37774	1.43034	57.90134	1.00535
19	0.18421	54.68814	78.28338	1.43145	54.95221	1.00483
20	0.17500	44.61663	63.95281	1.43339	44.76015	1.00322
21	0.16667	5.37765	8.33104	1.54920	5.37792	1.00005
22	0.15909	31.21918	44.52344	1.42616	31.26839	1.00158
23	0.15217	41.14495	58.63455	1.42507	41.25753	1.00274
24	0.14583	46.46945	66.22972	1.42523	46.63157	1.00349
25	0.14000	48.27741	68.83572	1.42584	48.45918	1.00377

in the definition of this indicator, see (7.31c). Nevertheless, for quadratic elements the effectivity index is much smaller that $\sqrt{2}$, which shows that in general this factor cannot be eliminated.

Figures 7.10–7.17 show plots of the effectivity index based on the data of the upper and lower estimators and demonstrate the very different behaviour of ${}^{1}\varepsilon_{\mathrm{Subd}}^{\mathrm{Neu}}$ and ${}^{2}\varepsilon_{\mathrm{Subd}}^{\mathrm{Neu}}$.

Moral: It is always better to use ${}^{2}\varepsilon_{\mathrm{Subd}}^{\mathrm{Neu}}$, than to use ${}^{1}\varepsilon_{\mathrm{Subd}}^{\mathrm{Neu}}$.

Table 7.6 Estimator 1 and Estimator 2 for Problem 2 using linear elements.

M	h	$\left\|e_{\Delta}^{[1]}\right\|_{\mathcal{U}}$	${}^{1}\mathcal{E}^{\text{Neu,U}}$	${}^{1}\varkappa_{\text{U}}$	${}^{2}\mathcal{E}^{\text{Neu,U}}$	${}^{2}\varkappa_{\text{U}}$
1	3.50000	217.93709	277.00573	1.27104	217.93709	1.00000
2	1.75000	181.84439	279.21391	1.53546	191.84057	1.05497
3	1.16667	160.38774	241.46890	1.50553	167.01964	1.04135
4	0.87500	138.94727	207.26869	1.49171	143.28106	1.03119
5	0.70000	110.62465	164.10688	1.48346	112.82409	1.01988
6	0.58333	58.00013	86.02243	1.48314	58.31949	1.00551
7	0.50000	52.35544	75.84381	1.44863	52.59046	1.00449
8	0.43750	70.59374	101.90278	1.44351	71.16865	1.00814
9	0.38889	78.08998	112.67128	1.44284	78.86744	1.00996
10	0.35000	78.39801	113.14597	1.44323	79.18467	1.01003
11	0.31818	70.93026	102.45507	1.44445	71.51339	1.00822
12	0.29167	49.40940	71.53629	1.44783	49.60693	1.00400
13	0.26923	29.33163	42.18053	1.43806	29.37302	1.00141
14	0.25000	47.89093	68.55452	1.43147	48.07081	1.00376
15	0.23333	56.67465	81.10682	1.43110	56.97255	1.00526
16	0.21875	59.75285	85.55005	1.43173	60.10187	1.00584
17	0.20588	57.09434	81.81700	1.43301	57.39889	1.00533
18	0.19444	45.41294	65.18684	1.43542	45.56635	1.00338
19	0.18421	14.89048	21.49716	1.44369	14.89590	1.00036
20	0.17500	35.91770	51.24583	1.42676	35.99365	1.00211
21	0.16667	45.27496	64.56802	1.42613	45.42698	1.00336
22	0.15909	49.55356	70.69222	1.42658	49.75281	1.00402
23	0.15217	49.23179	70.28207	1.42757	49.42719	1.00397
24	0.14583	42.46565	60.69563	1.42929	42.59111	1.00295
25	0.14000	17.59636	25.30163	1.43789	17.60530	1.00051

7.1.7 Averaging-based error indicators and estimators

The error estimators of Sections 7.1.1–7.1.6 were all based on residuals. We now consider a second class of estimators that are based on *averaging methods.*

The derivatives of the finite element solution $u_{\Delta}^{[p]}, p = 1, 2$ of (7.1b) are discontinuous polynomials of degree $p - 1$. The idea with averaging is to construct a continuous function $z^{[p]}(x) \in S_{\Delta}^{[p]}$ that approximates $\frac{du}{dx}$ better than does $\frac{d}{dx}u_{\Delta}^{[p]}$. This type of

Table 7.7 Estimator 1 and Estimator 2 for Problem 1 using quadratic elements.

M	h	$\left\Vert e_\Delta^{[1]}\right\Vert_{\mathcal{U}}$	${}^1\mathcal{E}^{\mathrm{Neu,U}}$	${}^1\varkappa_{\mathrm{U}}$	${}^2\mathcal{E}^{\mathrm{Neu,U}}$	${}^2\varkappa_{\mathrm{U}}$
1	3.50000	150.07728	200.99623	1.33928	150.07728	1.00000
2	1.75000	113.49891	146.57402	1.29141	15.86338	1.02083
3	1.16667	81.81327	96.51280	1.17967	82.69479	1.01077
4	0.87500	59.87796	66.47046	1.11010	60.22442	1.00579
5	0.70000	52.80359	59.72254	1.13103	53.04134	1.00450
6	0.58333	51.62551	63.13717	1.22298	51.84772	1.00430
7	0.50000	0.57013	0.74668	1.30967	0.57013	1.00000
8	0.43750	45.03753	60.62757	1.34616	45.18514	1.00328
9	0.38889	48.17046	60.30492	1.25191	48.35102	1.00375
10	0.35000	41.96278	48.45571	1.15473	42.08220	1.00285
11	0.31818	35.65381	39.24455	1.10071	35.72709	1.00206
12	0.29167	34.86646	39.43094	1.13091	34.93499	1.00197
13	0.26923	35.71794	43.64936	1.22206	35.79161	1.00206
14	0.25000	0.14359	0.18801	1.30940	0.14359	1.00000
15	0.23333	32.65075	43.85515	1.34316	32.70704	1.00172
16	0.21875	35.75140	44.61461	1.24791	35.82528	1.00207
17	0.20588	31.87992	36.69783	1.15113	31.93232	1.00164
18	0.19444	27.81545	30.57101	1.09907	27.85026	1.00125
19	0.18421	27.84095	31.47941	1.13069	27.87586	1.00125
20	0.17500	28.91428	35.32223	1.22162	28.95338	1.00135
21	0.16667	0.06390	0.08367	1.30935	0.06390	1.00000
22	0.15909	26.89422	36.09463	1.34210	26.92569	1.00117
23	0.15217	29.70249	37.02130	1.24640	29.74487	1.00143
24	0.14583	26.72993	30.73120	1.14969	26.76082	1.00116
25	0.14000	23.58272	25.90280	1.09838	23.60394	1.00090

function is often called *a recovered derivative function.* The error indicator associated with $z^{[p]}(x)$ has the form

$$\eta_q^2 = \int_{\tau_q} a \left(z^{[p]} - \frac{d}{dx} u_\Delta^{[p]} \right)^2 dx.$$

Because $z^{[p]} \in S_\Delta^{[p]}$ we determine first the values of z at the nodal points of Δ, and then interpolate to these. In order to do this we take again the subdomains $\bar{\omega}_q^\Delta$, where

Table 7.8 Estimator 1 and Estimator 2 for Problem 2 using quadratic elements.

M	h	$\left\Vert e_{\Delta}^{[1]} \right\Vert_{\mathcal{U}}$	${}^{1}\mathcal{E}^{\mathrm{Neu,U}}$	${}^{1}\varkappa_{\mathrm{U}}$	${}^{2}\mathcal{E}^{\mathrm{Neu,U}}$	${}^{2}\varkappa_{\mathrm{U}}$
1	3.50000	153.12224	203.66941	1.33011	153.12224	1.00000
2	1.75000	11.084412	140.74631	1.26977	113.07839	1.02016
3	1.16667	75.79268	87.28570	1.15164	76.50370	1.00938
4	0.87500	56.66197	62.79667	1.10827	56.95967	1.00525
5	0.70000	54.06973	63.61949	1.17662	54.32847	1.00479
6	0.58333	40.72964	54.24841	1.33191	40.84034	1.00272
7	0.50000	43.63929	59.47278	1.36283	43.77543	1.00312
8	0.43750	50.11609	63.22437	1.26156	50.32219	1.00411
9	0.38889	43.22327	49.84870	1.15328	43.35556	1.00306
10	0.35000	36.33897	39.97718	1.10012	36.41761	1.00216
11	0.31818	36.42009	41.85928	1.14935	36.49927	1.00217
12	0.29167	34.64543	44.10153	1.27294	34.71359	1.00197
13	0.26923	25.98129	35.94955	1.38367	26.01005	1.00111
14	0.25000	36.95891	47.62599	1.28862	37.04165	1.00224
15	0.23333	34.44553	40.55874	1.17747	34.51252	1.00194
16	0.21875	29.21950	32.21867	1.10264	29.26040	1.00140
17	0.20588	28.62700	32.25732	1.12681	28.66547	1.00134
18	0.19444	29.73566	36.52217	1.22823	29.77877	1.00145
19	0.18421	13.19781	18.53176	1.40415	13.20159	1.00029
20	0.17500	29.57928	38.96128	1.31718	29.62171	1.00143
21	0.16667	29.87524	36.04652	1.20657	29.91895	1.00146
22	0.15909	25.64767	28.58041	1.11435	25.67533	1.00108
23	0.15217	24.11422	26.76159	1.10978	24.13722	1.00095
24	0.14583	25.79313	30.75587	1.19241	25.82126	1.00109
25	0.14000	15.30749	20.98115	1.37065	15.31338	1.00038

$\bar{\omega}_1 = \bar{\tau}_1, \bar{\omega}_q \in \bar{\tau}_q \cup \bar{\tau}_{q-1}, q = 2, \ldots, M, \bar{\omega}_{M+1} = \bar{\tau}_M$, and in each subdomain $\bar{\omega}_q$ choose a set of *sample points* $y_{q,1}^{(k)} \in \bar{\tau}_{q-1}$ and $y_{q,2}^{(k)} \in \bar{\tau}_q, k = 1, \ldots, m$, where $m \geqq p, p = 1, 2$. We then determine the polynomial $\Psi_q^{[p]} \in \mathbb{P}^{[p]}(w_q)$ such that

$$\sum_{\substack{l=1,2 \\ k=1,\ldots,m}} a\left(y_{q,l}^{(k)}, k\right) \left(\Psi_q^{[p]}\left(y_{q,l}^{[k]}\right) - \frac{d}{dx} u_{\Delta}^{[p]}\left(y_{q,l}^{(k)}\right)\right)^2$$

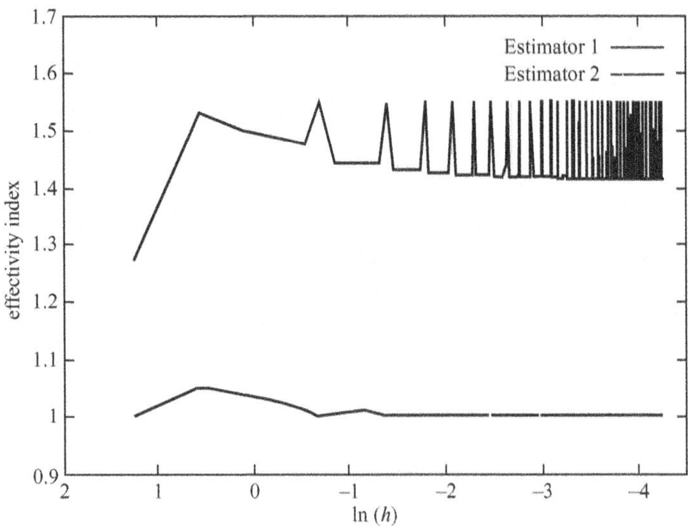

Fig. 7.10 Plot of effectivity index of the upper Neumann subdomain estimators ${}^{1}\varepsilon^{\mathrm{Neu,U}}$ and ${}^{2}\varepsilon^{\mathrm{Neu,U}}$ for Problem 1 using linear elements.

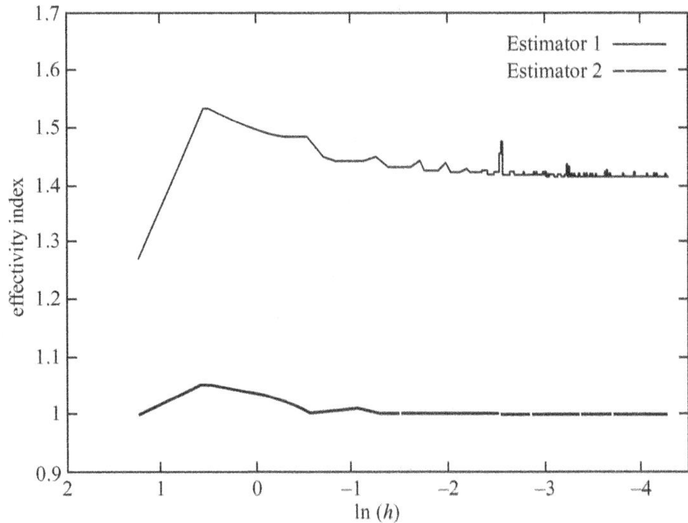

Fig. 7.11 Plot of effectivity index of the upper Neumann subdomain estimators ${}^{1}\varepsilon^{\mathrm{Neu,U}}$ and ${}^{2}\varepsilon^{\mathrm{Neu,U}}$ for Problem 2 using linear elements.

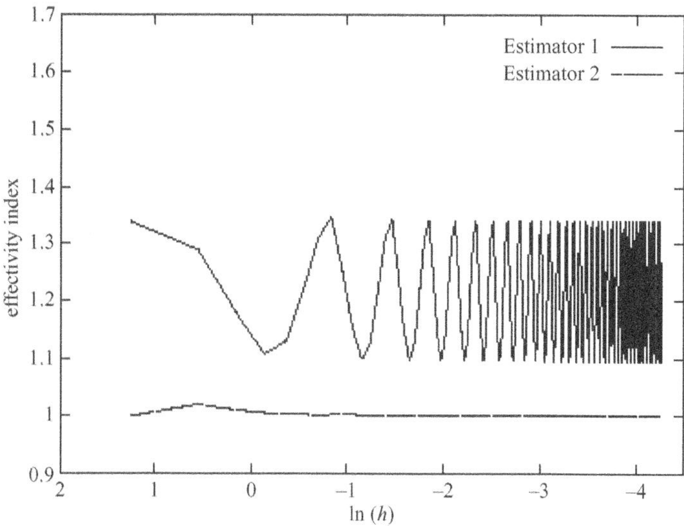

Fig. 7.12 Plot of effectivity index of the upper Neumann subdomain estimators ${}^{1}\varepsilon^{\mathrm{Neu,U}}$ and ${}^{2}\varepsilon^{\mathrm{Neu,U}}$ for Problem 1 using quadratic elements.

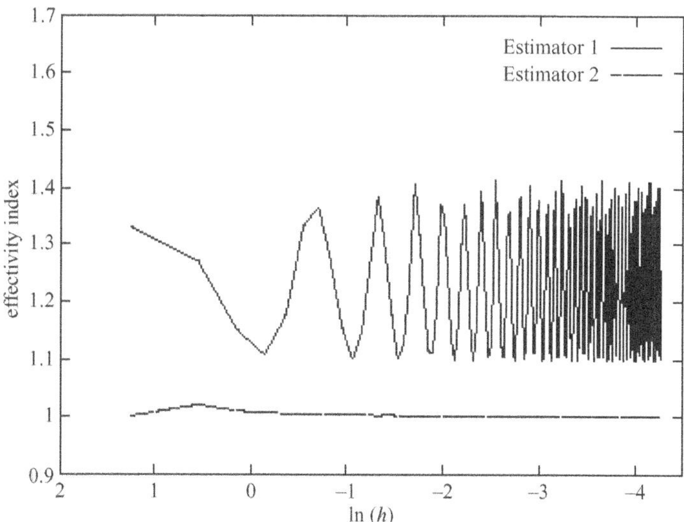

Fig. 7.13 Plot of effectivity index of the upper Neumann subdomain estimators ${}^{1}\varepsilon^{\mathrm{Neu,U}}$ and ${}^{2}\varepsilon^{\mathrm{Neu,U}}$ for Problem 2 using quadratic elements.

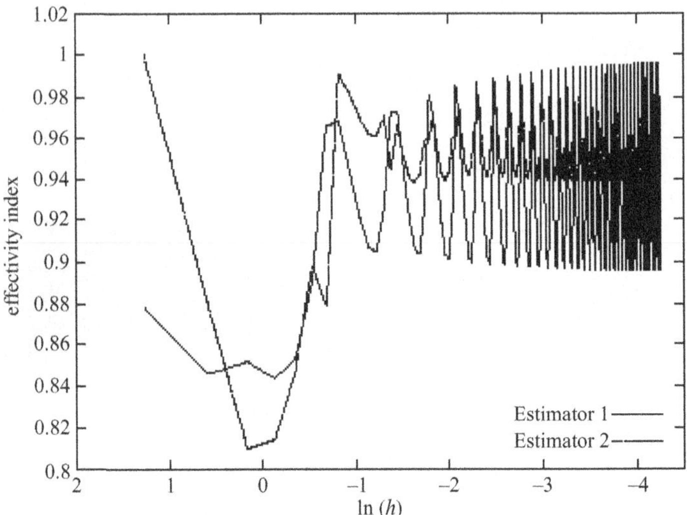

Fig. 7.14 Plot of effectivity index of the lower Neumann subdomain estimators ${}^{1}\varepsilon^{\mathrm{Neu,L}}$ and ${}^{2}\varepsilon^{\mathrm{Neu,L}}$ for Problem 1 using linear elements.

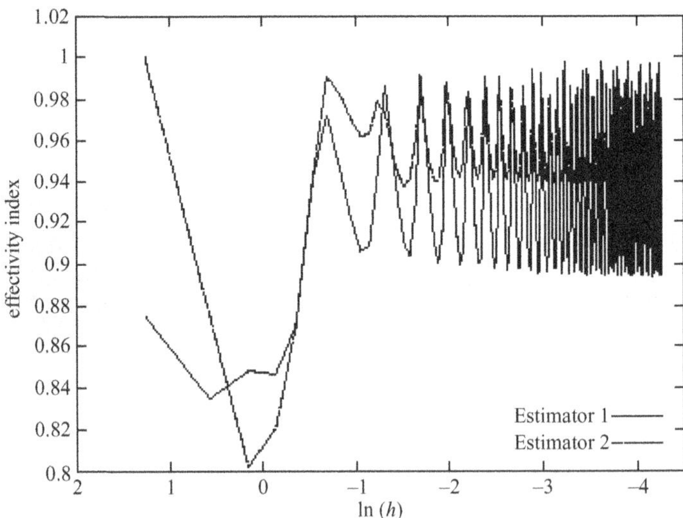

Fig. 7.15 Plot of effectivity index of the lower Neumann subdomain estimators ${}^{1}\varepsilon^{\mathrm{Neu,L}}$ and ${}^{2}\varepsilon^{\mathrm{Neu,L}}$ for Problem 2 using linear elements.

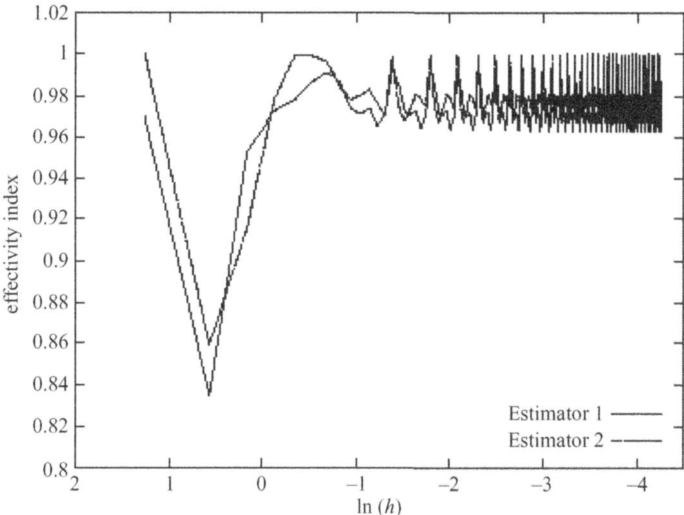

Fig. 7.16 Plot of effectivity index of the lower Neumann subdomain estimators ${}^{1}\varepsilon^{\mathrm{Neu,L}}$ and ${}^{2}\varepsilon^{\mathrm{Neu,L}}$ for Problem 1 using quadratic elements.

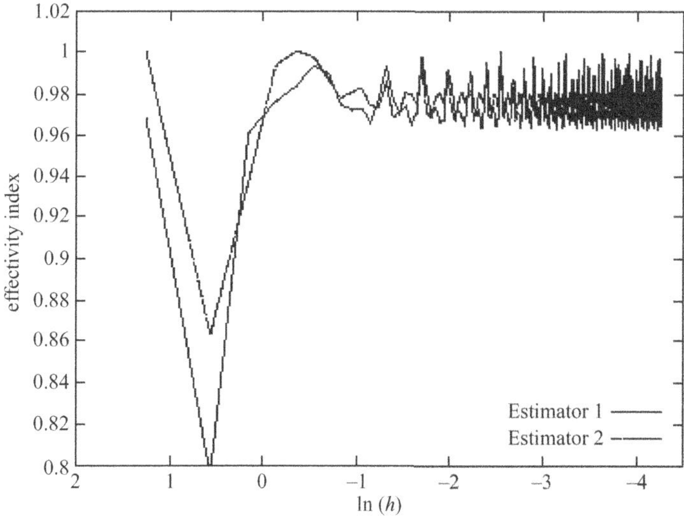

Fig. 7.17 Plot of effectivity index of the upper Neumann subdomain estimators for Problem 2 using quadratic elements.

is minimal. We then define $z^{[p]}(x) \in S_{\Delta}^{[p]}$ such that

$$z^{[p]}(x_q) \equiv \Psi_q^{[p]}(x_q),\ p = 1,$$

and for the case $p = 2$

$$z^{[2]}(x_{q+\frac{1}{2}}) \equiv \frac{1}{2}\left(\Psi_q(x_{q+\frac{1}{2}}) + \Psi_{q+1}(x_{q+\frac{1}{2}})\right).$$

At x_1 and x_{M+1}, and in the case $p = 2$ at $x_{1+\frac{1}{2}}$, and $x_{M+\frac{1}{2}}$, we determine the values of $\Psi^{[p]}$ using the patches ω_2 and ω_M. We note that for a problem of type (2.14) if we had a Neumann boundary condition prescribed at x_{M+1} we could use the exact value of the derivative.

The major task now is to select the sample points so that $\left|z^{[p]}(x) - \frac{du}{dx}(x)\right| \leq Ch^{\mu}, \mu > p$. The simplest way of doing this is to select as sample points the points of superconvergence of $u_{\Delta}^{[p]}$.

Theorem 7.8 *Consider problem (7.1a) and the finite element problem (7.1b), let $a \in C^2[I]$, $c \in C^1[I]$, $f \in C^1[I]$, let Δ be a quasiuniform mesh, and let $p = 1$. Then, using the sample points at the centres of the elements $y_{q,1}^{(1)} = (x_{q-1} + x_q)/2$, $y_{q,2}^{(1)} = (x_q + x_{q+1})/2$, $q = 1, \ldots, M$ we obtain*

$$\left\| z^{[1]} - \frac{du}{dx} \right\|_{L^{\infty}[I]} \leq Ch^2. \tag{7.33}$$

Proof In Theorem 6.5 we showed that

$$\left| \frac{d}{dx}\left(u - u_{\Delta}^{[1]}\right)(y_{q,i}) \right| \leq Ch^2. \tag{7.34}$$

We address first the case $q = 2, \ldots, M$; i.e. only interior nodes. The function $\Psi_q^{[1]}(x)$ is the linear interpolant to $\frac{du_{\Delta}^{[1]}}{dx}(y_{q,1}^{(1)})$ and $\frac{du_{\Delta}^{[1]}}{dx}(y_{q,2}^{(1)})$. We first show that

$$\left| \frac{du}{dx}(x_q^{\Delta}) - \Psi_q^{[1]}(x_q) \right| \leq Ch^2. \tag{7.35}$$

The assumptions on a, c and f give $\frac{du}{dx} \in C^2[I]$. Denoting by v_q the linear interpolant to $\frac{du}{dx}(y_{q,1}^{(1)})$ and $\frac{du}{dx}(y_{q,2}^{(2)})$, we find that

$$\left| v_q(x) - \frac{du}{dx}(x) \right| \leq Ch^2,\ x \in \omega_q, q = 2, \ldots, M, \tag{7.36a}$$

and so also

$$\left| v_q(x_q^{\Delta}) - \frac{du}{dx}(x_q^{\Delta}) \right| \leq Ch^2. \tag{7.36b}$$

Because $\Psi_q^{[1]}$ is the interpolant to $\frac{du_\Delta^{[1]}}{dx}(y_{q,1}^{(1)})$ and $\frac{du_\Delta^{[1]}}{dx}(y_{q,2}^{(1)})$, using (7.34), (7.35) and (7.36) we obtain

$$\left| z^{[1]}(x) - \frac{du}{dx}(x) \right| \leq Ch^2, \forall x \in \bar{\tau}_q,\ q = 2, \ldots, M.$$

We now have to address the boundary nodes x_1 and x_{M+1}. Using (7.36a) we define $z(x_1)$ as the linear interpolant of the values $\frac{du_\Delta^{(1)}}{dx}(y_{1,l}^{(1)}), l = 1, 2$, and similarly for $z(x_{M+1})$. If the Neumann boundary condition is prescribed at x_{M+1}, then $\frac{du}{dx}(x_{M+1})$ is known and we can use this value. In both cases we get

$$\left| z^{[1]}(x) - \frac{du}{dx}(x) \right| \leq Ch^2\ \forall x \in \bar{\tau}_1 \text{ and } x \in \bar{\tau}_M,$$

and (7.33) follows. □

Theorem 7.9 *Considering again (7.1a and b), letting $a \in C^3[I]$, $c \in C^2[I]$, $f \in C^2[I]$, let Δ be a quasiuniform mesh and $p = 2$, and using as sample points the superconvergence points $y_{q,l}^{(k)}$, $k = 1, 2$, $l = 1, 2$ where*

$$y_{q,1}^{(k)} = \frac{1}{2}(x_{q-1} + x_q) \pm \frac{h_{q-1}}{\sqrt{12}},\ y_{q,2}^{(k)} = \frac{1}{2}(x_q + x_{q+1}) \pm \frac{h_q}{\sqrt{12}},$$

we have that

$$\left\| z^{[2]}(x) - \frac{du}{dx}(x) \right\|_{L^\infty(I)} \leq Ch^3. \tag{7.37}$$

Proof This is analogous to the case $p = 1$, except that now $\Psi_q^{[2]}(x) \in \mathbb{P}^{[2]}(\omega_q)$ is not an interpolant but the least squares fit to $y_{q,l}^{(k)}$. From Theorem 6.5 we have that

$$\left| \frac{d}{dx}\left(u - u_\Delta^{[2]}\right)\left(y_{q,l}^{(k)}\right) \right| \leq Ch^3,\ q = 2, \ldots, M,\ k = 1, 2. \tag{7.38}$$

Denoting by v_q the best fit to the data $\frac{du}{dx}(y_{q,l}^{(k)})$, $q = 2, \ldots, M$, with $\frac{du}{dx} \in C^3[I]$ we obtain from (7.38)

$$\left\| v_q - \frac{du}{dx} \right\|_{L^\infty(\omega_q)} \leq Ch^3, q = 2, \ldots, M,$$

and also that

$$\left\| \Psi_q^{[2]} - \frac{du}{dx} \right\|_{L^\infty(\omega_q)} \leq Ch^3.$$

Now, in a manner analogous to the case $p = 1$, we obtain (7.37). For the values of the nodal points at the centre of the element we use the average of the values obtained from the patches that contain the element. □

Note that once more we do not use $\Psi_1^{[2]}$ and $\Psi_{M+1}^{[2]}$, but rather get the values of τ_1 and τ_M from $\Psi_2^{[2]}$ and $\Psi_M^{[2]}$.

If $\left|\dfrac{d^{p+1}u}{dx^{p+1}}\right| \geqq \gamma > 0$ on I, or on parts of I, then

$$\left\| \frac{du}{dx} - \frac{du_\Delta^{[p]}}{dx} \right\|_{L^\infty(I)} \geqq Ch^p.$$

On the other hand,

$$\left\| \frac{du}{dx} - z^{[p]} \right\|_{L^\infty(I)} \leqq Ch^{p+1}.$$

Hence,

$$\begin{aligned} \left\| \frac{du}{dx} - \frac{du_\Delta^{[p]}}{dx} \right\|_{L^\infty(I)} &\leqq \left\| \frac{du}{dx} - z^{[p]} \right\|_{L^\infty(I)} + \left\| \frac{du_\Delta^{[p]}}{dx} - z^{[p]} \right\|_{L^\infty(I)} \\ &= \left\| \frac{du_\Delta^{[p]}}{dx} - z^{[p]} \right\|_{L^\infty(I)} (1 + 0(h)) . \end{aligned}$$

Hence, we define the indicator

$$\eta_q^{ZZ} \equiv \left(\int_{\tau_q} a \left(\frac{du_\Delta^{[p]}}{dx} - z^{[p]} \right)^2 dx \right)^{\frac{1}{2}}, \tag{7.39a}$$

and the estimator

$$\varepsilon^{ZZ} \equiv \left(\sum_{q=1}^{M} \left(\eta_q^{zz}\right)^2 \right)^{\frac{1}{2}}. \tag{7.39b}$$

The indicator η_q^{ZZ} of (7.39a) involves $z^{[p]}x \in S_\Delta^{[p]}(I)$, which is the 'recovered' $\dfrac{du_\Delta}{dx}$. This function was constructed so that its values at the nodal points were superconvergent, and then $z^{[p]}(x)$ was the interpolant to these values. Correspondingly, in obtaining the superconvergent values at the nodal points we constructed the function $\Psi_q^{[p]}$ using the superconvergent points $y_{q,k}$. Nevertheless, $\Psi_q^{[p]}$ also had to be recovered from $\frac{du_\Delta}{dx}$ on each τ_q, so that it is possible to use $\Psi_q^{[p]}$ instead of $z^{[p]}$ in the indicator η_q^{ZZ}.

The notation ZZ is adopted because the indicator η_q^{ZZ} and estimator ε^{ZZ} were originally proposed by Zienkiewicz and Zhu 1987) and (1992). They are now universally known as the ZZ-indicator and estimator.

For sufficiently smooth input data

$$\left\| e_\Delta^{[p]} \right\|_{\mathcal{U}} \leqq \varepsilon^{ZZ}(1 + 0(h)).$$

In contrast to the Neumann estimator, the ZZ-estimator is not necessarily an upper estimate.

The error indicator and estimator (7.39a and b) do not take into consideration the terms $\int ce_\Delta^{[p]}$ and $\left(\frac{de_\Delta^{[p]}}{dx} - \alpha e_\Delta^{[p]}\right)^2$ of problem (7.1), see also (2.16)–(2.19), which are part of the total energy. We do not consider these here, because they are $0(h^2)$.

Remark 7.7 *In practice, in order to get good error estimators we need to construct recovered values of derivatives at nodal points; these need to be accurate but not necessarily superconvergent. For example, it is sufficient for the error in the recovered value to be, say, 10% of the original error.*

Remark 7.8 *In Theorems 7.7 and 7.8 above we have assumed that a is smooth. It is often only piecewise smooth and in this case, instead of recovering derivatives, we recover the flux $a\frac{du}{dx}$ because this is smooth. Denoting the recovered flux by $\hat{z}^{[p]}$, we define*

$$\eta_q^{zz} \equiv \left(\int_{\tau_q} \frac{1}{a} \left(\frac{du_\Delta^{[p]}}{dx} - \hat{z}^{[p]} \right)^2 dx \right)^{\frac{1}{2}},$$

which gives us the energy norm of the error. Of course it might be that we were in fact interested in the L_2-norms of the fluxes themselves.

Example 7.3 Let us consider again the problem of Example 7.1 and take a uniform mesh with $x_1 = 0$, $x_q^\Delta = (q-1)h$, and $M = 1/h$, and $p = 1$. Now, $\tau_q = (x_q, x_{q+1})$ and for $x \in \tau_q$, $q = 2, \ldots, M$ we have

$$u_h^{[1]} = \frac{4}{\pi^2}\left(\sin\left(\frac{\pi}{2}x_q\right) + \sin\left(\frac{\pi}{2}x_{q+1}\right) - \sin\left(\frac{\pi}{2}x_q\right)\right)\left(\frac{x - x_q}{h}\right).$$

Hence,

$$\frac{du_n^{[1]}}{dx}\left(x_{q+\frac{1}{2}}\right) = \frac{4}{h\pi^2}\left(\sin\left(\frac{\pi}{2}x_{q+1}\right) - \sin\left(\frac{\pi}{2}x_q\right)\right),$$

and

$$z\left(x_q\right) = \Psi\left(x_q\right) = \frac{2}{\pi^2}\frac{1}{h}\left(\sin\left(\frac{\pi}{2}x_{q+1}\right) - \sin\left(\frac{\pi}{2}x_{q-1}\right)\right)$$
$$= \frac{4}{h\pi^2}\sin\frac{\pi h}{2}\cos\frac{\pi}{2}x_q$$
$$= \frac{2}{\pi}\cos\frac{\pi}{2}x_q\left(1 + 0(h^2)\right).$$

Thus,

$$\frac{du}{dx}(x_q) = \frac{2}{\pi}\cos\frac{\pi}{2}x_q,$$

and

$$\left|\frac{du}{dx}(x_q) - z_q\right| \leqq 0(h^2).$$

Further,

$$z\left(0\right) = z(x_1) = z(x_2) - h\frac{z(x_3) - z(x_2)}{h}$$
$$= \frac{2}{\pi}\left(\cos\frac{\pi h}{2} - \left(\cos\frac{\pi}{2}2h - \cos\frac{\pi h}{2}\right)\right)\left(1 + 0(h^2)\right)$$
$$= \frac{2}{\pi}\left(1 + 0(h^2)\right),$$

and hence we have

$$\left|\frac{du}{dx}(x_1) - z(0)\right| \leqq 0(h^2),$$

so that for $x \in \tau_1$

$$\left|\frac{du}{dx}(x) - z(x)\right| \leqq 0(h^2).$$

In the above sections we have constructed various error indicators η_q and error estimators ε. The indicators approximate the energy norm of the error $e_\Delta^{[p]}$ on τ_q, whilst the estimators approximate the energy norm of the error on I. The estimators of the energy norm of the error perform successfuly even when there is pollution, whilst the error indicators do not indicate pollution, see Example 7.2. Although the respective indicators and estimators are closely related, they are used for different purposes. An error estimator is used as the *acceptance criterion* for a finite element solution, whilst an error indicator is usually used to determine where a mesh should be refined, providing a means of identifying where a calculated finite element solution is too inaccurate.

7.1.8 The performance of the ZZ-estimator for the one-dimensional engineering problems

Proceeding as for the previous estimators, we now illustrate the performance of the ZZ-estimator in the context of the 1D Eng Problems introduced in Section 2.1.5.

In the spirit of Remark 7.8, because the coefficient a is discontinuous in this problem we use the recovered fluxes (see Remark 7.8). As the ZZ-estimator does not include the term involving the Newton boundary condition, the norm of the error is computed without involving this term. The influence of this term can be assessed by comparing the energies illustrated in previous chapters. With linear elements for 1D Eng Problem 1 the energy of the error is 48.2774, whilst without the Newton term it is 48.2091.

Tables 7.9–7.12 show the errors and the effectivity indices for linear and quadratic elements for 1D Eng Problems 1 and 2. Values are given for:

(1) the energy norm of the error: $\left\|e_\Delta^{[p]}\right\|_{\mathcal{U}}$;
(2) the ZZ-estimator ε^{ZZ};
(3) the corresponding effectivity index $\varkappa_{ZZ}$.

Note that because the coefficient a in this problem is discontinuous, Remark 7.8 has been used in the computations.

It can be seen that the ZZ-estimator, whilst its quality is still very reasonable, is not necessarily the best. As before, for 1D Eng Problem 1 for M=7,14,21 the effectivity index is excellent.

Figures 7.18 and 7.19 show the true errors and the error obtained with averaging for linear elements. Note that here we show the derivative of the error and not the fluxes. The results obtained with quadratic elements exhibit similar behaviour.

Figures 7.20–7.23 give graphs of the effectivity indices for the ZZ-estimator. Comparing these with the plots for the Neumann estimator, we see large differences; (compare Figs. 7.1–7.9).

Moral: The ZZ-estimator is very easy to implement and is most effective for problems with smooth solutions. It is not so effective in the case of non-smooth solutions, where the Neumann estimator is much more robust.

7.1.9 The Richardson error estimator

In the previous sections of this chapter the error estimators that we have considered have been based on one computation on a single mesh. In this section we define estimators based on more than one mesh and employ the Richardson technique.

We know that, see (3.89a),

$$\left\|e_\Delta^{[p]}\right\|_{\mathcal{U}}^2 = 2\left(\Pi\left(u_\Delta^{[p]}\right) - \Pi(\mathrm{u})\right), \tag{7.40}$$

where Π is the total energy, and $\Pi\left(u_\Delta^{[p]}\right)$ is known, whilst $\Pi(\mathrm{u})$ is not. Hence, if we could somehow obtain $\Pi(\mathrm{u})$, we could compute the error using (7.40).

Table 7.9 ZZ-estimator for Problem 1 using linear elements.

M	h	$\left\|e_{\Delta}^{[1]}\right\|_{\mathcal{U}}$	ε^{ZZ}	$\varkappa_{ZZ}$
1	3.50000	205.79272	241.31104	1.17259
2	1.75000	173.69770	242.03132	1.39341
3	1.16667	155.78745	256.20119	1.64456
4	0.87500	139.20644	127.56608	0.91631
5	0.70000	119.44140	107.85285	0.90298
6	0.58333	90.45578	83.60417	0.92425
7	0.50000	16.08925	16.09853	1.00058
8	0.43750	54.47436	56.93990	1.04526
9	0.38889	68.49698	77.23211	1.12753
10	0.35000	74.72796	92.68906	1.24035
11	0.31818	75.43993	76.48012	1.01379
12	0.29167	70.42750	69.98174	0.99367
13	0.26923	56.68238	56.29699	0.99320
14	0.25000	8.06291	8.06407	1.00014
15	0.23333	38.33796	40.60245	1.05907
16	0.21875	49.88172	56.98342	1.14237
17	0.20588	55.78733	70.13586	1.25720
18	0.19444	57.47690	57.14667	0.99425
19	0.18421	54.58876	53.40549	0.97832
20	0.17500	44.56268	43.74907	0.98174
21	0.16667	5.37756	5.37790	1.00006
22	0.15909	31.20070	33.09377	1.06067
23	0.15217	41.10265	47.04066	1.14447
24	0.14583	46.40849	58.48582	1.26024
25	0.14000	48.20906	47.85344	0.99262

Let us consider computations using uniform meshes Δ_i with corresponding mesh sizes $h_i \equiv h(\Delta_i)$, where $h_{i+1} < h_i$. We now make the assumption that

$$\Pi\left(u_{\Delta}^{[p]}\right) \approx \Pi(\mathfrak{u}) + Ch^{2\beta}, \tag{7.41a}$$

Table 7.10 ZZ-estimator for Problem 1 using quadratic elements.

M	h	$\left\|e_{\Delta}^{[2]}\right\|_{\mathcal{U}}$	ε^{ZZ}	$\varkappa_{ZZ}$
1	3.50000	148.01116	229.60988	1.55130
2	1.75000	112.60788	118.30001	1.05055
3	1.16667	81.48017	81.08621	0.99516
4	0.87500	59.74750	63.79623	10.6776
5	0.70000	52.71414	77.32712	1.46691
6	0.58333	51.54292	46.34131	0.89910
7	0.50000	0.57013	0.57013	1.00000
8	0.43750	44.98204	59.46441	1.32196
9	0.38889	48.10256	54.63308	1.13576
10	0.35000	41.91790	42.76172	1.02013
11	0.31818	35.62629	38.30077	1.07507
12	0.29167	34.84072	51.36079	1.47416
13	0.26923	35.69027	35.14973	0.98485
14	0.25000	0.14359	0.14359	1.00000
15	0.23333	32.62961	40.86283	1.25232
16	0.21875	35.72365	38.60234	1.08058
17	0.20588	31.86025	31.83416	0.99918
18	0.19444	27.80239	29.66119	1.06686
19	0.18421	27.82785	40.61948	1.45967
20	0.17500	28.89960	27.53691	0.95285
21	0.16667	0.06390	0.06390	1.00000
22	0.15909	26.88241	33.64750	1.25165
23	0.15217	29.68658	32.08469	1.08078
24	0.14583	26.71833	26.68352	0.99870
25	0.14000	23.57476	25.13897	1.06635

where β is a constant (known or unknown) and $\mathcal{C}$ is independent of h. We can similarly use the number of degrees of freedom N of the computation to rewrite (7.41a) as

$$\Pi\left(u_{\Delta}^{[p]}\right) \approx \Pi(\mathrm{u}) + \mathcal{C}N^{-2\beta}, \tag{7.41b}$$

which has the advantage that we can use a non-uniform mesh. Note that the symbol $\approx$ means that the equality holds up to higher-order terms. Henceforth, we shall use $=$ rather than $\approx$.

Table 7.11 ZZ-estimator for Problem 2 using linear elements.

M	h	$\left\|e_{\Delta}^{[1]}\right\|_{\mathcal{U}}$	ε^{ZZ}	$\varkappa_{ZZ}$
1	3.50000	211.56025	251.80111	1.19021
2	1.75000	178.15684	257.48021	1.44524
3	1.16667	157.86335	276.05755	1.74871
4	0.87500	137.30921	123.93900	0.90263
5	0.70000	109.79978	99.82709	0.90917
6	0.58333	57.88156	55.36739	0.95656
7	0.50000	52.26825	53.81244	1.02954
8	0.43750	70.37986	78.61340	1.11699
9	0.38889	77.80037	96.54243	1.24090
10	0.35000	78.10496	79.06350	1.01227
11	0.31818	70.71331	70.29729	0.99412
12	0.29167	49.33612	49.28101	0.99888
13	0.26923	29.31630	30.01245	1.02375
14	0.25000	47.82420	52.84973	1.10508
15	0.23333	56.56403	68.92633	1.21855
16	0.21875	59.62320	59.62237	0.99999
17	0.20588	56.98125	55.79964	0.97926
18	0.19444	45.35606	44.58106	0.98291
19	0.18421	14.88847	14.98396	1.00641
20	0.17500	35.88956	38.80916	1.08135
21	0.16667	45.21859	53.57026	1.18470
22	0.15909	49.47964	49.98634	1.01024
23	0.15217	49.15930	48.19378	0.98036
24	0.14583	42.41914	41.48455	0.97797
25	0.14000	17.59305	17.55175	0.99765

If β is known, then we have in (7.41) the two unknowns $\mathfrak{u}$ and $\mathcal{C}$. Suppose that we compute using the two meshes Δ_1 and Δ_2, then for (7.41a)

$$\Pi\left(u_{\Delta_1}^{[p]}\right) = \Pi(\mathfrak{u}) + \mathcal{C}h_1^{2\beta},$$

$$\Pi\left(u_{\Delta_2}^{[p]}\right) = \Pi(\mathfrak{u}) + \mathcal{C}h_2^{2\beta},$$

Table 7.12 ZZ-estimator for Problem 2 using quadratic elements.

M	h	$\left\Vert e_{\Delta}^{[2]}\right\Vert_{\mathcal{U}}$	ε^{ZZ}	$\varkappa_{ZZ}$
1	3.50000	150.92716	245.45248	1.62630
2	1.75000	110.01431	113.31311	1.02999
3	1.16667	75.52792	74.97959	0.99274
4	0.87500	56.55144	68.73120	1.21538
5	0.70000	53.97369	49.99012	0.92619
6	0.58333	40.68860	38.30527	0.94143
7	0.50000	43.58881	53.98149	1.23843
8	0.43750	50.03962	57.56173	1.15032
9	0.38889	43.17423	43.98592	1.01880
10	0.35000	36.30983	40.39693	1.11256
11	0.31818	36.39076	58.68711	1.61269
12	0.29167	34.62018	33.55984	0.96937
13	0.26923	25.97064	28.57261	1.10019
14	0.25000	36.92825	41.72006	1.12976
15	0.23333	34.42071	35.01731	1.01733
16	0.21875	29.20435	30.12604	1.03156
17	0.20588	28.61276	40.94038	1.43084
18	0.19444	29.71970	28.25545	0.95073
19	0.18421	13.19642	13.60253	1.03077
20	0.17500	29.56357	40.23845	1.36108
21	0.16667	29.85905	31.13334	1.04268
22	0.15909	25.63743	25.54528	0.99641
23	0.15217	24.10571	30.90135	1.28191
24	0.14583	25.78271	25.01825	0.97035
25	0.14000	15.30532	14.97889	0.97867

and hence

$$\Pi^*(\mathrm{u}) \equiv \Pi\left(u_{\Delta_1}^{[p]}\right) - \frac{\Pi\left(u_{\Delta_1}^{[p]}\right) - \Pi\left(u_{\Delta_2}^{[p]}\right)}{1 - \left(\frac{h_2}{h_1}\right)^{2\beta}}$$

$$= \Pi\left(u_{\Delta_2}^{[p]}\right) - \frac{\Pi\left(u_{\Delta_1}^{[p]}\right) - \Pi\left(u_{\Delta_2}^{[p]}\right)}{\left(\frac{h_2}{h_1}\right)^{2\beta} - 1}. \tag{7.42}$$

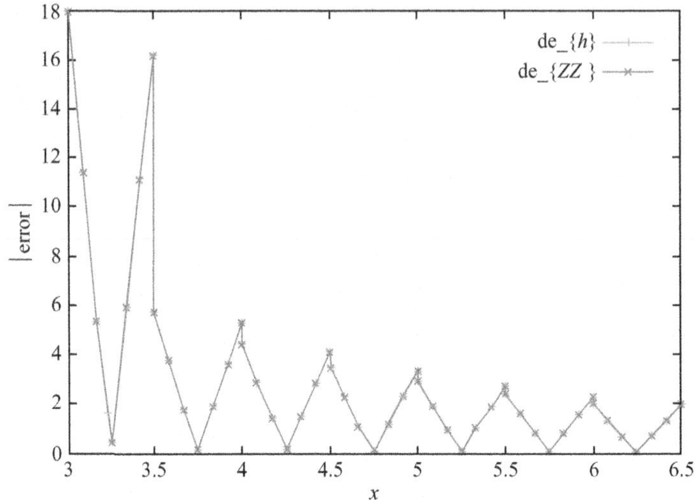

Fig. 7.18 Plot of $\left|e_h'(x)\right|$ and $\left|e_{ZZ}'(x)\right|$ for Problem 1 with number of elements = 7 using linear elements. Note that the interface coincides with a node.

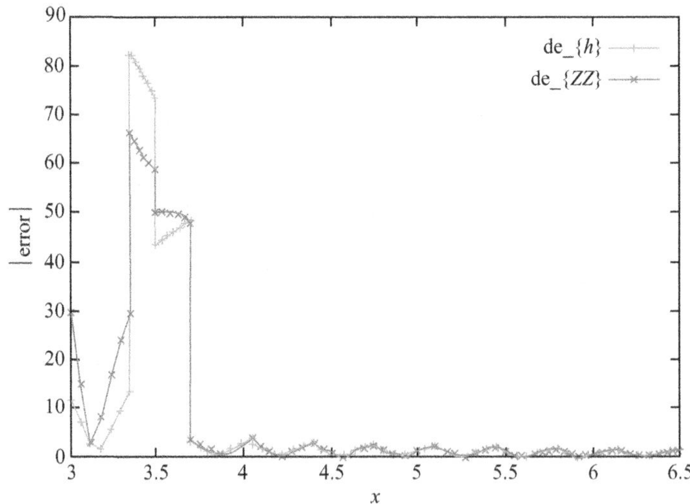

Fig. 7.19 Plot of $\left|e_h'(x)\right|$ and $\left|e_{ZZ}'(x)\right|$ for Problem 1 with number of elements = 10 using linear elements. Note that the interface does not coincide with a node.

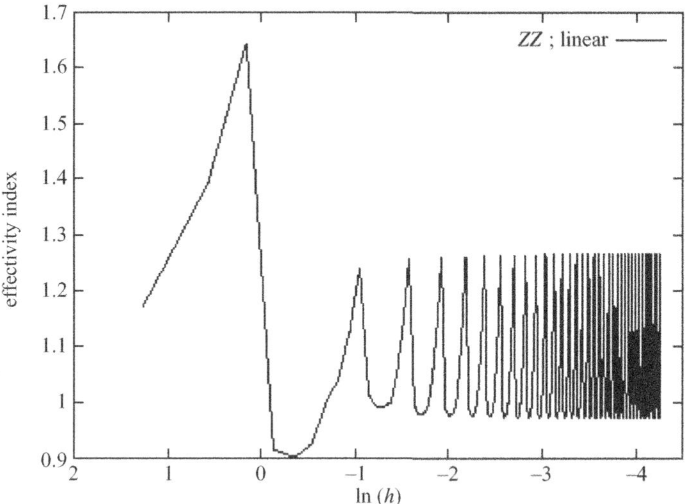

Fig. 7.20 Plot of effectivity index of the ZZ-estimator for Problem 1 using linear elements.

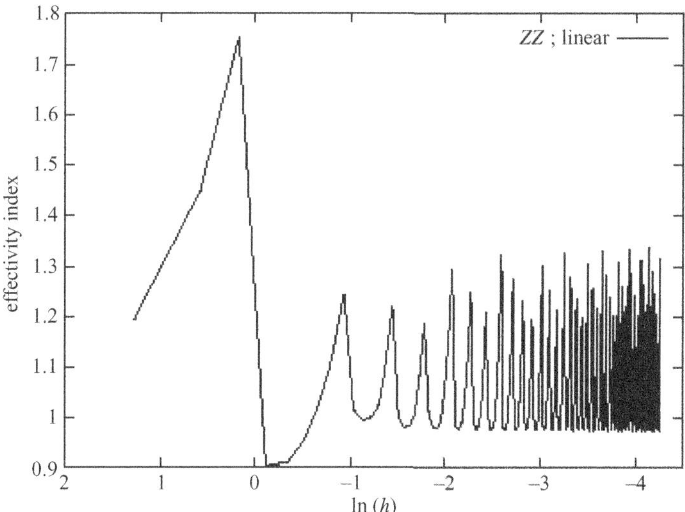

Fig. 7.21 Plot of effectivity index of the ZZ-estimator for Problem 2 using linear elements.

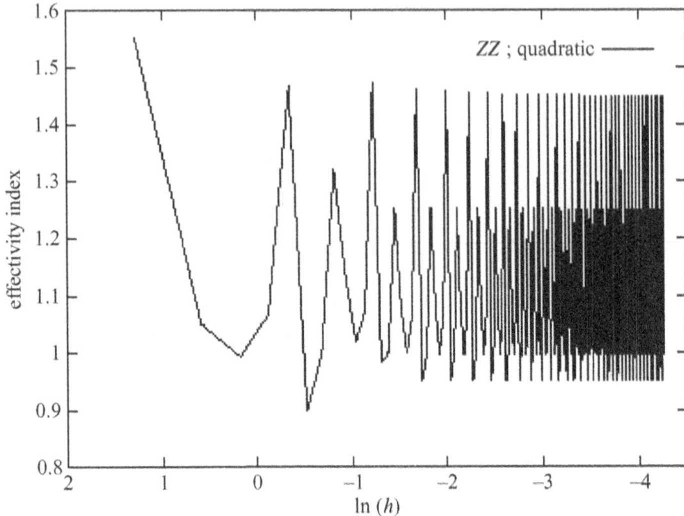

Fig. 7.22 Plot of effectivity index of the *ZZ*-estimator for Problem 1 using quadratic elements.

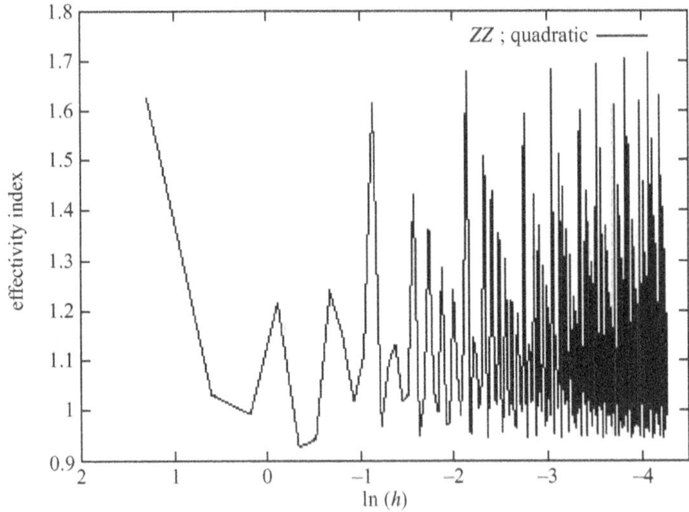

Fig. 7.23 Plot of effectivity index of the *ZZ*-estimator for Problem 2 using quadratic elements.

Note that $\Pi^*(\mathfrak{u})$ is an approximation to $\Pi(\mathfrak{u})$ and effectively it is assumed that $\Pi^*(\mathfrak{u}) = \Pi(\mathfrak{u})$.

In deriving (7.42) we have assumed that β is known. If this were not the case, we could compute with three meshes $\Delta_i, \Delta_{i+1}, \Delta_{i+2}$ and obtain

$$\frac{\Pi\left(u_{\Delta_1}^{[p]}\right) - \Pi(\mathfrak{u})}{\Pi\left(u_{\Delta_2}^{[p]}\right) - \Pi(\mathfrak{u})} = \left(\frac{h_2}{h_1}\right)^{2\beta}, \tag{7.43a}$$

and

$$\frac{\Pi\left(u_{\Delta_2}^{[p]}\right) - \Pi(\mathfrak{u})}{\Pi\left(u_{\Delta_3}^{[p]}\right) - \Pi(\mathfrak{u})} = \left(\frac{h_2}{h_3}\right)^{2\beta}, \tag{7.43b}$$

from which the β and $\Pi^*(\mathfrak{u})$ can be computed. In this general case this would necessitate the solving of a non-linear equation. However, in the special case that $h_{i+1} = \alpha h_i$ and $h_{i+2} = \alpha h_{i+1}$, we find that

$$\frac{\Pi\left(u_{\Delta_1}^{[p]}\right) - \Pi(\mathfrak{u})}{\Pi\left(u_{\Delta_2}^{[p]}\right) - \Pi(\mathfrak{u})} = \frac{\Pi\left(u_{\Delta_2}^{[p]}\right) - \Pi(\mathfrak{u})}{\Pi\left(u_{\Delta_3}^{[p]}\right) - \Pi(\mathfrak{u})},$$

so that we may define

$$\Pi^*(\mathfrak{u}) = \frac{\Pi\left(u_{\Delta_2}^{[p]}\right)^2 - \Pi\left(u_{\Delta_1}^{[p]}\right)\Pi\left(u_{\Delta_3}^{[p]}\right)}{2\Pi\left(u_{\Delta_2}^{[p]}\right) - \Pi\left(u_{\Delta_1}^{[p]}\right) - \Pi\left(u_{\Delta_3}^{[p]}\right)}. \tag{7.44}$$

Returning to the case where β in (7.41) is known, and using the two meshes Δ_i and Δ_{i+1} we can compute $\Pi^*(\mathfrak{u})$ from (7.42) using $u_{\Delta_i}^{[p]}$ and $u_{\Delta_{i+1}}^{[p]}$, and so define the estimators ε_i and ε_{i+1} where

$$\varepsilon_1^2 = 2\left(\Pi\left(u_{\Delta_1}^{[p]}\right) - \Pi^*(\mathfrak{u})\right) \approx \left\|e_{\Delta_1}^{[p]}\right\|_{\mathcal{U}}^2, \tag{7.45a}$$

$$\varepsilon_2^2 = 2\left(\Pi\left(u_{\Delta_2}^{[p]}\right) - \Pi^*(\mathfrak{u})\right) \approx \left\|e_{\Delta_2}^{[p]}\right\|_{\mathcal{U}}^2. \tag{7.45b}$$

Because $e_{\Delta_2}^{[p]}$ is more accurate than $e_{\Delta_1}^{[p]}$, it is usual to use ε_2. Note that assumption (7.41a) is essential for the estimators of (7.45), and we shall see the influence of this on the accuracy and the effectivity index for the estimator in the results that follow.

7.1.10 The performance of the Richardson estimator

Once again we use 1D Eng Problems 1 and 2 of Section 2.1.5 to demonstrate the use of an error estimator, in this case the Richardson estimator. We recall that the exact solution $\mathfrak{u}$, and hence $\Pi(\mathfrak{u})$, is known for each problem.

Recall again that in Problem 1 the interface between the two types of steel is $b = 3.5\,\mathrm{cm}$ and that for $M = 7, 14, 21$, where M is the number of elements of Δ, this is a nodal point of the meshes. The mesh size is as always h.

In Tables 7.13 to 7.19 we illustrate the accuracy of the Richardson estimator using (7.42) for $u_\Delta^{[1]}$ calculated using linear elements and for the values of M and h as shown. For Problems 1 and 2 values are given for:

(1) the total energy $\Pi\left(u_\Delta^{[1]}\right)$ of the finite element solution $u_\Delta^{[1]}$;
(2) the approximate value $\Pi^*(\mathrm{u})$ of the exact energy $\Pi(\mathrm{u})$ calculated using (7.42);
(3) $\varkappa_1$, the effectivity index of the estimator ε_1 when the finer mesh Δ_2 is used;
(4) $\varkappa_2$, the effectivity index of the estimator ε_2, when the coarser mesh Δ_1 is used;
(5) the relative error: $\mathrm{rel} = \left\| e_\Delta^{[1]} \right\|_{\mathcal{U}} / \|\mathrm{u}\|_{\mathcal{U}}$, as a percentage.

Table 7.13 Results for Problem 1, interface coinciding with a node, linear elements, $\beta = 1$.

M	h	$\Pi\left(u_\Delta^{[1]}\right)$	$\Pi^*(\mathrm{u})$	$\varkappa_1$	$\varkappa_2$	rel
7	0.5000	153 883.74584			0.999259	2.77622
14	0.2500	153 786.78088	153 754.45924	0.997133	0.999987	1.39160
28	0.1250	153 756.30761	153 757.27384	0.999793		

Table 7.14 Results for Problem 1, interface not coinciding with a node, linear elements, $\beta = 1$.

M	h	$\Pi\left(u_\Delta^{[1]}\right)$	$\Pi^*(\mathrm{u})$	$\varkappa_1$	$\varkappa_2$	rel
6	0.58333	157 886.86453			0.726842	15.681
12	0.29167	156 249.40605	155 703.58656	0.467707	0.869680	12.185
24	0.14583	154 833.97789	154 362.16850	0.661044		8.0157

Table 7.15 Results for Problem 1, interface not coinciding with a node, but located in same relative position in the element, linear elements, $\beta = 1/2$.

M	h	$\Pi\left(u_\Delta^{[1]}\right)$	$\Pi^*(\mathrm{u})$	$\varkappa_1$	$\varkappa_2$	rel
6	0.58333	157 886.86543			1.06414	15.68
13	0.26923	155 367.03481	153 206.74547	1.15732	1.04573	9.76
20	0.75000	154 749.59492	153 603.35329	1.07314		7.70

Table 7.16 Results for Problem 1, interface not coinciding with a node, but located in same relative position in the element, linear elements, $\beta = 1$.

M	h	$\Pi\left(u_\Delta^{[1]}\right)$	$\Pi^*(\mathrm{u})$	$\varkappa_1$	$\varkappa_2$	rel
6	0.58333	157 886.86543			0.88023	15.68
13	0.26923	155 367.03481	154 684.97562	0.65032	0.81420	9.76
20	0.75000	154 749.59492	154 297.87486	0.67367		7.70

Table 7.17 Results for Problem 1, interface not coinciding with a node, but located in same relative position in the element, linear elements, $\beta = 1/2$.

M	h	$\Pi\left(u_\Delta^{[1]}\right)$	$\Pi^*(\mathrm{u})$	$\varkappa_1$	$\varkappa_2$	rel
11	0.31818	156 619.80228			1.08317	13.06
18	0.19444	155 412.74885	153 515.95060	1.06944	1.03050	9.93
25	0.14000	154 919.62723	153 651.60021	1.04312		8.33

Table 7.18 Results for Problem 1, interface not coinciding with a node, but located in same relative position in the element, linear elements, $\beta = 1$.

M	h	$\Pi\left(u_\Delta^{[1]}\right)$	$\Pi^*(\mathrm{u})$	$\varkappa_1$	$\varkappa_2$	rel
11	0.31818	156 619.80228			0.81994	13.06
18	0.19444	155 412.74885	154 693.27370	0.65865	0.87863	9.93
25	0.14000	154 919.62723	154 388.82546	0.67490		8.33

Table 7.19 Results for Problem 2, linear elements, $\beta = 1$.

M	h	$\Pi\left(u_\Delta^{[1]}\right)$	$\Pi^*(\mathrm{u})$	$\varkappa_1$	$\varkappa_2$	rel
7	0.5000	153 027.67457			0.46668	9.09
14	0.2500	152 803.89889	152 729.30670	0.25503		8.32
28	0.1250	152 566.53978	149 154.086714	2.1265		14.34

We recall that u and $\Pi(\mathrm{u})$ are known for these problems, and that $\Pi^*(\mathrm{u}) - \Pi(\mathrm{u}) > 0$. Once again, values in the tables are given to far more significant figures that would appear sensible, because these are needed for the calculations that lead to the values in later columns.

Table 7.13 addresses for Problem 1 the case of linear elements where the interface coincides with a node. As the error in this case with linear elements is $0(h)$, we take $\beta = 1$ in (7.41a). Because, see Fig. 3.14, the errors are located virtually on the straight line with slope 1, the effectivity index $\varkappa$ must be close to 1; this is clearly visible from the table.

Table 7.14 addresses the case when the interface does not coincide with a node and, as would be expected, the effectivity index is not nearly so good as above. However, it is not a disaster, although (7.41) does not hold, because the ratio of successive mesh sizes is $\frac{1}{2}$, giving a relatively large difference in the number of elements used.

Returning to Fig. 3.14 we see that the envelope of the errors is nearly a straight line with slope $\frac{1}{2}$. Using meshes with $M = 6, 13, 20$ we locate the interface in the same relative position in the element in which it appears, so that we can expect an $0(h^{\frac{1}{2}})$ rate of convergence. Table 7.15 addresses this situation for Problem 1 with $\beta = \frac{1}{2}$ and we see that very reasonable values for $\varkappa$. Analogous results are shown in Tables 7.17 and 7.18 for $M = 11, 18, 25$.

Results for similar calculations for Problem 2 are presented in Table 7.19. Here, the position of the interface precludes it from coinciding with the mesh or from occurring in the same relative position in the elements that contain it. Hence, it is unreasonable to expect the Richardson technique to produce good effectivity indexes for any values of β.

Whilst we do not present any calculations here for cases when quadratic elements are used, we note that very similar behaviour will be obtained with $p = 2$ if we take $\beta = 2$.

We now consider the result of using the estimator based on (7.44), again with linear elements. Using meshes $\Delta_i, i = 1, 2, 3$ we can compute $u_{\Delta_i}^{[1]}, \Pi(u_{\Delta_i}^{[1]})$ and hence $\Pi^*(u)$ using (7.44). In Tables 7.20 and 7.21 we give values for $\Pi^*(u)$ and the effectivity index $\varkappa$ obtained for various values of M and h using linear elements, respectively, when the interface coincides and does not coincide with a node of the mesh. Note that with (7.44) we do not use β.

In Table 7.20 we see that in this case the effectivity index for the error is best for the coarsest mesh. The reason for this is that the error decays with M.

Again, we see from Table 7.21 that the value of the effectivity index with the coarsest mesh is the best; contrastingly the value with the fine mesh is very bad. In this case, the choice of meshes causes the relative position of the interface in an

Table 7.20 Results for Problem 1, interface coinciding with a node, linear elements.

M	h	$\Pi^*(u)$	$\varkappa$
14	0.2500	153 786.78088	0.99696
28	0.1250	153 762.40891	0.98787
56	0.0625	153 754.37180	0.95149

Table 7.21 Results for Problem 1, interface not coinciding with a node, linear elements.

M	h	$\Pi^*(u)$	$\varkappa$
10	0.3500	156 565.59777	0.78014
20	0.1750	154 749.59492	0.38619
30	0.0875	154 431.51879	0.09950

element to be M dependent. If the successive values of M were such that the relative positions remained the same, the values of $\varkappa$ would be better.

So far, in our use of Richardson techniques we have always had $h_i/h_{i+1} = 2$. If, contrastingly, we have $h_i \approx h_{i+1}$, then the estimators will not be good.

To show this we consider again Problem 1 and finite element solutions based on linear elements. We consider the accuracy of the solution for $h = 0.25$, i.e. $M = 14$ and refer to Table 7.13. There, we see that for $M = 7, 14$ with $h = 0.5$ and 0.25 the value of the effectivity index $\varkappa_1$ is $\varkappa_1 = 0.9971$.

Now, let us consider the meshes with $M = 13$ and 14 where correspondingly $h = 0.26923$ and 0.25. In this case the effectivity index has value $\varkappa_1 = 13.2831$, which is worthless. The values are given in Table 7.22.

The results of this section may be summarized as follows:

(1) The estimator is of very low quality in the case when $h_i \approx h_{i+1}$.
(2) If $h_i \approx 2h_{i+1}$ the effectivity index is not as bad as in (1).
(3) For some pairs h_i, h_{i+1} and corresponding N_i, N_{i+1} the effectivity index is very good. This depends on whether the error trend can be well approximated.
(4) Whilst the selection of the parameter β is important, even if the calculations are undertaken with an incorrect value of β the effectivity index will not be severely bad provided $h_i \geqq 2h_{i+1}$.
(5) The accuracy of the estimator depends on how well (7.41) is satisfied.

Moral: To obtain good results with the Richardson estimator one should take $h_{i+1} = \frac{1}{2}h_i$, otherwise the quality may be low. Assumption (7.41) is usually reasonable when the meshes are nested.

Table 7.22 Results for Problem 1 with meshes for which $h = 0.26923$ and 0.25, linear elements.

M	h	$\Pi(u_\Delta^{[1]})$	$\Pi^*(u)$	$\varkappa_1$
13	0.26923	155 367.03481		
14	0.25	153 786.78088	142 353.05	13.2831

7.2 Error indicators and estimators in two dimensions

7.2.1 The Dirichlet element-based error estimator

As for the one-dimensional context in the two-dimensional context, we start by discussing residual estimators. Let us consider the weak problem (5.51) and the finite element problem (5.55) so that we have for $p = 1, 2$

$$\text{find } \mathfrak{u} \in \mathcal{U}_0 \text{ such that } B(\mathfrak{u}, v) = F(v)\ \forall v \in \mathcal{U}_0, \tag{7.46a}$$

and

$$\text{find } u_\Delta^{[p]} \in S_{\Delta,0}^{[p]} \text{ such that } B\left(u_\Delta^{[p]}, v\right) = F(v)\ \forall v \in S_{\Delta,0}^{[p]}, \tag{7.46b}$$

where the $B(\mathfrak{u}, v)$ and $F(v)$ are, respectively, as in (5.52) and (5.53). Again, defining the error $e_\Delta^{[p]} \equiv \mathfrak{u} - u_\Delta^{[p]}, p = 1, 2$ we have, as in Section 7.1.1,

$$B\left(e_\Delta^{[p]}, v\right) = B\left(\mathfrak{u} - u_\Delta^{[p]}, v\right) = F(v) - B\left(u_\Delta^{[p]}, v\right) \equiv \Re(v), \tag{7.47}$$

where $\Re(v)$ is now the residual for the two-dimensional case. Once more, for simplicity we shall assume that $A \equiv \{a_{ij}\} \in C^1(\Omega),\ c \in C^0(\Omega), \alpha \in C^0(\Gamma_N),\ f \in L^2(\Omega)$ and $G \in L_2(\Gamma_N)$.

As before, we denote by τ_q an element of the triangular mesh Δ over Ω. Then, considering this element we have

$$B_{\tau_q}(u_\Delta^{[p]}, v) = -\int_{\tau_q} \left(\mathcal{D}u_\Delta^{[p]}, v\right) d\mathbf{x} + \int_{\partial\tau_q} \aleph_\tau^{[p]} \cdot \mathbf{n}_\tau v ds + \int_{\partial\tau\cap\Gamma_N} \alpha u_\Delta^{[p]} v ds,$$

where

$$\mathcal{D}u_\Delta^{[p]} \equiv \sum_{i,j=1}^{2} \frac{\partial}{\partial x_i} a_{i,j} \frac{\partial}{\partial x_j} u_\Delta^{[p]} - c u_\Delta^{[p]},$$

and

$$\aleph_\tau^{[p]} \cdot \mathbf{n}_\tau = \sum_{i,j=1}^{2} n_i a_{ij} \frac{\partial u_\Delta^{[p]}}{\partial x_j},$$

with $\mathbf{n}_\tau = (n_1, n_2)$ the outward normal to $\partial\tau_q$. Then, defining on the element τ_q the *interior residual* r_q, where

$$r_q = f + \mathcal{D}u_\Delta^{[p]}, \tag{7.48}$$

analogously to (7.2a), we may define the *residual*

$$\Re(v) = \sum_{\tau_q\in\Delta} \left(\int_{\tau_q} r_q v d\mathbf{x} - \int_{\partial\tau_q} \left(\aleph_\tau^{[p]} \cdot \mathbf{n}_\tau\right) v ds - \int_{\tau_q\cap\Gamma_N} \alpha u_\Delta^{]p[} v ds \right). \tag{7.49}$$

Further, rearranging the terms on the boundary of the elements, we obtain

$$-\sum_{\tau_q\in\Delta}\int_{\partial\tau_q}\left(\aleph_{\tau_q}^{[p]}\cdot\mathbf{n}_{\tau_q}\right)vds+\int_{\partial\tau_q\cap\Gamma_N}\alpha u_{\Delta}^{]p[}vds=+\sum_{\varepsilon\in\varepsilon(\Delta)}\int_{\varepsilon}\mathcal{J}_{\varepsilon}\left(n\left(\tau_q\right)\right)vds, \tag{7.50a}$$

where for an adjacent element τ_q^*

$$\mathcal{J}_{\varepsilon}\left(n\left(\tau_q\right)\right)=\begin{cases}\left(\aleph_{\tau_q}^{[p]}-\aleph_{\tau_q^*}^{[p]}\right)\cdot\mathbf{n}_{\tau_q}, & \varepsilon=\partial\tau_q\cap\partial\tau_q^*\not\in\partial\Omega,\\ G-\alpha u_{\Delta}^{[p]}-\aleph_{\tau_q}^{[p]}\cdot\mathbf{n}_{\tau_q}, & \varepsilon\subset\Gamma_N\\ 0, & \varepsilon\subset\Gamma_D\end{cases} \tag{7.50b}$$

in which $\mathcal{J}_\varepsilon$ is the *jump* on the edge of ε of τ_q. We have used the notation $\mathcal{J}_\varepsilon\left(n\left(\tau_q\right)\right)$ because of the normal orientation. We have $\mathcal{J}_\varepsilon\left(n\left(\tau_q^*\right)\right)=-\mathcal{J}_\varepsilon\left(n\left(\tau_q\right)\right)$. Note that we denote by $\varepsilon(\Delta)$ the set of all element edges of the mesh Δ.

Again, analogously to the one-dimensional case we may write

$$\Re(v)=\sum_{\tau_q\in\Delta}\int_{\tau_q}r_q v d\mathbf{x}+\sum_{\varepsilon\in\varepsilon(\Delta)}\int_{\varepsilon}\mathcal{J}_{\varepsilon}vds. \tag{7.51}$$

The exact meaning of the second term in (7.51) requires some explanation. Every $\varepsilon=\bar{\tau}_q\cap\bar{\tau}_q^*$ and the jump across ε is $\mathcal{J}_\varepsilon\left(n\left(\tau_q\right)\right)$. The integration on ε is in the direction that is 90° counterclockwise related to $n(\tau_q)$ and so the value of the integral is independent of the selection of n at ε, i.e. $n(\tau)$ or $n(\tau^*)$.

We now give a sequence of theorems, which are analogous to those of the one-dimensional case.

Theorem 7.10 *Let* $\mathrm{u}\in\mathcal{U}_0$ *and* $u_{\Delta}^{[p]}\in S_{\Delta,0}^{[p]}$ *be the solutions of problems (5.51) and (5.55), respectively. Then, for* $\Re(v), v\in\mathcal{U}_0$, *as in (7.36)*

$$\left\|e_{\Delta}^{[p]}\right\|_{\mathcal{U}}=\sup_{v\in\mathcal{U}_0}\frac{\Re(v)}{\|v\|_{\mathcal{U}}},\quad p=1,2. \tag{7.52}$$

Proof The proof is identical to that of Theorem 7.1. □

We now define the two-dimensional error indicator, η_q^{Dir}, the error estimator $\varepsilon^{\mathrm{Dir}}$, and the effectivity index $\varkappa$ as,

$$\eta_q^{\mathrm{Dir}}\equiv\left\|\hat{e}_{\Delta}^{[p]}\right\|_{\mathcal{U}(\tau_q)}, \tag{7.53a}$$

$$\varepsilon^{\mathrm{Dir}}-\left(\sum_{\tau_q\in\Delta}\left(\eta_q^{\mathrm{Dir}}\right)^2\right)^{\frac{1}{2}}, \tag{7.53b}$$

$$\varkappa^{\mathrm{Dir}}=\frac{\varepsilon^{\mathrm{Dir}}}{\left\|e_{\Delta}^{[p]}\right\|,\mathcal{U}} \tag{7.53c}$$

where

$$\hat{e}_{\Delta}^{[p]} \in \mathcal{U}_0(\tau_q) = \{v \mid v \in \mathcal{U}(\tau_q),\ v = 0 \text{ on } \partial\tau_q\}, \tag{7.53d}$$

and

$$B_{\tau}\left(\hat{e}_{\Delta}^{[p]}, v\right) = \int_{\tau_q} r_q v d\mathbf{x},\ \forall v \in \mathcal{U}_0(\tau_q). \tag{7.53e}$$

Theorem 7.11 *Let* $u \in \mathcal{U}_0$ *and* $u_{\Delta}^{[p]} \in S_{\Delta,0}^{[p]}$ *be as in Theorem 7.7. Then,*

$$\varepsilon^{\text{Dir}} \leqq \left\| e_{\Delta}^{[p]} \right\|_{\mathcal{U}_0}. \tag{7.54}$$

Proof This is completely analogous to that of Theorem 7.1. □

We note that in contrast to the one-dimensional case ε^{Dir} for two dimensions can be very inaccurate. The next example demonstrates this.

Example 7.4 Let us consider the problem

$$-\Delta u(\mathbf{x}) = 0,\quad x \in \Omega$$
$$\frac{\partial u}{\partial n}(x) = G,\quad x \in \Gamma_N = \Gamma,$$

where $\int_\Gamma G dx = 0$.

The weak formulation of this problem is to find $u \in \mathcal{U}$ such that

$$\int_\Omega \nabla u \nabla v d\mathbf{x} = \int_\Gamma v G ds \ \ \forall v \in \mathcal{U}.$$

Let Δ be a triangular mesh partition of Ω and take $p = 1$. Then, on every $\tau_q \in \Delta$ we have that $\Delta u_{\Delta}^{[1]} = 0$, and

$$\int_{\tau_q} \nabla u_{\Delta}^{[1]} \nabla v = 0 \ \ \forall v \in \mathcal{U}_0(\tau_q).$$

Hence, we see that $\hat{e}_q^{\text{Dir}} = 0$ and hence $\eta_q^{\text{Dir}} = 0$.

Remark 7.9 *We showed for the problem of Example 7.4 that* $\eta_q^{\text{Dir}} = 0$ *for* $p = 1$. *It is often the case that this estimator also gives reasonable results when* $p = 2$.

7.2.2 The Neumann element-based error estimator

The main idea for this two-dimensional estimator is the same as that in the one-dimensional case, although the implemention is more complicated.

Let $\tau_q \in \Delta$ be an element with vertices XYZ, see Fig. 7.24, with edges $\varepsilon_1 \equiv \overline{XY}, \varepsilon_2 \equiv \overline{YZ}$ and $\varepsilon_3 \equiv \overline{ZX}$ and outward normals $n(\tau_q)$ as indicated. We differentiate between the case of an interior element, where neither the vertices nor the sides coincide with part of the boundary Γ, and a boundary element where a vertex lies on Γ.

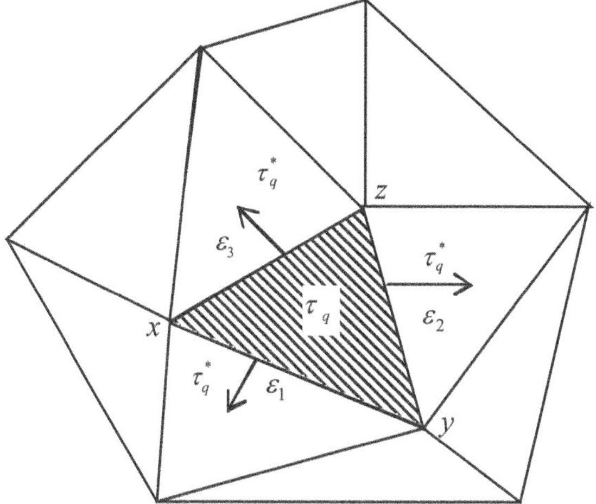

Fig. 7.24 Interior element τ_q of the mesh Δ and associated patch elements.

In the case where τ_q is an interior element let Φ_X, Φ_Y, Φ_Z be the three hat functions associated with the vertices X, Y, Z, and let $\omega_X, \omega_Y, \omega_Z$ be their respective supports; the ω's are called *patches* and clearly τ_q belongs to all three patches.

The jumps $\mathcal{J}_\varepsilon(n(\tau_q))$ for $\varepsilon_i = \tau_q \cap \tau_q^*$, $q = 1, 2, \ldots, M$, where τ_q and τ_q^* are adjacent elements, with $n(\tau_q)$ being the outward normal to τ_q on ε, are defined for every interior edge ε_i of the mesh Δ. For the element τ_q we split the jumps $\mathcal{J}_\varepsilon(n(\tau_q))$ defined on each ε_i into classes; $\mathcal{J}_\varepsilon(n(\tau_q))$ associated with τ_q and $\mathcal{J}_\varepsilon(n(\tau_q^*))$ associated with the elements τ_q^* in such a way that

$$\mathcal{J}_{\varepsilon_i}^{(1)}(n(\tau_q)) = \frac{1}{2}\mathcal{J}_{\varepsilon_i}(n(\tau_q)) = -\mathcal{J}_{\varepsilon_i}^{(1)}(n(\tau_q^*)).$$

The negative sign in the last expression arises from the fact that $n(\tau_q^*) = -n(\tau_q)$. In this way, we split the jumps associated with τ_q and the surrounding elements τ_q^*, as shown in Fig. 7.24.

However, for this splitting we have that

$$\int_{\tau_q} r_q d\mathbf{x} + \sum_{i=1}^{3} \int_{\varepsilon_i} \mathcal{J}_{\varepsilon_i}^{(1)}(n(\tau_q))ds \tag{7.55}$$

is not equal to zero, so that we have to make *corrections* $\Theta\left(\varepsilon_i(n(\tau_q))\right)$ in $\mathcal{J}_{\varepsilon_i}^{(1)}(n(\tau_q))$ on each edge ε_i of τ_q, with similar corrections $\Theta\left(\varepsilon_i(n(\tau_q^*))\right)$ for τ_q^*. We then define

$$\widehat{\mathcal{J}}_{\varepsilon_i}^{(1)}(n(\tau_q)) = \mathcal{J}_{\varepsilon_i}^{(1)}(n(\tau_q)) + \Theta\left(\varepsilon_i(n(\tau_q))\right),$$
$$-\widehat{\mathcal{J}}_{\varepsilon_i}^{(1)}(n(\tau_q^*)) = -\mathcal{J}_{\varepsilon_i}^{(1)}(n(\tau_q^*)) + \Theta\left(\varepsilon_i^*(n(\tau_q^*))\right),$$

so that

$$\widehat{\mathcal{J}}^{(1)}_{\varepsilon_i}(n(\tau_q)) - \mathcal{J}^{(1)}_{\varepsilon_i}(n(\tau_q^*)) = \mathcal{J}_\varepsilon(n(\tau_q));$$

again the signs are related to the fact that $n(\tau_q) = -\, n(\tau_q^*)$. Hence, we take

$$\Theta\left(\varepsilon_i(n(\tau_q))\right) = \Theta\left(\varepsilon_i(n(\tau_q^*))\right),$$

and these corrections are made such that now

$$\int_{\tau_q} r_q d\mathbf{x} + \sum_{i=1}^{3} \int_{\varepsilon_i} \widehat{\mathcal{J}}^{(1)}_{\varepsilon_i}(n(\tau_q)) ds = 0. \tag{7.56}$$

We now define $\hat{e}_q^{[p]} \in \mathcal{U}(\tau_q)$ such that

$$B_{\tau_q}\left(\hat{e}_q^{[p]}, v\right) = \int_{\tau_q} r_q v d\mathbf{x} + \sum_{i=1}^{3} \int_{\varepsilon_i} \widehat{\mathcal{J}}^{(1)}_{\varepsilon_i}(n(\tau_q)) v ds$$
$$\forall v \in \mathcal{U}(\tau_q), \tag{7.57}$$

which is analogous to (7.14).

The function $\hat{e}_q^{[p]}$ exists because of (7.56). If $c \neq 0$, then $\hat{e}_q^{[p]}$ is unique, whilst if $c = 0$ it is unique up to an additional constant that is determined so that $\int_{\tau_q} \hat{e}_q^{[p]} d\mathbf{x} = 0$.

We now define the Neumann element-based error indicator and estimator

$$\eta_q^{\text{Neu,U}} = \left\| \hat{e}_q^{[p]} \right\|_{\mathcal{U}(\tau_q)}, \tag{7.58}$$

and

$$\varepsilon^{\text{Neu,U}} = \left(\sum_{q=1}^{M} \left(\eta_q^{\text{Neu,U}}\right)^2 \right)^{\frac{1}{2}}. \tag{7.59}$$

In a manner completely analogous to the one-dimensional case we obtain

$$\left\| e_\Delta^{[p]} \right\|_{\mathcal{U}} \leqq \varepsilon^{\text{Neu,U}},$$

under the assumption that either the $\hat{e}_q^{[p]}$ are determined exactly, or that an upper estimate of $\left\| \hat{e}_q^{[p]} \right\|_{\mathcal{U}(\tau_q)}$ is made.

The next task is to compute the corrections Θ, which is now relatively easy because condition (7.56) has to be satisfied on every element. Whilst this is at first sight a global problem, it can in fact be done locally using the fact that the residual on the patch ω_X is orthogonal to the linear shape function Φ, so that

$$B_{\omega_X}\left(e_\Delta^{[p]}, \Phi_X\right) = \int_{\omega_X} r_X \Phi_X d\mathbf{x} + \sum_i \int_{\varepsilon_{XX^i}} \mathcal{J}_{XX^i} \Phi_X ds = 0, \tag{7.60}$$

where ε_{XX^i} is the edge connecting the vertex X with the neighbouring vertex X^i of the patch, as shown in Fig. 7.24, and $\mathcal{J}_{XX^i}$ is the jump on the edge ε_{XX^i}, and the patch based on X has m vertices. Denoting $|\varepsilon_{XX^i}|$ to be the length of the edge ε_{XX^i} we now define on these edges functions

$$\Psi_i^X = \frac{2}{|\varepsilon_{XX^i}|}\left(2\Phi_X - \Phi_{X^i}\right),$$

where the Φ_X and the Φ_{X^i} are, respectively, the (hat) shape functions associated with the vertices X and X^i. The corrections $\Theta_{\varepsilon_{XX^i}}(\eta(\tau_q))$ can then be defined as

$$\Theta_{\varepsilon_{XX^i}}(\eta(\tau_q)) = \sum_{i=1}^{m} a_{XX^i}\Psi_i^X, \tag{7.61}$$

where the coefficients a_{XX^i} satisfy the systems of equations

$$\begin{bmatrix} 1\,0\,0\,0 & & -1 \\ -1\,1\,0\,0 & & 0 \\ & & \\ & & \\ 0\,0\,0\,0 & -1 & 1 \end{bmatrix} \begin{bmatrix} a_{XX^1} \\ a_{XX^2} \\ \\ \\ a_{XX^m} \end{bmatrix} = \begin{bmatrix} -\mathcal{R}_1(\Phi_X) \\ -\mathcal{R}_2(\Phi_X) \\ \\ \\ -\mathcal{R}_m(\Phi_X) \end{bmatrix}, \tag{7.62}$$

where for $i = 1, 2, \ldots, m$

$$\mathcal{R}_i(\Phi_X) = \int_{\tau_i} r_{\tau_i}\Phi_X d\mathbf{x} + \frac{1}{2}\left(\int_X^{X_i} \mathcal{J}_{\varepsilon_{XX^i}}(\eta(\tau_q))\,\Phi_X\,|_{\varepsilon_{XX^i}}\,ds\right)$$
$$+\frac{1}{2}\left(\int_X^{X_{i+1}} \mathcal{J}_{\varepsilon_{XX^{i+1}}}(\eta(\tau_q))\,\Phi_X\,|_{\varepsilon_{XX^{i+1}}}\,ds\right).$$

Equation (7.60) ensures that $\sum_{i=1}^{m}\mathcal{R}_i(\Phi_X) = 0$. This condition is necessary for the solvability of the system (7.59), because the matrix on the left-hand side has the vector $(1\ 1, \ldots, 1)^T$ as the only eigenvector associated with the zero eigenvalue. Hence, the vector $\{a_{XX^i}\}$ exists and is determined up to a constant C.

Various authors have implemented the Neumann element-based estimator in different ways. In Bank and Weiser (1985), BW, the constant C is taken such that $a_{XX^1} = 0$, whilst for the estimator L (Ladeveze and Leguillon, 1983) the constant is determined so that it leads to the minimum of $\sum_{i=1}^{m}\rho_i\,|a_{XX^1}|^2$, where either (L1), $\rho_i = 1$, or (L2), $\rho_i = |\varepsilon_{XX^i}|^{-2}$.

In this section, we have so far discussed only the case when τ_q is an interior element. If, however, τ_q has a side that coincides with part of Γ_{D}, then we take $\hat{e}_\Delta^{[p]} = 0$ on this side and split the jumps $\mathcal{J}$ only on the other two sides of τ_q. The procedure is then analogous.

We have here given only the main ideas for the BW and L estimators. For greater detail the reader should refer to the papers above and to Babuška and Strouboulis (2001). The orthogonality condition (7.60) is valid for shape functions of degree p. If this condition is utilized for shape functions of degree $q \leqq p$, then we shall speak of equilibration of order q.

7.2.3 The performance of the Neumann element-based estimator

We consider once again Benchmark Problem 1 of Section 3.4.3, and for this apply finite element methods using the meshes of Fig. 3.22. In the tables that follow we show the effectivity indices for the Bank–Weiser (BW) and Ladevese (L1 and L2) estimators. As was already said the errors $\hat{e}_{\Delta}^{[p]}$ have to be computed numerically. This is done using polynomials of degree $p+s$, for $p=1$ (linear elements) and $p=2$ (quadratic elements) and for values of s as shown. Equilibration of orders $q=1,2$ is used, respectively, for $p=1,2$. The results for the effectivity indices for the three estimators are given in Tables 7.23–7.28 and, as might be expected, lead to different values for the different indices. We note the following features of the performance of the estimators:

(1) The indices grow with s. This is related to the fact that $\left\|\hat{e}_{\Delta}^{[p]}\right\|_{\mathcal{U}(\tau_q)}$ grows with increasing p. As we cannot compute the estimator exactly, the values presented are lower than the exact ones and cannot be guaranteed as upper bounds.
(2) For $p=1$ and $q=1$ it is impossible to say which estimator is superior in all cases, although it appears that the BW one is inferior.
(3) For $p=q=2$ the L2 estimator seems to perform the best.
(4) The L1 and L2 estimators produce very reasonable effectivity indices in all cases.
(5) The L2 estimator is more robust than the BW and L1 estimators with respect to the character of the mesh.

We note that the performance of the estimators as above was for a problem with a smooth solution and a regular mesh. For the performance with an irregular mesh, see Section 7.4.1 and the 2D Eng Problem.

7.2.4 The Dirichlet subdomain (patch) estimator

The construction of this estimator follows very closely that for the one-dimensional case of Section 7.1.4.

Let ω_X be the patch of elements associated with any vertex X of Δ; ω_X is of course the support of the hat function Φ_X located at X. We introduce the subdomain residual error indicator function $\hat{e}_{\omega_X}^{[p]}$ where

$$\hat{e}_{\omega_X}^{[p]} \in U_0(\omega_X) = \left\{u \in H^1(\omega_X) \mid u = 0 \text{ on } \partial\omega_X - \Gamma_N\right\},$$

which, for every $v \in U_0(\omega_X)$, satisfies

$$B_{\omega_X}\left(\hat{e}_{\omega_X}^{[p]}, v\right) - \mathcal{R}_{\omega_X}(v) = \int_{\omega_X} f d\mathbf{x} - B\left(u_{\Delta}^{[p]}, v\right).$$

The function $\hat{e}_{\omega_X}^{[p]}$ exists and is unique and with this we can define the Dirichlet subdomain indicator

$$\eta_{\omega_X}^{\text{Dir}} \equiv \left\|\hat{e}_{\omega_X}^{[p]}\right\|_{\mathcal{U}(\omega_X)}, \tag{7.63a}$$

Table 7.23 Effectivity index for the BW estimator for Benchmark Problem 1 with linear elements $p = 1$ and $q = 1$.

h	s	Regular	Chevron	Union Jack	Criss-Cross
	1	1.55635	1.43568	1.61314	1.24460
0.5	2	1.63707	1.48342	1.68910	1.26150
	3	1.69579	1.51466	1.73902	1.27406
	1	1.27104	1.55720	2.33472	1.33795
0.25	2	1.29312	1.68679	2.67483	1.34336
	3	1.30824	1.77410	2.89625	1.34726
	1	1.20397	1.63413	2.71792	1.40470
0.125	2	1.21095	1.80824	3.15549	1.40627
	3	1.21491	1.92575	3.44131	1.40740
	1	1.19721	1.66484	2.89111	1.44253
0.0625	2	1.19965	1.85692	3.36838	1.44295
	3	1.20074	1.98663	3.68095	1.44325
	1	1.20026	1.67983	2.97147	1.46253
0.03125	2	1.20123	1.88017	3.46634	1.46264
	3	1.20158	2.01549	3.79090	1.46271
	1	1.20319	1.68756	3.00994	1.47279
0.015625	2	1.20362	1.89193	3.51304	1.47282
	3	1.20375	2.02998	3.84324	1.47284

and the corresponding Dirichlet subdomain estimator

$$\varepsilon_{\text{Subd}}^{\text{Dir}} \equiv \left(\sum_{x \in \Delta} \left(\eta_{\omega x}^{Dir} \right)^2 \right)^{\frac{1}{2}}. \tag{7.63b}$$

When the $\hat{e}_{\omega x}^{[p]}$ are computed exactly, the estimator $\varepsilon_{\text{Subd}}^{\text{Dir}}$ is simultaneously both an upper and a lower estimator, up to a reasonable multiplicative constant. In practice of course, the $\hat{e}_{\omega x}^{[p]}$ have to be computed numerically using elements of a higher degree on a finer mesh over the patches. Then, as in Section 7.1.4, we can compute $w \in \mathcal{U}_0$ where

$$w(\mathbf{x}) \equiv \sum_{x \in \Delta} \hat{e}_{\omega x}^{[p]}(\mathbf{x}), \tag{7.64}$$

Table 7.24 Effectivity index for the L1 estimator for Benchmark Problem 1 using linear elements $p = 1$ and $q = 1$.

h	s	Regular	Chevron	Union Jack	Criss-Cross
	1	1.18427	1.29519	1.28962	1.08653
0.5	2	1.23655	1.34688	1.36492	1.09415
	3	1.24136	1.36626	1.38704	1.09575
	1	1.38207	1.45701	1.42675	1.11476
0.25	2	1.41418	1.48606	1.45050	1.11683
	3	1.41905	1.49040	1.45544	1.11705
	1	1.48850	1.54965	1.50901	1.12814
0.125	2	1.50557	1.57296	1.51914	1.12895
	3	1.50841	1.57512	1.52093	1.12905
	1	1.54215	1.59301	1.54677	1.13429
0.0625	2	1.55080	1.61377	1.55160	1.13465
	3	1.55228	1.61511	1.55241	1.13470
	1	1.56871	1.61348	1.56440	1.13726
0.03125	2	1.57306	1.36600	1.56678	1.13742
	3	1.57381	1.63394	1.56717	1.13745
	1	1.58183	1.62336	1.57288	1.13872
0.015625	2	1.58401	1.64226	1.57406	1.13880
	3	1.58439	1.64301	1.57426	1.13882

where it is understood that $\hat{e}^{[p]}_{\omega_X}(\mathbf{x}) = 0, \mathbf{x} \notin \omega_X$. Using w in Theorem (7.2) we obtain a guaranteed lower estimator for any numerical computations. Again, as in Section 7.1.4, we can define the explicit Dirichlet estimator $\varepsilon_\omega^{\mathrm{Dir,E}}$, where

$$\varepsilon_\omega^{\mathrm{Dir,E}} \equiv \left(\sum_{\tau_q \in \Delta} |\tau_q| \, \|r_q\|^2_{L^2(\tau_q)} + \sum_{e_q \in E} |e_q| \, \left|\mathcal{J}_{\varepsilon_q}\right|^2_{L^2(\omega_X)} \right)^{\frac{1}{2}}.$$

This estimator is also both an upper and lower bound for $\left\| e^{[p]}_\Delta \right\|_{\mathcal{U}}$ up to a multiplicative constant. This constant is not easy to control and usually this estimator is not very accurate.

Table 7.25 Effectivity index for the L2 estimator for Benchmark Problem 1 using linear elements $p = 1$ and $q = 1$.

h	s	Regular	Chevron	Union Jack	Criss-Cross
	1	1.16870	1.31455	1.19534	1.08647
0.5	2	1.22279	1.34827	1.23912	1.09093
	3	1.22760	1.35309	1.24673	1.09110
	1	1.39872	1.48135	1.42882	1.11266
0.25	2	1.43451	1.51161	1.45012	1.11400
	3	1.43923	1.51505	1.45356	1.11408
	1	1.50512	1.56329	1.51584	1.12639
0.125	2	1.52498	1.58855	1.52659	1.12685
	3	1.52788	1.59070	1.52822	1.12689
	1	1.55271	1.60001	1.55109	1.13328
0.0625	2	1.56303	1.62203	1.55648	1.13346
	3	1.56461	1.62342	1.55727	1.13348
	1	1.57456	1.61699	1.56673	1.13673
0.03125	2	1.57981	1.63720	1.56944	1.13680
	3	1.58062	1.63818	1.56983	1.13682
	1	1.58490	1.62512	1.57409	1.13845
0.015625	2	1.58754	1.64437	1.57544	1.13849
	3	1.58796	1.64514	1.57563	1.13849

Again, we do not give here numerical results for this estimator. We refer the interested reader to Babuška and Strouboulis (2001), Fig. 54.1.

7.2.5 The Neumann subdomain (patch) estimator

The construction of this estimator again follows closely that for the one-dimensional case, of Section 7.1.5. Let $\Phi_X^{[1]}$ be the linear hat function located at X with support ω_X. We now construct functions $\hat{\overline{\overline{e}}}_{\omega_X} \in \mathcal{U}(\omega_X)$ such that $\forall v \in \mathcal{U}(\omega_X)$

$$B_{\omega_X}\left(\hat{\overline{\overline{e}}}_{\omega_X}, v\right) = \int_{\omega_X} f\Phi_X v d\mathbf{x} - B\left(u_\Delta^{[p]}, \Phi_X v\right). \tag{7.65}$$

Table 7.26 Effectivity index of the BW estimator for Benchmark Problem 1 using quadratic elements $p = 2$ and $q = 2$.

h	s	Regular	Chevron	Union Jack	Criss-Cross
	1	1.90674	2.12835	1.95610	2.93822
0.5	2	2.10712	2.37364	2.15037	3.20243
	3	2.25217	2.52486	2.24816	3.32782
	1	2.09463	2.43653	1.83784	3.38292
0.25	2	2.23870	2.59891	1.97423	3.66902
	3	2.33653	2.69483	2.03553	3.81297
	1	2.28464	2.68492	1.87299	3.60162
0.125	2	2.41517	2.82204	1.97367	3.90349
	3	2.49293	2.89827	2.00963	4.06471
	1	2.39884	2.82862	1.92221	3.70265
0.0625	2	2.52957	2.96103	2.01083	4.01286
	3	2.60242	3.03316	2.03888	4.18366
	1	2.46014	2.90352	1.95503	3.75033
0.03125	2	2.59280	3.03554	2.03971	4.06476
	3	2.66484	3.10716	2.06565	4.24046
	1	2.49178	2.94147	1.97348	3.77355
0.015625	2	2.62585	3.07379	2.05678	4.09020
	3	2.69796	3.14552	2.08216	4.26846

The function $\overset{=}{\hat{e}}_{\omega_x}$ exists and is unique, and with this we can define the Neumann subdomain indicators

$$ {}^1\varepsilon^{\text{Neu}}_{\text{Subd}} \equiv \left(\sum_{x \in \Delta} \left({}^1\eta^{\text{Neu}}_{\tau_q} \right)^2 \right)^{\frac{1}{2}}, \tag{7.66a} $$

$$ {}^2\varepsilon^{\text{Neu}}_{\text{Subd}} \equiv \left(\sum_{x \in \Delta} \left({}^2\eta^{\text{Neu}}_{\tau_q} \right)^2 \right)^{\frac{1}{2}}, \tag{7.66b} $$

where

$$ {}^1\eta^{\text{Neu}}_{\omega_X} = \sqrt{3} \left\| \overset{=[p]}{\hat{e}}_{\omega_X} \right\|_{\mathcal{U}(\tau_q)}, $$

Table 7.27 Effectivity index for the L1 estimator for Benchmark Problem 1 using quadratic elements $p = 2$ and $q = 2$.

h	s	Regular	Chevron	Union Jack	Criss-Cross
	1	1.38163	1.32729	1.36408	1.30194
0.5	2	1.46623	1.41224	1.44458	1.32131
	3	1.48432	1.43979	1.47525	1.32923
	1	1.37206	1.34938	1.35293	1.26117
0.25	2	1.47702	1.44418	1.44865	1.28089
	3	1.49865	1.46363	1.47129	1.28359
	1	1.38625	1.36592	1.35468	1.24290
0.125	2	1.50643	1.48060	1.47039	1.26734
	3	1.53172	1.50055	1.48693	1.26879
	1	1.39647	1.37582	1.35837	1.23397
0.0625	2	1.52520	1.50327	1.48760	1.26177
	3	1.55241	1.52378	1.50133	1.26281
	1	1.40198	1.38100	1.36092	1.22957
0.03125	2	1.53512	1.51531	1.49764	1.25924
	3	1.56323	1.53606	1.51009	1.26010
	1	1.40489	1.38372	1.36246	1.22769
0.015625	2	1.54022	1.52150	1.50303	1.25825
	3	1.56872	1.54232	1.51485	1.25902

and

$$ {}^2\eta^{\mathrm{Neu}}_{\tau_q} = \left\| \sum_{\tau_q \in \omega_e} \overset{=[p]}{\hat{e}}_{\omega_X} \right\|_{\mathcal{U}(\tau_q)} . \tag{7.67} $$

We can obtain a lower estimate analogously to the one-dimension case.

7.2.6 The performance of the Neumann subdomain (patch) estimator

The Neumann subdomain estimator is applied to Benchmark Problem 1 and, in particular, results for the effectivity indices of the estimator ${}^2\varepsilon^{\mathrm{Neu}}_{\mathrm{Subd}}$ for $p = 1, 2$, see (7.32b), for various values of h are given in Tables 7.29 and 7.30, where the functions $\overset{=}{\hat{e}}{}^{[p]}_{\omega_X}$ were computed on the patches ω_X using in each element of the patch elements of

Table 7.28 Effectivity index of the L2 estimator for Benchmark Problem 1 using quadratic elements $p = 2$ and $q = 2$.

h	s	Regular	Chevron	Union Jack	Criss Cross
	1	0.97543	0.96629	0.95807	0.95625
0.5	2	0.97790	0.97171	0.96549	0.98036
	3	0.97861	0.97317	0.96693	0.98347
	1	0.99094	0.98032	0.97406	0.96312
0.25	2	0.99210	0.98586	0.98392	0.98726
	3	0.99240	0.98673	0.98526	0.99022
	1	0.99648	0.98365	0.97871	0.96427
0.125	2	0.99705	0.98892	0.98995	0.98843
	3	0.99717	0.98962	0.99135	0.99135
	1	0.99852	0.98484	0.98022	0.96434
0.0625	2	0.99880	0.98997	0.99216	0.98849
	3	0.99885	0.99061	0.99362	0.99140
	1	0.99935	0.98536	0.98076	0.96434
0.03125	2	0.99949	0.99043	0.99306	0.98845
	3	0.99951	0.99105	0.99455	0.99137
	1	0.99979	0.98568	0.98104	0.96461
0.015625	2	0.99986	0.99071	0.99351	0.98866
	3	0.99987	0.99133	0.99503	0.99161

Table 7.29 Effectivity indices for the Neumann subdomain estimator ${}^{2}\varepsilon_{\mathrm{Subd}}^{\mathrm{Neu}}$ for linear elements (Benchmark Problem 1).

h	Regular	Chevron	Union Jack	Criss-Cross
0.5	1.40166	1.32182	1.34282	1.39330
0.25	1.27475	1.23780	1.23182	1.22958
0.125	1.22125	1.19295	1.17416	1.13545
0.25	1.19472	1.16597	1.13827	1.08327
0.03125	1.18091	1.15079	1.11782	1.05555
0.015625	1.17365	1.14274	1.10687	1.04123

Table 7.30 Effectivity indices for the Neumann subdomain estimator ${}^{2}\varepsilon_{\text{Subd}}^{\text{Neu}}$ for quadratic elements (Benchmark Problem 1).

h	Regular	Chevron	Union Jack	Criss-Cross
0.5	1.14349	1.13751	1.13905	1.28062
0.25	1.11239	1.10037	1.07976	1.13685
0.125	1.09745	1.08066	1.05215	1.05183
0.25	1.08990	1.07001	1.03902	1.00613
0.03125	1.08611	1.06445	1.03258	0.98252
0.015625	1.08426	1.06166	1.02944	0.97073

degree $p+k, k=2$. This estimator would be an upper estimator on $\left\|e_{\Delta}^{[p]}\right\|_{\mathcal{U}}$ if $\hat{\bar{\bar{e}}}_{\omega_X}^{[p]}$ were computed exactly. However, as this is not the case and it is computed numerically, a smaller value for the estimator is produced, which is not necessarily an upper bound. The estimator ${}^{2}\varepsilon_{\text{Subd}}^{\text{Neu}}$ is chosen because for Benchmark Problem 1 it produces better results than ${}^{1}\varepsilon_{\text{Subd}}^{\text{Neu}}$, as in the one-dimensional case.

As has been said, this estimator is an upper estimator only if the functions $\hat{\bar{\bar{e}}}$ are computed exactly. Numerical computation of $\hat{\bar{\bar{e}}}$ produces underestimation of the error. This underestimation is visible for the effectivity indexes for the cross-cross mesh in Table 3.30.

7.2.7 Averaging-based indicators and estimators (ZZ)

Once again we proceed in a manner similar to that for the one-dimensional case of Section 7.1.7, and, using averaging methods, we seek continuous functions $z_{x_1}^{[p]}(\mathbf{x})$, $z_{x_2}^{[p]}(\mathbf{x})$ that are better approximations to $\frac{\partial u}{\partial x_1}(\mathbf{x})$ and $\frac{\partial u}{\partial x_2}(\mathbf{x})$ than are the derivatives of $u_{\Delta}^{[p]}$. Again, for each vertex $X \in \Delta$ we consider a patch ω_X of elements, as in Fig. 7.25, with sampling points $s_i, i=1,2,\ldots,$ as shown. Using finite element solutions values of the derivatives of $u_{\Delta}^{[p]}$ at the sampling points are calculated and then polynomials $\Psi_{x_1}, \Psi_{x_2} \in \mathbb{P}^{[p]}(\omega_X)$ are determined by least squares fitting to these derivative values. For each ω_X these polynomials determine recovered values for the derivatives at $X \in \Delta$ and the functions $z_{x_1}^{[p]}$ and $z_{x_2}^{[p]}$ are produced by interpolating to these values at the nodal points in each element $\tau_q \in \Delta$.

For the case $p=1$ the sampling points $s_1, s_2, \ldots s_6$ are the centroids of each triangle of the patch, see Fig. 7.25(a), whilst for $p=2$ derivative values of the solutions in two adjacent elements are taken at the midpoints of the common side, see Fig. 7.25(b), giving 18 sampling points.

The situation when a vertex point is located on the boundary Γ is analogous to that in one dimension, and we again use extrapolation techniques.

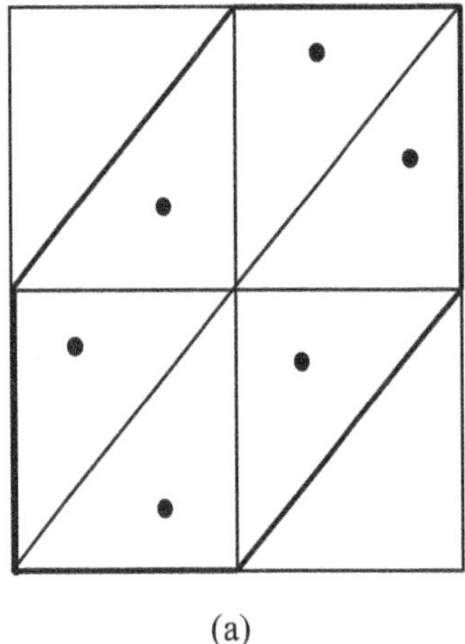
(a)

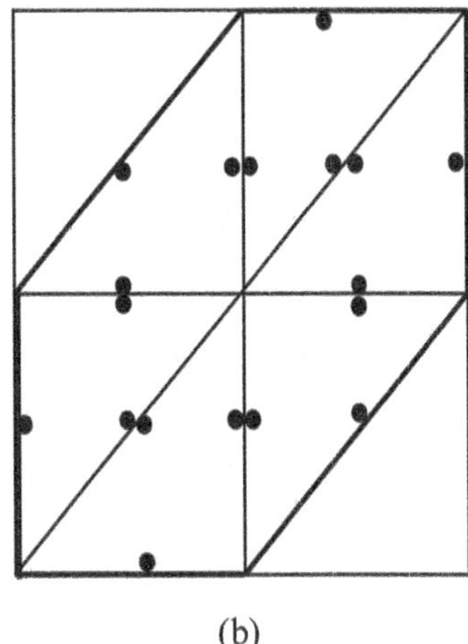
(b)

Fig. 7.25 Positions of the sampling points for the patch ω_X for $p = 1$, (a) and for $p = 2$ (b).

The indicator and estimator in this two-dimensional context are defined as

$$\eta_\tau^{ZZ} = \left(\int_\tau a \left(\left(\frac{\partial u_\Delta^{[p]}}{\partial x_1} - z_{x_1}^{[p]}(x) \right)^2 + \left(\frac{\partial u_\Delta^{[p]}}{\partial x_2} - z_{x_2}^{[p]}(x) \right)^2 \right) d\mathbf{x} \right)^{\frac{1}{2}}, \tag{7.68a}$$

$$\varepsilon^{ZZ} \equiv \left(\sum_{\tau \in \Delta} \left(\eta_q^{ZZ} \right)^2 \right)^{\frac{1}{2}}, \tag{7.68b}$$

where again the ZZ relates to Ziekiewicz and Zhu, see (1987) and (1992).

7.2.8 The performance of the ZZ-estimator

Continuing as before we apply the ZZ-estimator to Benchmark Problem 1 with the meshes of Fig. 3.22. The effectivity indices for ε^{ZZ} for $p = 1$ and 2 are shown in Tables 7.31 and 7.32.

We see that the effectivity indices are very reasonable, although they are mesh dependent. Nevertheless, the quality is good, especially noting the simplicity of implementation.

7.2.9 The Richardson error estimator and its performance

Just as in the one-dimensional case of Section 7.1.9 we may employ the Richardson technique to define estimators based on more than one mesh. As the analysis for two dimensions is identical to that of Section 7.1.9, we do not repeat it.

The Richardson estimator is applied to Benchmark Problem 1, which of course has a very smooth solution, for the four uniform nested meshes of Fig. 3.22. The effectivity indices obtained are excellent, exhibiting only minor differences for the different mesh patterns. This high quality also arises because the meshes are obtained by successive halving of the mesh size.

Table 7.31 Effectivity index for the ε^{ZZ} estimator using linear elements, $p = 1$, Benchmark Problem 1.

h	Regular	Chevron	Union Jack	Criss-Cross
0.5	1.05950	1.11360	0.96935	1.24522
0.25	1.28569	1.21106	1.32522	1.08485
0.125	1.25018	1.13640	1.19521	1.03246
0.065	1.15819	1.06351	1.11157	1.01391
0.015625	1.04707	0.98926	1.03129	1.00306

Table 7.32 Effectivity index for the ε^{ZZ} estimator using quadratic elements, $p = 2$ Benchmark Problem 1.

h	Regular	Chevron	Union Jack	Criss-Cross
0.5	1.06433	1.01432	1.32099	1.56196
0.25	1.21332	1.00038	1.23672	1.96822
0.125	1.28828	1.00789	1.58601	2.09271
0.25	1.31414	1.01439	2.00222	2.12321
0.03125	1.32251	1.01816	2.59929	2.12979
0.015625	1.32542	1.02015	3.48921	2.13086

Values for $\varkappa_1$ and $\varkappa_2$ the effectivity indices for Richardson estimators ε_1 and ε_2, see (7.45), together with the relative errors rel $= \left\| e_{\Delta}^{[p]} \right\|_{\mathcal{U}} / \|\mathrm{u}\|_{\mathcal{U}}$ as percentages are given in Tables 7.33–7.40.

Table 7.33 Effectivity indices $\varkappa_1$ and $\varkappa_2$ for the Richardson estimators ε_1 and ε_2 and rel with linear elements for Regular pattern meshes, Benchmark Problem 1.

h	$\varkappa_1$	$\varkappa_2$	rel
0.5		0.97364	30.2737
0.25	0.9058	0.99260	16.2715
0.125	0.97136	0.99787	8.3135
0.0625	0.99159	0.99940	4.1831
0.03125	0.99761	0.99982	2.0953
0.015625	0.99930		1.0482

Table 7.34 Effectivity indices $\varkappa_1$ and $\varkappa_2$ for the Richardson estimators ε_1 and ε_2 and rel with linear elements for Chevron pattern meshes, Benchmark Problem 1.

h	$\varkappa_1$	$\varkappa_2$	rel
0.5		0.98574	36.5752
0.25	0.94635	0.99409	19.0485
0.125	0.97699	0.99757	9.6910
0.0625	0.99039	0.99897	4.8806
0.03125	0.99591	0.99953	2.4478
0.015625	0.99815		1.2256

Table 7.35 Effectivity indices $\varkappa_1$ and $\varkappa_2$ for the Richardson estimators ε_1 and ε_2 and rel with linear elements for Union Jack pattern meshes, Benchmark Problem 1.

h	$\varkappa_1$	$\varkappa_2$	rel
0.5		0.98199	35.9046
0.25	0.96956	0.99612	18.3700
0.125	0.99231	0.99903	9.2382
0.0625	0.99807	0.99976	4.6258
0.03125	0.99953	0.99991	2.3137
0.015625	0.99982		1.1570

Table 7.36 Effectivity indices $\varkappa_1$ and $\varkappa_2$ for the Richardson estimators ε_1 and ε_2 and rel with linear elements for Criss-Cross pattern meshes, Benchmark Problem 1.

h	$\varkappa_1$	$\varkappa_2$	rel
0.5		0.99962	20.6660
0.25	0.99846	0.99990	10.3446
0.125	0.99961	0.99997	5.1738
0.0625	0.99989	0.99999	2.5871
0.03125	0.99994	1.00000	1.2936
0.015625	0.99999		0.6468

Table 7.37 Effectivity indices $\varkappa_1$ and $\varkappa_2$ for the Richardson estimators ε_1 and ε_2 and rel with quadratic elements for Regular pattern meshes, Benchmark Problem 1.

h	$\varkappa_1$	$\varkappa_2$	rel
0.5		0.99838	2.7719
0.25	0.97461	0.99933	1.7342
0.125	0.98939	0.99973	0.4379
0.0625	0.99587	0.99996	0.1099
0.03125	0.99902	1.00000	0.0275
0.015625	0.99981		0.0069

Table 7.38 Effectivity indices $\varkappa_1$ and $\varkappa_2$ for the Richardson estimators ε_1 and ε_2 and rel with quadratic elements for Chevron pattern meshes, Benchmark Problem 1.

h	$\varkappa_1$	$\varkappa_2$	rel
0.5		0.99776	5.6224
0.25	0.96594	0.99940	1.4519
0.125	0.99528	0.99981	0.3662
0.0625	0.99709	0.99992	0.0918
0.03125	0.99823	1.00000	0.0230
0.015625	0.99961		0.0057

Table 7.39 Effectivity indices $\varkappa_1$ and $\varkappa_2$ for the Richardson estimators ε_1 and ε_2 and rel with quadratic elements for Union Jack pattern meshes, Benchmark Problem 1.

h	$\varkappa_1$	$\varkappa_2$	rel
0.5		0.99542	5.5604
0.25	0.96514	0.99882	1.4370
0.125	0.99673	0.99971	0.3624
0.0625	0.99531	0.99998	0.0908
0.03125	0.99912	1.00000	0.0227
0.015625	0.99996		0.0057

Table 7.40 Effectivity indices $\varkappa_1$ and $\varkappa_2$ for the Richardson estimators ε_1 and ε_2 and rel with quadratic elements for Criss-Cross pattern meshes, Benchmark Problem 1.

h	$\varkappa_1$	$\varkappa_2$	rel
0.5		0.99903	2.0804
0.25	0.98482	0.99979	0.5276
0.125	0.99677	0.99994	0.1323
0.0625	0.99919	0.99999	0.6331
0.03125	0.99993	1.00000	0.0083
0.015625	1.00000		0.0021

7.3 Comparison of the various error estimates

We have in Sections 7.1 and 7.2 for one and two dimensions addressed various error estimators of finite element solutions measured in the energy norm. As a result of this, we now make the following observations.

7.3.1 The Neumann element error estimator

(1) This estimator is robust and gives very reasonable effectivity indices.
(2) The estimator offers a guaranteed bound if the local problems are solved exactly. In practice, this estimator does not underestimate the error.
(3) Although the implementation of this estimator is not completely simple, it has the advantages that the local problems are solved on individual elements and only a single mesh is required.

7.3.2 The Neumann subdomain error estimator

(1) This estimator is very robust and leads to upper bounds on the energy norm of the error provided the local problems are solved exactly. Lower bounds can also be obtained.
(2) The estimator has the disadvantage that the local problems have to be solved on patches and not on single elments.
(3) This estimator is easy to implement and only a single mesh is required.

7.3.3 Averaging-based error estimators

(1) These estimators usually give reasonable effectivity indices, provided the elements of the mesh are not distorted. The estimators are sensitive to the character of the mesh.
(2) With these estimators the effectivity indices are better with linear elements than with quadratic, although both indices are reasonable.

(3) These estimators are not robust with respect to the smoothness of the solutions of the problems, and do not produce guaranteed upper and lower bounds.
(4) These estimators are very easy to implement and require only a single mesh.

7.3.4 The Richardson error estimator

(1) Richardson estimators perform very well for problems with smooth solutions when uniform meshes are used, provided the finite element space associated with the coarser mesh is a subspace of that associated with the finer mesh. If the finer mesh consists of refinements of every element of the coarser mesh, the effectivity index is not too bad, even if the problem solution is not smooth, giving a robust estimator.
(2) The estimator is not robust if the above inclusion property is not present.
(3) The selection of the parameter β, see (7.41), is important.
(4) The major disadvantage of this estimator is that one has to work with two meshes and to solve the two associated finite element problems. The course mesh must be reasonable, and the solution of the problem on the fine mesh can be costly.

The effectiveness of the various error indicators and estimators in general depends on factors such as the meshes and the smoothness of the solutions of the problems being solved. As has been said, the quality of an estimator can be measured by its asymptotic behaviour. The overriding wish of anyone using an error estimator is that it should be robust and reasonably simple to implement.

7.4 *a posteriori* error estimations for the 2D engineering problem

In the previous section we addressed various *a posteriori* error estimators for Benchmark Problem 1. In this section we now apply various estimators to the 2D Eng Problem, following on from the computations of Sections 5.2.3 and 6.2.4.

There is a major difference between the computations for Benchmark Problem 1 and those for the 2D Eng Problem. The meshes used in the computational analysis of the 2D Eng Problem have elements with large aspect ratios; i.e. small minimal angles, although the maximal angle is always far from π (see Remark 4.10 on the maximal angle condition). As in Section 6.2.4 we consider a sequence of meshes obtained by halving the crude original mesh as shown in Fig. 3.27.

7.4.1 The Neumann element-based estimator

The error estimator for $\left\|e_{\Delta}^{[p]}\right\|_{\mathcal{U}}$ has to be computed numerically, and for this we shall as before use polynomials of degree $p+s$. We now show tables analogous to those shown for Benchmark Problem 1 in the previous sections, see Tables 7.41–7.47.

In Tables 7.41–7.47 the estimators were computed numerically using full polynomials of degree $p+s$. Alternatively, we can augment the polynomial of degree p using only bubble functions of degree $p=s$. In this case, the error estimator will be an underestimate. This is illustrated for the BW estimator with $p=q=2$, and we note that for the other estimators the effect is very similar.

Table 7.41 Effectivity index of estimated error using q-order equilibration and full $(p+s)$ degree solution with Bank–Weiser equilibration where, $p=1$ and $q=1$, 2D Eng Problem.

		Pattern			
Refinement	s	Regular	Chevron	Union Jack	Criss-Cross
	1	1.5364	1.4488	1.4240	2.1602
1	2	1.7162	1.6423	1.6601	2.2669
	3	1.7475	1.6843	1.7045	2.2702
	1	1.6923	1.5767	1.6908	2.1789
2	2	1.8739	1.7787	1.8772	2.2874
	3	1.9071	1.8208	1.9168	2.2906
	1	1.7359	1.7604	1.7746	2.1465
3	2	1.9554	2.0050	1.9650	2.2738
	3	1.9991	2.0583	2.0069	2.2776
	1	1.7507	1.6607	1.8118	2.1220
4	2	1.9921	1.9043	2.0038	2.2572
	3	2.0423	1.9592	2.0469	2.2612
	1	1.7573	1.6738	1.8292	2.1082
5	2	2.0100	1.9278	2.0216	2.2466
	3	2.0640	1.9866	2.0653	2.2505
	1	1.7605	1.6807	1.8382	2.0981
6	2	2.0197	1.9411	2.0304	2.2376
	3	2.0759	2.0024	2.0742	2.2415
	1	1.7578	1.6754	1.8427	2.0941
7	2	2.0202	1.9384	2.0344	2.2339
	3	2.0777	2.0011	2.0783	2.2380
	1	1.7593	1.6707	1.8447	2.0806
8	2	2.0242	1.9352	2.0358	2.2203
	3	2.0826	1.9986	2.0797	2.2241

Table 7.42 Effectivity index of estimated error using q-order equilibration and full $(p+s)$ degree solution with Ladeveze 1 equilibration, where $p = 1$ and $q = 1$, 2D Eng Problem.

		Pattern			
Refinement	s	Regular	Chevron	Union Jack	Criss-Cross
	1	3.6436	3.6014	4.1951	5.9654
1	2	4.0415	3.8521	4.6240	6.2566
	3	4.3414	3.9917	4.9339	6.3990
	1	2.9160	3.1368	2.4587	4.8812
2	2	3.2586	3.3734	2.6830	5.1239
	3	3.4129	3.4733	2.7544	5.1820
	1	2.8560	3.2182	2.2942	4.6357
3	2	3.1886	3.4562	2.5013	4.8753
	3	3.3285	3.5459	2.5553	4.9180
	1	2.8235	2.8884	2.2594	4.5336
4	2	3.1492	3.1015	2.4589	4.7763
	3	3.2816	3.1763	2.5091	4.8115
	1	2.8040	2.8224	2.2567	4.5245
5	2	3.1261	3.0321	2.4524	4.7738
	3	3.2539	3.1025	2.5012	4.8048
	1	2.7990	2.7837	2.2721	4.6040
6	2	3.1191	2.9921	2.4655	4.8642
	3	3.2440	3.0604	2.5139	4.8925
	1	2.8023	2.7547	2.3022	5.0636
7	2	3.1205	2.9618	2.4944	5.3600
	3	3.2433	3.0288	2.5429	5.3847
	1	2.8301	2.7518	2.3472	4.9421
8	2	3.1474	2.9584	2.5387	5.2355
	3	3.2691	3.0249	2.5878	5.2602

Table 7.43 Effectivity index of estimated error using q-order equilibration and full $(p+s)$ degree solution with Ladeveze 2 equilibration, where $p=1$ and $q=1$, 2D Eng Problem.

		Pattern			
Refinement	s	Regular	Chevron	Union Jack	Criss-Cross
	1	1.5586	1.4685	1.4255	2.1587
1	2	1.7408	1.6640	1.6636	2.2668
	3	1.7726	1.7063	1.7084	2.2694
	1	1.6018	1.5107	1.6723	2.1145
2	2	1.7961	1.7310	1.8409	2.2335
	3	1.8316	1.7746	1.8750	2.2366
	1	1.6391	1.6749	1.7450	2.0647
3	2	1.8311	1.9151	1.9154	2.1888
	3	1.8666	1.9610	1.9496	2.1920
	1	1.6581	1.5803	1.7796	2.0340
4	2	1.8469	1.8024	1.9496	2.1612
	3	1.8819	1.8439	1.9834	2.1644
	1	1.6675	1.5932	1.7968	2.0163
5	2	1.8550	1.8145	1.9668	2.1455
	3	1.8897	1.8553	2.004	2.1488
	1	1.6722	1.5995	1.8058	2.0041
6	2	1.8592	1.8201	1.9758	2.1345
	3	1.8938	1.8606	2.0092	2.1378
	1	1.6706	1.5943	1.8104	1.9995
7	2	1.8570	1.8132	1.9802	2.1304
	3	1.8913	1.8531	2.0135	2.1339
	1	1.6726	1.5895	1.8124	1.9853
8	2	1.8590	1.8072	1.9821	2.1164
	3	1.8933	1.8468	2.0153	2.1198

Table 7.44 Effectivity index of estimated error using q-order equilibration and full $(p+s)$ degree solution with Bank–Weiser equilibration, where $p=2$ and $q=2$, 2D Eng Problem.

		Pattern			
Refinement	s	Regular	Chevron	Union Jack	Criss-Cross
	1	1.4845	1.3438	1.3058	1.8694
1	2	1.8837	1.7731	1.7430	2.2781
	3	2.0056	1.8980	1.8862	2.3373
	1	1.4751	1.1940	1.5562	1.8608
2	2	1.8463	1.5674	1.9260	2.2677
	3	1.9580	1.6721	2.0223	2.3258
	1	1.4682	1.3147	1.5635	1.8547
3	2	1.8263	1.7198	1.9297	2.2615
	3	1.9317	1.8315	2.0264	2.3195
	1	1.4649	1.3106	1.5679	1.8536
4	2	1.8193	1.7145	1.9350	2.2613
	3	1.9230	1.8253	2.0325	2.3195
	1	1.4607	1.3050	1.5756	1.8573
5	2	1.8131	1.7074	1.9448	2.2652
	3	1.9160	1.8176	2.0431	2.3235
	1	1.4512	1.2974	1.5952	1.8669
6	2	1.8009	1.6972	1.9690	2.2741
	3	1.9029	1.8066	2.0686	2.3323
	1	1.4424	1.2908	1.6125	1.9216
7	2	1.7897	1.6880	1.9900	2.3246
	3	1.8910	1.7967	2.0906	2.3823

7.4.2 Performance of the Neumann estimator

From the tables in Section 7.4.1 we can draw the following conclusions:

(1) An estimator that is computed numerically grows with increasing s. The guaranteed error is obtained only as $s \to \infty$, although the estimator overestimates the error in all cases. The use of bubbles rather than complete polynomials for the augmentation has a large effect.

Table 7.45 Effectivity index of estimated error using q-order equilibration and full $(p+s)$ degree solution with Ladeveze 1 equilibration, where $p = 2$ and $q = 2$, 2D Eng Problem.

		Pattern			
Refinement	s	Regular	Chevron	Union Jack	Criss-Cross
	1	5.0323	2.9334	5.0840	5.3379
1	2	5.6365	3.3322	5.6856	5.7929
	3	5.9971	3.4757	6.0668	5.9139
	1	2.8323	2.2147	3.1885	5.0685
2	2	3.5237	2.6180	3.6267	5.5022
	3	3.7775	2.7459	3.7714	5.5637
	1	2.9303	2.4804	3.1811	5.4078
3	2	3.5997	2.9203	3.6115	5.8212
	3	3.8423	3.0574	3.7501	5.8755
	1	3.0941	2.6235	3.2544	5.9083
4	2	3.7435	3.0493	3.6762	6.2967
	3	3.9791	3.1834	3.8111	6.3474
	1	3.2889	2.8249	3.4039	6.5793
5	2	3.9163	3.2335	3.8175	6.9405
	3	4.1453	3.3650	3.9504	6.9893
	1	3.5043	3.0666	3.6369	7.5142
6	2	4.1067	3.4579	4.0452	7.8449
	3	4.3287	3.5872	4.1783	7.8919
	1	3.7571	3.3503	3.9199	9.9973
7	2	4.3351	3.7264	4.3228	10.2637
	3	4.5507	3.8544	4.4569	10.3029

(2) Comparison of the performance of the estimators when used for Benchmark Problem 1 and for the 2D Eng Problem shows that the L1 estimator is very inferior when used for the 2D Eng Problem, in contrast to when it is used for Benchmark Problem 1. For the 2D Eng Problem the BW estimator is clearly better, and is of about the same quality as the L2 estimator. The results indicate that the L1 estimator is not robust with respect to the aspect ratio of the elements used in the meshes. Robustness of an estimator is essential if it is to be used with confidence in practice.

Table 7.46 Effectivity index of estimated error using q-order equilibration and full $(p+s)$ degree solution with Ladeveze 2 equilibration, where $p=2$ and $q=2$, 2D Eng Problem.

		Pattern			
Refinement	s	Regular	Chevron	Union Jack	Criss-Cross
	1	1.4838	1.3436	1.2982	1.8335
1	2	1.8844	1.7746	1.7408	2.1979
	3	2.0065	1.9000	1.8845	2.2328
	1	1.4687	1.1915	1.5556	1.8272
2	2	1.8399	1.5681	1.9226	2.1943
	3	1.9533	1.6752	2.0166	2.2295
	1	1.4610	1.3110	1.5623	1.8192
3	2	1.8231	1.7236	1.9264	2.1876
	3	1.9333	1.8404	2.0202	2.2227
	1	1.4570	1.3070	1.5664	1.8173
4	2	1.8168	1.7197	1.9323	2.1865
	3	1.9258	1.8361	2.0269	2.2215
	1	1.4526	1.3015	1.5742	1.8205
5	2	1.8107	1.7130	1.9425	2.1900
	3	1.9191	1.8289	2.0378	2.2249
	1	1.4431	1.2940	1.5938	1.8301
6	2	1.7986	1.7029	1.9668	2.1988
	3	1.9061	1.8181	2.0635	2.2336
	1	1.4344	1.2874	1.6111	1.8854
7	2	1.7875	1.6937	1.9879	2.2500
	3	1.8942	1.8082	2.0856	2.2843

7.4.3 Performance of the ZZ-estimator

Throughout this book we have consistently computed the energy norm of the error $e_\Delta^{[p]} = u - u_\Delta^{[p]}$, which is

$$\left\|e_\Delta^{[p]}\right\|_{\mathcal{U}} = \left(\int_\Omega \alpha \left|\nabla e_\Delta^{[p]}\right|^2 d\Omega + \int_\Gamma \beta \left(e_\Delta^{[p]}\right)^2 ds\right)^{\frac{1}{2}}.$$

Table 7.47 Effectivity index of estimated error using q-order equilibration and $(p+s)$ degree bubble solution with Bank–Weiser equilibration, where $p = 2$ and $q = 2$, 2D Eng Problem.

		Pattern			
Refinement	s	Regular	Chevron	Union Jack	Criss-Cross
	1	1.1191	1.0606	1.0181	1.1391
1	2	1.2392	1.1904	1.1716	1.2561
	3	1.2721	1.2310	1.2087	1.2706
	1	1.1295	0.9537	1.1827	1.1368
2	2	1.2441	1.0739	1.2818	1.2518
	3	1.2771	1.1114	1.3206	1.2657
	1	1.1309	1.0549	1.1833	1.1352
3	2	1.2441	1.1887	1.2815	1.2499
	3	1.2784	1.2307	1.3205	1.2636
	1	1.1303	1.0528	1.1842	1.1351
4	2	1.2427	1.1868	1.2823	1.2498
	3	1.2772	1.2290	1.3216	1.2636
	1	1.1277	1.0481	1.1886	1.1356
5	2	1.2394	1.1819	1.2871	1.2505
	3	1.2741	1.2240	1.3266	1.2642
	1	1.1206	1.0411	1.2022	1.1363
6	2	1.2314	1.1742	1.3019	1.2513
	3	1.2659	1.2161	1.3420	1.2650
	1	1.1138	1.0346	1.2139	1.1432
7	2	1.2238	1.1669	1.3146	1.2590
	3	1.2581	1.2085	1.3552	1.2727

For the 2D Eng Problem Γ is the outside boundary of the pipe, see Fig. 2.13. As was said in Sections 7.1.7 and 7.1.8 the ZZ-estimator neglects the second term on the right-hand side of the above equation and estimates only $\int_\Omega a \left|\nabla e_\Delta^{[p]}\right|^2 d\Omega$.

Because in the 2D Eng Problem a is only piecewise smooth we estimate the fluxes in a manner analogous to that 1D Eng Problems using Remark 7.8. Results shown the effectivity indices of the ZZ-estimator applied to the 2D Eng Problem of Fig. 2.15 are given in Table 7.48 for $p = 1$ and Table 7.49 for $p = 2$.

From these, we see that the quality of the estimator is high, in spite of the elements being very disturbed as in the meshes of Fig. 3.27. This estimator is not too sensitive to the character of the mesh, and is neither an upper nor a lower estimator.

7.4.4 Performance of the Dirichlet subdomain estimator

The Dirichlet subdomain indicator, $\eta_{\omega_X}^{\text{Dir}} \equiv \left\| \hat{e}_{\omega_X}^{[p]} \right\|_{\mathcal{U}(\omega_X)},$ was defined in (7.63a). Using polynomials of degree $p+s$, $p = 1, 2$, and for various s, this has been applied to the

Table 7.48 Effectivity indices for the ZZ -estimator with $p = 1$ for the 2D Eng Problem.

		Pattern		
Refinement	Regular	Chevron	Union Jack	Criss-Cross
1	1.0939	0.9121	0.8414	0.6982
2	1.0416	0.9816	0.9185	0.8172
3	0.9925	0.9724	0.9185	0.8424
4	0.9851	0.9815	0.9360	0.8536
5	0.9903	0.9916	0.9515	0.8600
6	0.9975	0.9996	0.9628	0.8639
7	1.0021	1.0041	0.9702	0.8695
8	1.0075	1.0105	0.9750	0.8695

Table 7.49 Effectivity indices for the ZZ -estimator with $p = 2$ for the 2D Eng Problem.

		Pattern		
Refinement	Regular	Chevron	Union Jack	Criss-Cross
1	0.8316	0.9421	1.1190	0.8733
2	0.8369	0.9300	1.1586	0.8667
3	0.8522	0.9512	1.1903	0.8818
4	0.8661	0.9669	1.1289	0.8948
5	0.8746	0.9755	1.2356	0.9026
6	0.8784	0.9789	1.2551	0.9053
7	0.8825	0.9822	1.2781	0.9063

2D Eng Problem and results showing the effectivity indices of the estimator $\varepsilon_{\text{Subd}}^{\text{Dir}}$, see (7.63b), are given in Tables 7.50 and 7.51.

The value of the effectivity index has to increase with s and it converges to a limit as $s \to \infty$; this convergence is very rapid. For this estimator, we have that $C_{\text{L}}\varepsilon_{\text{Subd}}^{\text{Dir}} \leqq \left\| e_{\Delta}^{[p]} \right\|_{\mathcal{U}} \leqq C_{\text{U}}\varepsilon_{\text{Subd}}^{\text{Dir}}$, where the constants C_{L} and C_{U} depend on characteristics of the mesh; the maximum number of elements having a common vertex, the minimal angles of the triangles, and the coefficients of the governing differential equation. These constants do not depend on the mesh size nor the solution $\mathfrak{u}$. The range of the constants is visible from the values of the effectivity indices given in Tables 7.50 and 7.51.

7.4.5 Performance of the Richardson estimator

In Section 3.4.4, computed solutions with a sequence of meshes obtained by successive halving of the mesh were presented. It was observed from Figs. 3.36 and 3.37 that the computed values for the total energy Q_3 converged at a rate of ≈ 0.66 with respect to the degrees of freedom.

As explained in Section 7.1.9 various versions of Richardson extrapolation can be used; for example using two terms assuming that the rate of convergence is known or with three terms without it. Values of the effectivity indices using Π^* as in (7.44) with meshes Δ_{i-1}, Δ_i and Δ_{i+1} to compute the error $\left\| e_{\Delta_i}^{[p]} \right\|_{\mathcal{U}}$ are given in Tables 7.52 and 7.53, respectively for $p = 1$ and 2.

From the tables, it can be seen that Richardson extrapolation gives excellent error estimation. However, it must be acknowledged that it is very expensive because of the need to compute each time three successive solutions. This can of course be mitigated to a certain extent by starting with a crude mesh.

The idea behind Richardson extrapolation is very simple, the method can be used to estimate the error in any computed quantity of interest and it is very reliable and easy to implement. This generality allows for 'quality control' (verification) of computations in many circumstances.

In fact, it is not too strong to say that ideally the Richardson approach should always be used as a control for convergence of a computation. The approach can also give insight into the formulation of a problem, because possible instability of a computation could be a manifestation of instability of the mathematical problem or of the nature of the quantity of interest.

7.4.6 The performance of the *a posteriori* error estimators

In this chapter we have presented theory and results showing the performance of various estimators of the error in the finite elements solutions measured in energy norms for various problems. The numerical results were illustrated particularly in the form of effectivity indices, which of course can be computed only if the exact solutions of the problems are known. In the problems considered the solutions were either known analytically or were determined by the use of 'overkill' computations. The numerical results were deliberately presented first for the simple benchmark problem and then

Table 7.50 Effectivity indices for the Dirichlet subdomain estimator with $p = 1$, 2D Eng Problem.

		Pattern			
Refinement	s	Regular	Chevron	Union Jack	Criss-Cross
	1	0.6466	0.5251	0.5466	0.4844
1	2	0.7120	0.5978	0.6175	0.5576
	3	0.7321	0.6224	0.6410	0.5680
	1	0.8653	0.8095	0.8004	0.7085
2	2	0.9169	0.8652	0.8579	0.7660
	3	0.9287	0.8784	0.8718	0.7732
	1	0.9743	1.0038	0.9153	0.7915
3	2	1.0159	1.0509	0.9584	0.8444
	3	1.0235	1.0603	0.9672	0.8509
	1	1.0323	0.9950	0.9721	0.8309
4	2	1.0689	1.0334	1.0078	0.8816
	3	1.0748	1.0403	1.0146	0.8880
	1	1.0642	1.0290	1.0026	0.8518
5	2	1.0979	1.0645	1.0343	0.9012
	3	1.1030	1.0706	1.0401	0.9075
	1	1.0828	1.0481	1.0199	0.8624
6	2	1.1148	1.0817	1.0491	0.9109
	3	1.1194	1.0873	1.0544	0.9172
	1	1.0914	1.0536	1.0300	0.8672
7	2	1.1222	1.0858	1.0575	0.9149
	3	1.1264	1.0911	1.0624	0.9212
	1	1.0987	1.0559	1.0357	0.8685
8	2	1.1288	1.0872	1.0622	0.9157
	3	1.1329	1.0923	1.0669	0.9220

also for the more demanding engineering problems that display various features found in practice for more complicated problems; for example using non-uniform meshes with elongated elements and problems with nonsmooth solutions. We now compare the results obtained with the various estimators, acknowledging that the computed

Table 7.51 Effectivity indices for the Dirichlet subdomain estimator with $p = 2$, 2D Eng Problem.

		Pattern			
Refinement	s	Regular	Chevron	Union Jack	Criss-Cross
	1	0.8923	0.8602	0.8943	0.6972
1	2	0.9390	0.9000	0.9430	0.7222
	3	0.9470	0.9076	0.9521	0.7263
	1	0.9080	0.7858	0.8795	0.7025
2	2	0.9505	0.8204	0.9116	0.7307
	3	0.9576	0.8266	0.9167	0.7349
	1	0.9146	0.8792	0.8826	0.7130
3	2	0.9575	0.9182	0.9149	0.7417
	3	0.9646	0.9252	0.9201	0.7460
	1	0.9198	0.8855	0.8883	0.7212
4	2	0.9634	0.9254	0.9211	0.7503
	3	0.9707	0.9325	0.9264	0.7546
	1	0.9214	0.8863	0.8949	0.7266
5	2	0.9653	0.9266	0.9283	0.7559
	3	0.9726	0.9337	0.9335	0.7603
	1	0.9174	0.8829	0.9070	0.7297
6	2	0.9613	0.9232	0.9409	0.7591
	3	0.9686	0.9303	0.9463	0.7636
	1	0.9127	0.8786	0.9168	0.7355
7	2	0.9565	0.9187	0.9511	0.7652
	3	0.9637	0.9258	0.9566	0.7697

indices may be mildly influenced by various 'crimes' such as numerical integration, round-off error and the accuracy of the overkilled solutions.

As a result of the above we now suggest some guidelines for assessing which estimator is preferable.

1. Any estimator should lead to reasonable effectivity indices. We would like the values of these to be in the range (0.6, 2.0) for crude meshes with a tighter range for finer meshes.
2. Estimators that are guaranteed upper estimators are preferable.

Table 7.52 Effectivity indices for the Richardson estimator with $p = 1$, 2D Eng Problem.

	Pattern			
Refinement	Regular	Chevron	Union Jack	Criss-Cross
2	0.9974	1.0044	0.9656	0.9916
3	1.0225	1.0357	1.0377	0.9905
4	1.0134	1.0221	1.0226	0.9979
5	1.0078	1.0123	1.0124	0.9935
6	1.0047	1.0063	1.0078	1.0011
7	1.0013	1.0024	1.0014	1.0029

Table 7.53 Effectivity indices for the Richardson estimator $p = 2$, 2D Eng Problem.

Refinement	Regular	Chevron	Union Jack	Criss-Cross
2	0.9172	0.9066	0.9444	0.9284
3	0.8865	0.8899	0.8829	0.9024
4	1.0061	1.0056	1.0085	0.9795
5	1.0073	1.0039	1.0066	1.0044
6	1.0065	1.0074	1.0118	1.0033

3. An error estimator should be robust in that it performs more or less uniformly well for a large class of mesh patterns and sizes, and solutions (smooth and non-smooth).
4. The implementation of an estimator should not be too complicated and the cost of implementation should not be high; we consider two different costs, the direct cost of the estimator and the indirect cost related to its accuracy. Overestimation of the error would lead to the use of an unnecessarily fine mesh, whilst underestimation could be misleading.
5. The results presented here have been for one- and two-dimensional problems. Any estimator must be such that it can be generalized to three dimensions and more complicated situations such as systems of elliptic equations as are found in elasticity. The complexity of such generalisations could be a determining factor in the assessment of an estimator. It could be that estimators for other measures than the energy norm are needed.

We now attempt to assess the estimators in the light of the above five factors.

1. **The reasonableness of the values of the effectivity indices**. The tables of results presented here indicate that in the contexts considered all the estimators

lead to reasonable effectivity indices, with the exception of the Neumann-based element estimator L1. We emphasize that the ZZ-estimator does not estimate the error in terms of the full energy norm as it neglects the energy arising from the Newton/Robin boundary conditions. This part of the energy is negligible, or not relevant in most practical cases, although this is not invariably the case.

2. **Upper estimators.** The Neumann element-based and the Neumann subdomain estimators are guaranteed estimators when the local element problems are solved exactly and hence the effectivity indices have value greater than or equal to one. However, when computed numerically these estimators decrease and are not guaranteed upper estimators. Notwithstanding this, in practical terms they are upper estimators, as can been seen from the numerical results. None of the other estimators is a guaranteed upper estimator and their effectivity indices have in some cases values below one, although not very much below one.
3. **Robustness.** The robustness of all the estimators has been shown to be very reasonable, with again the exception of the estimator L1. This exhibited good indices in the one-dimensional setting but poor indices in the two-dimensional settings when the elements had elongated geometry. The robustness of the ZZ-estimator may in fact come as a surprise. The Richardson estimator is robust only when a proper sequence of meshes is used, for example when they are obtained by successive halving.
4. **Implementation**. The simplest estimators to use are those of ZZ and Richardson. However, as has been said, the Richardson estimator can be computationally very expensive because of the necessity to use double or triple meshes. Subdomain estimators are also relatively simple to implement, although of course one has in this case to solve local problems on patches rather than on single elements. The Neumann element-based estimator is slightly more complex because of the necessity of programming the equilibrations, although of course the local problems are in this case solved only on single elements.
5. **Generalization to three dimensions and to more complicated equations.** With the exception of the Neumann-based estimators the generalization of the subdomain estimators to higher dimensions is straightforward. The generalization to different norms, such as the L_∞ norm of the solution or of its gradient in either the entire domain or part of it is also important. However, this type of estimator is quite different from what we have considered and cannot be obtained by easy generalization of the estimators discussed above. We therefore do not comment further on this.

7.4.7 Recommendations for approaching error estimation

We feel that at the end of this long chapter on *a posteriori* error estimation, in which the performance of the various estimators has been demonstrated and discussed, we are honour bound to try to give 'advice' to anyone who is faced with the question of which estimator to use in a specific context. Our opinions are of course subjective.

Our first choice would be to use the ZZ-estimator, particularly because it is cheap and easy to program. Experience shows that it performs well in a wide variety of contexts although, as has been said above, it is neither an upper nor lower estimator, and it does not lead to full energy norm estimation. Our second preference would be to use the Neumann subdomain estimator with patch computations using refined meshes on the patches. This estimator is an upper estimator, provided that the patch computation is exact. A third possibility is to use the Richardson estimator on a suitable mesh, then on the mesh obtained by halving the original, and finally on the mesh obtained by halving again. This is a simple way of applying the method.

We stress again that the above suggestions are subjective and are based upon our experience and that of colleagues in estimating error.

7.5 *a posteriori* estimation of errors in the functional

In Sections 6.1.1 and 6.2.1 for the functional $\Phi(\mathfrak{u})$ we introduced the error $e_{\Phi}^{[p]}$ in the finite element solution $\Phi\left(u_{\Delta}^{[p]}\right)$,

$$e_{\Phi}^{[p]} = \Phi(\mathfrak{u}) - \Phi\left(u_{\Delta}^{[p]}\right).$$

Further, in Theorem 6.1 we introduced the functions $\Psi_{\Phi} \in \mathcal{U}_0$ and $\Psi_{\Phi,\Delta} \in S_{\Delta,0}^{[p]}$ that satisfy, respectively,

$$B\left(\Psi_{\Phi,\Delta}, v\right) = \Phi(v)\ \forall v \in \mathcal{U}_0,$$

and

$$B\left(\Psi_{\Phi,\Delta}, v\right) = \Phi(v)\ \forall v \in S_{\Delta,0}^{[p]}.$$

By showing that

$$e_{\Phi}^{[p]} = B\left(\Psi_{\Phi} - \Psi_{\Phi,\Delta}^{[p]}, \mathfrak{u} - u_{\Delta}^{[p]}\right) = B\left(e_{\Phi,\Delta}^{[p]}, e_{\Delta}^{[p]}\right), \tag{7.69}$$

we obtained, see (6.3c)

$$\left|e_{\Phi}^{[p]}\right| \leqq \left\|e_{\Phi,\Delta}^{[p]}\right\|_{\mathcal{U}} \left\|e_{\Delta}^{[p]}\right\|_{\mathcal{U}}. \tag{7.70}$$

This result indicates how we can estimate the error $\left|e_{\Phi}^{[p]}\right|$. If we compute $\Psi_{\Phi,\Delta}^{[p]}$ and $u_{\Delta}^{[p]}$ then we have

$$\left|e_{\Phi}^{[p]}\right|_1 \leqq \varepsilon\left(u_{\Delta}^{[p]}\right) \cdot \varepsilon\left(\Psi_{\Phi,\Delta}^{[p]}\right), \tag{7.71}$$

where $\varepsilon\left(u_{\Delta}^{[p]}\right)$ and $\varepsilon\left(\Psi_{\Phi,\Delta}^{[p]}\right)$ are any of the *a posteriori* error estimations of the previous sections. Equality (7.69) can also be written in the form

$$e_{\Phi}^{[p]} = \sum_{\tau_q} B_{\tau_q}\left(e_{\Phi,\Delta}^{[p]}, e_{\Delta}^{[p]}\right), \tag{7.72}$$

from which, using any of the elementwise estimators,

$$\left| e_{\Phi}^{[p]} \right|_2 \leqq \sum_{\tau_q} \eta_{\tau_q} \left(\Psi_{\Phi,\Delta}^{[p]} \right) \cdot \eta_{\tau_q} u_{\Delta}^{[p]}. \tag{7.73}$$

Analogues of (7.73) can be obtained for any of the subdomain estimators.

If, for the element estimators we have on every element τ_q that $e_{\Delta}^{[p]} \approx \hat{e}_{\Delta}^{[p]}$ and $e_{\Phi,\Delta}^{[p]} \approx \hat{e}_{\Phi,\Delta}^{[p]}$, then we can write

$$\left| e_{\Phi}^{[p]} \right|_3 \leqq \left| \sum_{\tau_q} B_{\tau_q} \left(\hat{e}_{\Phi,\Delta}^{[p]}, \hat{e}_{\Delta}^{[p]} \right) \right|. \tag{7.74}$$

The major difference between (7.74) and (7.73) is that the terms in (7.74) may have different signs and that cancellation takes place. Obviously (7.74) leads to a smaller estimate than (7.71) and (7.73) but it could grossly underestimate the error.

Exercise 7.3 Show that $\left| e_{\Phi}^{[p]} \right|_1 \geqq \left| e_{\Phi}^{[p]} \right|_2 \geqq \left| e_{\Phi}^{[p]} \right|_3$.

Remark 7.10 *The functions* $\mathfrak{u}$ *and* Ψ_{Φ} *can have different regularity; often* Ψ_{Φ} *is less regular that* $\mathfrak{u}$. *In this case, the estimate* $\left| e_{\Phi}^{[p]} \right|_1$ *would indicate that we should use a finer mesh for* $\Psi_{\Phi,\Delta}^{[p]}$ *than was used for* $u_{\Delta}^{[p]}$. *It is necessary to realize that the estimate has utilized Galerkin orthogonality. Denoting by* $S_{\Delta,0}^{[p]}$, *the space for* $u_{\Delta}^{[p]}$ *and by* $S_{\Phi\Delta,0}^{[p]}$ *the space for* $\Psi_{\Phi,\Delta}^{[p]}$ *it is necessary that* $S_{\Delta,0}^{[p]} \subset S_{\Phi,0}^{[p]}$.

7.6 Adaptive finite element methods

In Section 5.1.1 we introduced the concept of an *equilibrated* mesh, which is a mesh that produces errors of approximately the same size in every element. A mesh of this type is optimal. The process of constructing such a mesh is one of *adaptivity*, which can be described as follows.

Given a mesh Δ_n, we compute indicators $\eta(\tau)$ and $\max_{\tau \in \Delta_n} (\eta(\tau))$ is identified. Those elements for which $\eta(\tau) \geqq \beta \max_{\tau \in \Delta_n} (\eta(\tau))$, where $\beta \leqq 1$,(usually $0.5 \leqq \beta \leqq 0.8$), are then refined, thus producing a new mesh Δ_{n+1} with which a new finite element solution is computed. The same refinement principle is used for the mesh $\Delta_{n+2.}$ In this way, a sequence of meshes $\Delta_m, m = 1, 2, \dots$ and corresponding spaces $S_{\Delta_m}^{[p]}$ are constucted together with the estimators ε_m and the energy norm $\left\| u_{\Delta_m}^{[p]} \right\|_{\mathcal{U}(\Omega)}$.

When the relative error $\varepsilon_m / \|\mathfrak{u}\|_{\mathcal{U}} < \alpha$, where α is an *a priori* chosen tolerence, the mesh and the corresponding solution are accepted.

Note that the selection of β is particularly arbitrary. One could take $\beta = 1$, but this would require more problems in the sequence to be solved. If, alternatively, β is selected to be too small, many elements will be refined unnecessarily and the spaces $S_{\Delta_m}^{[p]}$ will be of larger dimension than needed. The choice of $0.5 \leqq \beta \leqq 0.8$ is thus a compromise, and experience shows that the refinement process is not highly sensitive for β in this range.

In any refinement strategy one seeks to have the situation where all the meshes have approximately the same minimum angles. It is also desirable that $S^{[p]}_{\Delta_{m+1}} \supset S^{[p]}_{\Delta_m}$, so that the error decreases monotonically with m. The adaptive procedure described above leads to approximately equilibrated meshes.

The effectiveness of any adaptive procedure should always be tested. One way of doing this is to apply it to a benchmark problem, such as one of our problems in an L-shaped domain. If the solution of this problem were smooth, then the error would be $0(h^p) = 0\left(N^{-p/2}\right)$. If an adaptive procedure is applied in the case when the problem contains a boundary singularity, then it is effective only when $e^{[p]}_{\Delta} \approx 0\left(N^{-p/2}\right)$. We saw that in this situation use of a uniform mesh leads to a convergence rate of $0\left(N^{-\beta/2}\right)$. However, the $0\left(N^{-p/2}\right)$ rate can be achieved a suitable adapted mesh.

Clearly, the form of the meshes Δ_m depends on the solution u of the problem. If we are interested in problem arising from a differential equation with different right-hand sides, and we let uj, $j = 1, \ldots k$ be the solutions, then for a given mesh we compute the error indicators $\eta j(\tau)$, $j = 1, 2, \ldots, k$, and we control the refinement using a norm of the vector $\left(\eta^1(r), \ldots, \eta^k(\tau)\right)$.

The adaptive procedure for the computation of a functional is the same as above. For every element we compute $\eta_{\Delta_m}(\tau)$ and $\eta_{\Phi,\Delta_m}(\tau)$ and the refinement strategy is based now on equilibration of the product $\eta_{\Delta_m}(\tau) \cdot \eta_{\Phi,\Delta_m}(\tau)$ ·

Mesh adaptivity is a huge subject in itself, and has been the subject of many books. We have not sought here to define mesh-refinement strategies. Our aim has simply been to point out some ideas behind underlying adaptive techniques.

A note on verification

As we discussed in Chapter 1, verification is an essential part of finite element computation and addresses the question of whether the finite element technique leads to accurate approximate solutions. Verification essentially has two parts, verification of the method and verification of the code implementing the method. In this book, we have addressed verification of the method using theory and the evidence of numerical experiments. In these it has been assumed that all numerical integrations are sufficiently accurate, that the systems of linear equations are solved accurately and that round-off error is negligible. Although we have not addressed it, we have tacitly assumed that there is no error in the codes used in the computations.

So, how should verification be done in general? Typically, it can be approached using the 'manufactured solution process' in which a series of problems with known exact solutions are solved. One can identify two classes of such problems.

For the first of these the solution is chosen such that if it is linear the finite element solution is exact for elements of degree 1, if it is quadratic the finite element solution is exact for elements of degree 2, and it also satisfies the various boundary conditions, different right-hand sides of the equations and different coefficients in the equations. If the finite element solutions are not exact in all these contexts, then an investigation has to be undertaken.

The second class of problems is that in which the exact solution is available but the finite element solution is only an approximation. In this case solutions are computed for various meshes, errors are computed in the energy norm, and the finite element energy norm and the exact energy norm are compared. The relation between these two energies is then considered in the light of the theory as presented and *a posteriori* errors are estimated using specific estimators. In the case of the Neumann subdomain estimator, a check is made that this is an upper estimator, or almost an upper estimator because the local problem is not solved exactly, and the effectivity indices are analyzed. In this whole process, problems are chosen that have both smooth and singular solution behaviour. If everything has worked satisfactorily for these classes of problems, then the user has confidence in the finite element method for solving more general problems. If not, then further investigations have to be made. We emphasize that we have used the word confidence here, because theoretically one can never be absolutely sure that the computation of a practical problem will not show some form of anomaly.

Epilogue

At the beginning of this book in both the Preface and in the introductory first chapter we sought to explain our reasons for writing yet another book on finite elements and to set out our vision of what we were seeking to achieve. It is fitting at the end of the book that we should examine whether our goals have been met, and what we feel we have achieved.

Our goals were to explain in as simple a manner as possible the main ideas of the finite element method and the analysis of finite element error. We positioned this in the context of reliability, giving as our purpose the need that any modeller has to be confident that the computed finite element results are reliable in the sense that they are close to the exact solution of the problem; this is verification. We asked how one could build up such confidence.

Our approach to this has been to develop theory in each chapter, leading up to theoretical results on error estimation. One of the essential checks for the correctness of a computation is that it should be possible to explain the results in the terms of theory. Simultaneously to giving theory we have therefore presented a large number of computations both for benchmark-type problems and also for more demanding 'engineering problems' in one and two space dimensions. The computational results have illustrated the theoretical ones, and explanations have been given, all contributing to the building of confidence in the methods. Advice has been given throughout to those wishing to use finite element methods in these contexts in the form of 'morals'.

The sequence of the material has led up to the defining of 'calculable' *a posteriori* error estimators in the final chapter. Many different estimators have been defined, analyzed theoretically and their use illustrated and compared computationally. It is not possible to say which estimator is the 'best' under all circumstances. However, in illustrating the performance of the estimators in multiple contexts, including our engineering problems, we have tried to indicate how to assess the quality of an estimator and when one can hope to use it successfully. Finally, in Section 7.4.7 we have made 'recommendations'.

So, have we fulfilled our objectives? We hope that this book will enable students and others to gain insight into the finite element method and the analysis of its error, as well as the pitfalls in finite element computation, and that it will enable those who are interested to proceed to study the more advanced literature.

We finish with the observation that computation of the type discussed in this book is a somewhat risky business, in a manner similar to that in which the entirety of life is a challenging and risky business. In life one makes decisions based on experience and well-established confidence. Similarly, confidence in computations that are based on results of this book is in our opinion a 'well-founded' confidence.

It can also be asserted that verification suffers from the presence of ‘guilt’, in that in any computation there is error and it is the task of the practitioner to prove ‘innocence’, i.e. that there is in fact not an error. This is the reverse of most judicial systems, where in a court of law innocence is automatically assumed and guilt has to be proven!

Appendix A

A.1 Linear spaces, normed linear spaces, linear functionals, bilinear forms

We shall deal here only with spaces of real numbers.

A.1.1 Linear space

A linear space $\mathcal{U}$ is a family of elements $u, v, w, \ldots$ with the following properties: if $u, v, w, \ldots, \in \mathcal{U}$ and $\alpha, \beta \in \mathbb{R}$ are real numbers then,

(1) $(u+v) \in \mathcal{U}$;
(2) $u+v = v+u$, $u+(v+w) = (u+v)+w$;
(3) $\alpha u \in \mathcal{U}$;
(4) there is a unique element of $\mathcal{U}$, denoted by 0, such that $u+0 = u$ for any $u \in \mathcal{U}$;
(5) with every element $u \in \mathcal{U}$ is associated a unique element $-u \in \mathcal{U}$ such that $u+(-u) = 0$;
(6) $\alpha(u+v) = \alpha u + \alpha v$;
(7) $(\alpha+\beta)u = \alpha u + \beta u$;
(8) $\alpha(\beta u) = (\alpha\beta)u$;
(9) $1 \cdot u = u$;
(10) $0 \cdot u = 0$.

A.1.2 Normed linear space

In a normal linear space with every $u \in \mathcal{U}$ is associated a real number $\|u\|_{\mathcal{U}}$, called the norm, such that for any $u, v \in \mathcal{U}$

(1) $\|u\|_{\mathcal{U}} \geqq 0$ and $= 0$ if and only if $u = 0$;
(2) $\|\alpha u\|_{\mathcal{U}} = |\alpha| \, \|u\|_{\mathcal{U}}$ for any α;
(3) $\|u+v\|_{\mathcal{U}} \leqq \|u\|_{\mathcal{U}} + \|v\|_{\mathcal{U}}$.

A.1.3 Inner product spaces

The study of linear problems invariably involves the use of inner product spaces, in which a norm can be defined through the inner product. An inner product $(\cdot,\cdot)$ is a function from $\mathcal{U} \times \mathcal{U}$ to $\mathbb{R}$ (real numbers) such that if $u, v, w, \in \mathcal{U}$, and $\alpha, \beta \in \mathbb{R}$, then

(1) $(u,u) \geqq 0$ and $= 0$ if and only if $u = 0$;
(2) $(u,v) = (v,u)$;
(3) $(\alpha u + \beta v, w) = \alpha(u,v) + \beta(v,w)$.

In this case, the space $\mathcal{U}$ with the inner product $(\cdot,\cdot)$ is an inner product space.

Based on the properties of inner products and norms as above, it is clear that an inner product $(\cdot,\cdot)$ produces a norm through the relation

$$\|u\| = (u,u)^{\frac{1}{2}},\ u \in \mathcal{U}.$$

In this case, the inner product space is a normed linear space. In order to verify the triangle inequality using the two definitions, the Schwarz inequality is required.

A.1.4 Schwarz inequality

If $u, v \in \mathcal{U}$, an inner product space, then

$$|(u,v)| \leqq ((u,u)\,(v,v))^{\frac{1}{2}},$$

and equality holds if and only if u, v are linearly dependent.

Proof The result clearly holds when either u or $v = 0$. When this is not the case, so that $u, v \neq 0$, we note that

$$(u + \lambda v, u + \lambda v) \geqq 0,\ \in \mathbb{R},$$

so that

$$(u,u) + 2\lambda\,(u,v) + \lambda^2\,(v,v) \geqq 0,\ \ \lambda \in \mathbb{R}.$$

This inequality involves a quadratic expression for λ that is non-negative, so that its discriminant

$$[2\,(u,v)]^2 - 4\,(u,u)\,(v,v) \leqq 0,$$

and the Schwarz inequality holds. □

A.2 Convergence, completeness and Hilbert spaces

A.2.1 Convergence

A sequence of functions $u_n \in \mathcal{U}$, $n = 1, 2, \ldots,$ converges in the normed space $\mathcal{U}$ to $u \in \mathcal{U}$ if for every $\varepsilon > 0$ there is a number $N(\varepsilon)$ such that for any $n > N(\varepsilon)$

$$\|u - u_n\|_{\mathcal{U}} < \varepsilon.$$

A.2.2 Cauchy sequence

Let $u_{1,}u_2, \ldots, u_n \in \mathcal{U}$ be a sequence. If for any $\varepsilon > 0$ there exists a number $N(\varepsilon)$. If the convergence is to an element of $\mathcal{U}$ then the sequence is a Cauchy sequence such that $\|u_n - u_m\| < \varepsilon$ for $n, m \geqq N(\varepsilon)$, then the sequence is called a Cauchy sequence.

A.2.3 Hilbert space

If $\mathcal{U}$ is a complete inner product space it is called a Hilbert space. Clearly, it is in this case a complete normed space, where $\|u\|_{\mathcal{U}}^2 = (\cdot,\cdot)$.

A.3 Linear functionals and bilinear forms

A.3.1 Linear functionals

Let $\mathcal{U}$ be a normed linear space and F(u) be a process that associates with every $u \in \mathcal{U}$ a real number $\mathcal{F}(u)$. Then, $\mathcal{F}(u)$ is a linear functional on $\mathcal{U}$ if

(1) $\mathcal{F}(u_1 + u_2) = \mathcal{F}(u_1) + \mathcal{F}(u_2)$;
(2) $\mathcal{F}(\alpha u) = \alpha\mathcal{F}(u)$;
(3) $|\mathcal{F}(u)| \leqq \mathcal{C}\, \|u\|_{\mathcal{U}}$ with $\mathcal{C}$ independent of u.

A.3.2 Bilinear forms

Let $\mathcal{U}$ and $\mathcal{V}$ be normed linear spaces and $B(u,v)$ be a process that associates with every $u \in \mathcal{U}$ and $v \in \mathcal{U}$ a real number $B(u,v)$. Then, $B(u,v)$ is a bilinear form on $\mathcal{U} \times \mathcal{V}$ if

(1) $B(u_1 + u_2, v) = B(u_1, v) + B(u_2, v)$;
(2) $B(u, v_1 + v_2) = B(u, v_1) + B(u, v_2)$;
(3) $B(\alpha u, v) = \alpha B(u, v)$;
(4) $B(u, \alpha v) = \alpha B(u, v)$;
(5) $|B(u,v)| \leqq \mathcal{C}\, \|u\|_{\mathcal{U}}\, \|v\|_{\mathcal{V}}$, with $\mathcal{C}$ independent of u and v.

Note that $B(u,v)$ is not necessarily symmetric.

A.3.3 The Lax–Milgram lemma

Let $\mathcal{U}$ be a Hilbert space with inner product $B(u,v)$, $u, v \in \mathcal{U}$, and let there exist two constants $\mathcal{C}_1$ and $\mathcal{C}_2$ independent of u and v such that for every $u, v \in \mathcal{U}$

$$|B(u,v)| \leqq \mathcal{C}_1\, \|u\|_{\mathcal{U}}\, \|v\|_{\mathcal{U}},$$
$$B(u,u) \geqq \mathcal{C}_2\, \|u\|_{\mathcal{U}}^2.$$

Then, every bounded linear functional $\mathcal{F}$ on $\mathcal{U}$ can be expressed in the form

$$\mathcal{F}(v) = B(\mathfrak{u}, v)\ \forall v \in \mathcal{U},$$

with $\mathfrak{u}$ uniquely determined by this functional; see, e.g., Ciarlet (1979) and Whiteman (1975).

Bibliography

Abramowitz M. and Stegun I. A. Handbook of Mathematical Functions. Dover, New York, 1972.

Ainsworth M. and Oden J. T. A-posteriori Error Estimation in Finite Element Analysis. John Wiley and Sons Inc., New York. 2000.

Babuška I. and Aziz A. K. On the Angle Condition in the Finite Element Method. SIAM J. Numer. Anal. 13, 214–226, 1976.

Babuška I., Nobile F. and Tempone R. Reliability in Computational Science. Number. Meth. Partial Diff. Equations 23, 753–784, 2007.

Babuška I. and Rheinboldt W. C. A-posteriori Error Estimates for the Finite Element Method. Int. J. Numer. Meth. Eng. 12, 1597–1615, 1978.

Babuška I. and Rheinboldt, W. C. Error Estimates for Adaptive Finite Element Computation. SIAM J. Numer. Anal. 15, 736–754, 1978.

Babuška I. and Silvia R., Numerical Treatment of Engineering Problems with Uncertainties. The Fuzzy set Approach and its Application to the Heat Exchanger Problem. Int. J. Numer. Meth. Eng. (to appear). 2010.

Babuška I. and Strouboulis T. The Finite Element Method and its Reliability. Oxford University Press, Oxford, 2001.

Babuška I., Strouboulis T., Upadhjay A. S. and Gangaraj S. K. Computer-based Proof of the Existence of Superconvergence Points in the Finite Element Method. Numer Meth. Partial Diff. Equations 12, 347–392, 1996.

Bank E. and Weiser A. Some *A Posteriori* Error Estimators for Elliptic Partial Differential Equations. Math. Comp. 44, 283–301, 1985.

Bathe K-J. Finite Element Procedures in Engineering Analysis, Prentice Hall, NJ, 1996.

Becker E., Carey G. F. and Oden J. T. Finite Elements, An Introduction Vol 1. Prentice Hall, NJ, 1981.

Ben-Aki M. The Bug that Destroyed a Rocket. J. Comp. Sci. Educ. 2, 15–17, 1997.

Braess D. Finite Elements. Cambridge University Press, Cambridge, 1997.

Brenner S. C. and Scott L. R. The Mathematical Theory of Finite Element Methods. Springer-Verlag, New York, 1994.

Ciarlet P. G. The Finite Element Method for Elliptic Problems. North Holland, Amsterdam, 1979.

Demkowicz L. Computing with $h-p$ Adaptive Finite Elements I, One- and Two-Dimensional Elliptic and Maxwell Problems. Chapman and Hall, New York, 2006.

Ern A. and Goermond J. L. Theory and Practice of Finite Elements. Springer-Verlag, Berlin, 2004.

Goodsell G. and Whiteman J. R. A Unified Treatment of Superconvergent Recovered Gradient Functions for Piecewise Linear Finite Element Approximations. Int. J. Numer. Meth. Eng. 27, 469–481, 1989.

Grisvard P. Elliptic Problems in Unsmooth Domains. Pitman, London, 1985.

Hinton E. and Owen D. R. J. Finite Element Programming. Academic Press, London, 1977.

Hughes T. J. R. The Finite Element Method. Prentice Hall, NJ, 1987.

Ladeveze P. and Leguillon D. Error Estimate Procedure in the Finite Element Method and Applications. SIAM J. Numer. Anal. 20, 485–509, 1983.

Ladeveze P. and Pelle J. Mastering Calculations in Linear and Nonlinear Mechanics. Translated from French by T. Strouboulis, Springer-Verlag, Berlin, 2005.

Mackerle J. Finite Element Modelling, a Bibliography, 1994–2004, website.

Martin H. C. and Carey G. F. Introduction to Finite Element Analysis, Theory and Applications. McGraw Hill, New York, 1973.

Nazarov S. A. and Plamenevsky B. A. Elliptic Problems in Domains with Smooth Boundaries. Walter de Gruyer, Berlin, 1994.

Nitsche J. A. and Schatz A. H. Interior Estimates for Ritz–Galerkin Methods. Math. Comp. 28, 937–968, 1974.

Oden J. T. Finite Elements of Nonlinear Continua. McGraw-Hill, New York, 1972.

Oden J. T. and Prudhomme S., Estimation of Modeling Error in Computational Mechanics. J. Comput. Phys. 182, 496–515, 2002.

Oden J. T. and Reddy J. N. An Introduction to the Mathematical Theory of Finite Elements. Wiley-Interscience, New York, 1976.

Pares N., Diez P. and Huerta A. Subdomain-based Flux-free *A Posteriori* Error Estimation. Comp. Meth. Appl. Mech. Eng. 195, 297–323, 2006.

Rectorys K. Survey of Applicable Mathematics. 2nd edn. Kluwer Academic, Vol. 1, p.387, Dordrecht, 1994.

Reddy J. N. Energy and Variational Methods in Applied Mechanics. Wiley-Interscience, New York, 1984.

Richardson, L. F. The Approximate Arithmetical Solution by Finite Differences of Physical Problems Involving Differential Equations, with Application to Stresses in a Masonry Dam. Proc Roy Soc London, Ser A, 210, 307–357, 1910.

Schatz A. H. and Wahlbin L. B. Interior Maximum Norm Estimates for the Finite Element Method. Math. Comp. 31, 414–442, 1977.

Schatz A. H. and Wahlbin L. B. Interior Maximum Norm Estimates for the Finite Element Method, II. Math. Comp. 64, 907–928. 1995.

Schwarz H. R. Finite Element Methods. Academic Press, London, 1988.

Shaw S., Warby M. K. and Whiteman J. R. Discretization Error and Modelling Error in the Context of the Rapid Inflation of Hyperelastic Membranes. IMA J. Numer. Anal. Vol 30, 302–333, 2010.

Szabo B. and Babuška I. Finite Element Analysis. Wiley, New York, 1991.

Wahlbin L. B. Superconvergence in Galerkin Finite Element Methods. Springer-Verlag, New York, 1995.

Wait R and Mitchell A. R. Finite Element Analysis and Applications. John Wiley and Sons, Chichester, 1985.

Wheeler M. F. and Whiteman J. R. Superconvergent Recovery of Gradients on Subdomains for Piecewise Linear Finite Element Approximations. Numer. Meth. Partial Diff. Equations, 65–82, 1987.

Whiteman J. R. (ed.). The Mathematics of Finite Elements and Applications (7 volumes). Academic Press, London, 1972, 1975, 1978, 1981, 1984, 1987, 1990.

Whiteman J. R. (ed.) The Mathematics of Finite Elements and Applications, Highlights 1993 and 1996. Wiley, Chichester, 1993, 1996.

Whiteman J. R. (ed.) The Mathematics of Finite Elements and Applications X. Elsevier, Amsterdam, 1999.

Whiteman J. R. and Goodsell G. Superconvergent Recovery of Stresses from Finite Element Approximations on Subdomains for Planar Problems of Linear Elasticity. p 28–53 of J. R. Whiteman (ed.) The Mathematics of Finite Elements and Applications VI, Academic Press, London, 1988.

Zienkiewicz O. C. The Finite Element Method. McGraw Hill, Maidenhead, 1977.

Zienkiewicz O. C. and Taylor R. L. The Finite Element Method in Solid and Structural Mechanics. McGraw Hill, Maidenhead, 2005.

Zienkiewicz O.C and Zhu J. Z. A Simple Error Estimator and the Adaptive Procedure for Simple Finite Engineering Analysis. Int. J. Numer. Meth. Eng. 24, 337–357, 1987.

Zienkiewicz O. C. and Zhu J. Z. The Superconvergent Patch Recovery and *A-posteriori* Error Estimates. Int. J. Numer. Meth. Eng. 33, 1331–1382, 1992.

Index

Note: page numbers in *italics* refer to Figures and Tables.

The manufacturer's authorised representative in the EU for product safety is Oxford University Press España S.A. of El Parque Empresarial San Fernando de Henares, Avenida de Castilla, 2 - 28830 Madrid (www.oup.es/en or product.safety@oup.com). OUP España S.A. also acts as importer into Spain of products made by the manufacturer.
Printed and bound by CPI Group (UK) Ltd, Croydon, CR0 4YY
06/07/2026
02157632-0014